D0070379

The Art of Managing Technical Projects

PERSONAL
MANAGEMENT
STYLE:
THE INDIVIDUAL
SYSTEM

PEOPLE MANAGEMENT:
THE PROJECT SYSTEM

PROJECT OPERATIONS:
THE PROJECT SYSTEM

MEASURING:
THE ORGANIZATIONAL SYSTEM

FORECASTING:
THE ORGANIZATIONAL SYSTEM

FOUNDATION: PROJECT MANAGER'S AUTHORITY

MELVIN SILVERMAN

Atrium Associates, Inc.

Prentice-Hall, Inc., Englewood Cliffs, New Jersey 07632

Library of Congress Cataloging-in-Publication Data
SILVERMAN, MELVIN.
 The art of managing technical projects.

 Includes bibliographies and index.
  1. Industrial project management. 2. Executives.
3. Industrial management. I. Title.
HD69.P75S54 1987   658.4′04   87-1750
ISBN 0-13-047010-4

Editorial/production supervision
 and interior design: **PATRICK WALSH**
Cover design: **BEN SANTORA**
Manufacturing buyer: **RHETT CONKLIN**

© 1987 by Prentice-Hall, Inc.
A Division of Simon & Schuster
Englewood Cliffs, New Jersey 07632

Printed in the United States of America

10 9 8 7 6 5 4 3 2 1

ISBN 0-13-047010-4 025

Prentice-Hall International (UK) Limited, *London*
Prentice-Hall of Australia Pty. Limited, *Sydney*
Prentice-Hall Canada Inc., *Toronto*
Prentice-Hall Hispanoamericana, S.A., *Mexico*
Prentice-Hall of India Private Limited, *New Delhi*
Prentice-Hall of Japan, Inc., *Tokyo*
Prentice-Hall of Southeast Asia Pte. Ltd., *Singapore*
Editora Prentice-Hall do Brasil, Ltda., *Rio de Janeiro*

Contents

PERSONAL MANAGEMENT STYLE: THE INDIVIDUAL SYSTEM

PEOPLE MANAGEMENT: THE PROJECT SYSTEM

PROJECT OPERATIONS: THE PROJECT SYSTEM

MEASURING: THE ORGANIZATIONAL SYSTEM

FORECASTING: THE ORGANIZATIONAL SYSTEM

FOUNDATION: PROJECT MANAGER'S AUTHORITY

Chapter 1: Definitions: The Important Beginning   15

Chapter 2: Forecasting: Starting with the Charter   46

Chapter 3: Forecasting: Techniques   66

Chapter 4: Forecasting: Tieing it All Together   100

Chapter 5: Measuring: The Second System   122

Part II: Project-Oriented Systems   145

Preface

PERSONAL MANAGEMENT STYLE: THE INDIVIDUAL SYSTEM

PEOPLE MANAGEMENT: THE PROJECT SYSTEM

PROJECT OPERATIONS: THE PROJECT SYSTEM

MEASURING: THE ORGANIZATIONAL SYSTEM

FORECASTING: THE ORGANIZATIONAL SYSTEM

FOUNDATION: PROJECT MANAGER'S AUTHORITY

WHY?

I suppose that this book was started over 20 years ago when I first became a project manager but, of course, I didn't know that at the time. I had been a manufacturing plant manager for several years, and the job had been fascinating because there was so much to do, at first. But eventually, the work force became well trained and responsive, quality rose to more than adequate levels, and in general things were running very smoothly. One day a management recruiter called and said he had heard about the positive things that had been done at my present company. He offered me a wonderful opportunity as a project manager in a much larger industrial firm in another city. From his description, the position seemed too good to be true. It was.

The interview and the preliminary discussions with the new company's managers became very vague when it came to describing what my duties as a project manager would be. But that was all right, because I thought I knew what had to be done. After all, I had been managing complex engineering and manufacturing operations for years, and this couldn't be any more difficult that that. I was wrong.

After joining the firm, I learned some hard truths, such as, project management is entirely different from line management. I quickly found out that

1. I had control over neither the technical objectives nor the selection of the project team members. The specifications had already been agreed upon, and top management had assigned the team members to the project. Those team members still reported to their original bosses.

2. I had no control over the length of time for the project. The contract defining that had been signed just before I arrived.

3. I had no control over the costs. The Accounting department allocated funds and measured expenditures.

4. But I was responsible for it all! When project goals were not achieved or budgets were exceeded, I was supposed to explain and correct the problems! Those were interesting days.

It took quite awhile, but I finally *did* get control over that project, and it even eventually succeeded. In retrospect, it took more pain and work than it should have. Managing subsequent projects was easier. I was able to analyze the foundations of my past successes, determine those that each new project needed, and use the applicable systems that I had previously developed. It was really a painful, unstructured kind of on-the-job training. I suspected then, and I'm sure now, that the pain is not a necessary part of project management. This conclusion has been reinforced over the years as I compared my experience with many other project managers, gained more experience in general management through consulting with many industrial organizations, and lectured in countless seminars on the subject.

There are ways to set up and manage projects that minimize most of the classic, repetitive problems. These are dependent on the repetititve systems that can provide the project manager with a basis for achieving success and a unique management style to deal with unforeseen organizational situations. This book is intended to be a guide in developing both those systems and one's own personal style of managing technical projects successfully, that is, without the pain. Success, however, is a moving target. Therefore the redesign of project systems and improvement of personal style always goes on.

SOME PRELIMINARY IDEAS: SCIENCE AND ART

Managers and would-be managers have searched for success as if it were a badge of excellence or an honored medal to be proudly worn or displayed as a trophy on the office wall. That search is in vain since management success isn't an end goal. It's a never-ending, ever-changing process. It varies with each organization and with each person in that organization.

Management fads that promise inevitable success come and go. Their adherents fill the literature of the day with their implied promises of universal solutions. Eventually, however, new and unexpected problems arise, and the "universal solution" of the past fails. Then a new technique is born, promising to overcome the, by now, familiar problems that those older solutions couldn't solve. It's just another promise of the "universal solution," and universal solutions don't really apply to management. They can apply to a physical science with its clearly defined rules and procedures. But management is not a science; it's more like an art. And a successful artist is always at least slightly different from everyone else. The physical sciences have open experiments and procedures that are independent of the experimenter or the manager and that are replicable by others. Management doesn't follow this model because the experimenter (manager) is part of the process. In management, "experiments" are done in the mind of the experimenter or manager. Those experiments can't be observed. Even the procedures are not replicable, because time changes the situation. The company's costs, sales, product designs, and so forth are always changing. For example, we all know that when a different manager occupies a management position, the situation changes. Additionally, management depends upon people, and each person is unique. Management is, therefore, still very much an art form, and each job often requires different skills and knowledge. I don't believe that a successful manager in one situation can be equally successful in all others, unless there is a major personal change to fit the situation.

Depending on the position, there may be differing requirements for knowledge of economics, finance, psychology, physics, chemistry, and so forth. But even when those requirements are satisfied, there is often more required. In the words of the logician, it's necessary but incomplete. Knowledge may be applied differently now. There are often additional needs for unique solutions that come from the manager's ability to use his or her mental and physical resources optimally. But even the combination of knowledge and creativity may not be sufficient to grasp success. There may be external, uncontrollable variables that are not within the manager's control, such as changes in technology (e.g., buggy whips became obsolete with the auto and semaphore signaling with the telegraph) or changes in consumer demand (e.g., fuel price increases that affect demand for home insulation). There is always a complex, multivariable interaction between the manager and the situation. Managing this changing situation well with oneself as only one variable requires the highest degree of artistry. The universal solution that solves all problems and assures the inevitable success simply does not apply.

Learning how to be a successful manager, therefore, is different from learning to be a successful engineer, physicist, or chemist. The deterministic, relentlessly logical way of thinking in which there *must* be a "best" answer to a particular problem is not very useful in management, in my opinion. There are too many variables in us, as human beings, and in the situation for any neat equations to solve. There can be many answers, some better than others. There is another factor to consider. Because managers are part of the problem as well as the solution, managers are never sure whether they have considered all the variables or if they have chosen the one best answer. It's interactive. Managers can't extricate themselves from a situation and go back to try again as if nothing happened. There cannot be a truly "controlled" experiment. It's an ongoing process that cannot be stopped while they search for the best answer. Time is not an elastic resource. Managers often have to choose a solution *now* and go with it. The search for the "best" answer that is typical of science is therefore impossible in management.

Project managers who were trained in the physical sciences must use a different way of thinking. They can start with some basic solutions to common problems that are relatively predictable, which is also done in the physical sciences. They can learn to handle the classical problems of management from the many texts published every year that explain how to "plan, organize, direct, and control," or "optimize operations," or "achieve excellence," or whatever the current management watchwords are. These texts deal with the basics. However, when the organizational situation begins to deviate from the "classical," when the situation is no longer congruent with that latest "universal" solution (and very often, it's not), the manager has few documented guidelines that can help. The creative management "artist" now needs to develop the unique, not "universal," solution to fit the needs of the specific situation. That is partially a self-development process that depends on a basic knowledge of business (i.e., organization structure, individual motivation, finance, etc.) and the creativity to solve unique problems. But how is this done?

Just as the successful artist develops a unique style of drawing, composing, sculpting, based on systems of colors, shapes, forms, and so forth, managers can develop their own styles of management based on systems of forecasting, measuring, and so forth. In other words, managers can design practical systems as foundations upon which to build the unique processes that really lead to success in a particular organizational situation. For example, after the artist learns color mixing, perspective development, and form drawing, he or she may go on to paint in an impressionist or a cubist style resulting in unique works of art. Similarly, managers learn forecasting, how to write software, measuring, and cost accounting systems before becoming project managers in the creative environment of, for instance, a computer company. The management systems, like the painter's colors, may be similar, but the computer designs that are the end results of the project, like the paintings, are unique. It never ends. Rarely does one painting or one design make the artist's or the manager's reputation. Continual, high-quality output is the important thing.

THE ARTIST AS A PROFESSIONAL

But there might be a middle ground between the relatively tight logic of the scientist and the comparatively free creativity of the artist. I think that this ground is occupied by the "professional," whose characteristics might be listed as follows:

1. A specialized body of knowledge: There is an extensive training program that results in the acquisition of a specialized body of knowledge. It includes basic data, techniques for applying that data, and the ethics to consider in applying those techniques.

2. Competency: After some period of apprenticeship under the direction of established practitioners, competency is established by passing an examina-

tion or a test of some kind. The examination itself is developed by those who have already demonstrated that competency.

3. Peer review: Because of the high level of skill required, the receiver of the professional's services is unable to evaluate them adequately. Therefore, peer review is used because personal judgement is always a factor.

An obvious example of the professional is the physician. A lengthy academic and hospital training program provides basic knowledge, followed by internship under the supervision of the permanent teaching hospital staff. During this internship, there are oral and written examinations that test the knowledge and the clinical skills of the student physician. If the internship is successfully completed, there are further examinations by the various governmental licensing boards before the physician is allowed to offer professional services to the public. Finally, if there is any dispute about any actions the physician has taken, there are peer review boards responsible for evaluating those actions. These, of course, cover only the barest of essentials. The physician, like the artist, also has to deal creatively with the situation. No two patients are exactly alike, and the interaction between the physician and the patient obviously affects progress (although some physicians may not agree with me).

Another example of the professional closer to our subject of technical projects is the engineer. Professional engineers follow a path similar to physicians. The three elements of specialized knowledge, demonstration of competency, and peer review are alike in many respects. Similarly, the professional engineer must also develop subjective judgments and decisions, such as, for example, when judging safety factors in designing a structure and the ability of an end user to safely apply a product.

Now let us consider managers. A case might be made that they are professionals but the discipline of our simplified, three-fold criteria is rarely considered to apply uniformly to managers. For example, Andrew Carnegie, who developed a major steel-producing complex in the United States in the nineteenth century, was understood to be an excellent manager. But, he never had much formal schooling, and no one approved his competency or reviewed his actions. We can contrast him with Robert MacNamara, who, as Secretary of Defense, was also considered to be an excellent manager. In contrast to Andrew Carnegie, MacNamara is the product of the best schooling, his competency has been approved by governments, and his peers have vouched for almost all of his managerial decisions. Both men achieved much as managers. Yet in these examples there could be arguments against my suggestion that both were professionals. Carnegie didn't consider the needs of some of the clients (i.e., his steel workers) as much as his customers and therefore had his labor strikes. MacNamara wasn't free of negative criticism either.

If these examples are typical, it's reasonable to suggest that management hasn't even reached a stage of development that is defined as professional. Perhaps the best that I can do is stand squarely in the quicksand of the middle of the road and say that managers are "professional artists." They possess some of the attributes of both professionals and artists. They train either in an institution of higher learning or informally through self-learning programs. They pass an examination either in academia and through on-the-job achievements. Their peer review board is their regular assessment by the company's top management. That assessment determines future progress in that company. But there is no common body of specialized learning, no objective examination, and no objective peer review. And since each organizational situation is unique, managers must be as creative as artists in their problem solving and decision making. They also must be prolific as professional artists, since it's impossible to rest on yesterday's achievements.

The new product designs, the productivity improvement, and the financial accounts tell only of today's successes. New products are eventually superseded, and today's improvements or profits may fade away in the face of competition. The manager must produce a steady stream of achievements in many situations to be considered a "success." Success is a moving target requiring equally changing skills. Professional management who are successful artists obtain many of their skills by first learning the basics, then embarking on a never-ending personal improvement program. Good managers never stop their education. This book is intended to help this educational process.

STARTING TO BECOME A PROFESSIONAL ARTIST, OR AN ARTISTIC PROFESSIONAL: HOW TO PROCEED

To become a successful manager, one must first concentrate on the foundations or the practical, tested systems that support an ongoing learning and skill development process. These systems are not intended to produce "universal" answers. Nor do I claim that the research supporting them or my experience concerning them exhausts all the potential answers. It might be a case of selective attention or of "satisficing," but I've found that these systems *do* work in technical projects, most of the time. "Satisficing" is defined as "the discovery and selection of satisfactory alternatives," not optimal alternatives. "An example is the difference between searching a haystack to find the sharpest needle in it and finding one to sew with." (March & Simon, 1958, p. 140)

These systems provide the project manager (or the professional artist) with the basic tools. Then those basic tools have to be applied or implemented. We follow this sequence in system development: first definition and description, then design and implementation. We'll move from basics to complexities, and from generalizations to specifics. These systems do work. They are to the manager as the needle is to the tailor. Both are necessary for success.

WHAT ABOUT PROJECT MANAGEMENT AS PART OF MANAGEMENT IN GENERAL?

During much of the design and implementation discussions, we'll compare two general categories of managers: functional managers and project managers. They have different goals in the organization. *Functional managers* are responsible for operating the company and optimizing the use of its resources on an overall basis. They are the managers whose jobs go on "forever." Typically, they are the chief engineers, the controllers, the sales managers, and the production managers, who supervise the daily and continuing operations of any company. (I was a functional manager when I was a manufacturing plant manager). *Project managers,* on the other hand, have a specific goal, a specific time, and a specific budget. When the project is finished, their job is over. They are also concerned with optimizing resources but only as applied to their projects. Sometimes there are clashes between these two managerial species because their definitions of *"optimization"* may differ.

In my opinion, project management requires more "artistry" because it is much more complex than functional management. Project goals are almost always novel. They have to be achieved within closer constraints than those usually imposed on functions. Projects are established either to create something new or solve a problem that is larger or more complex than the regular organization can handle. Some project goals are not even clearly defined initially. They may change as the project progresses, changing resource needs drastically. This is not like functional management, which has more predictability and in which drastic changes are almost regarded as failures rather than as learning experiences or progressions.

The project team members may be new to one another and may have to learn to work together quickly. Functional personnel work together for long periods of time. Finally, when the project accomplishes some acceptable goal, it is dissolved. Functional groups are rarely dissolved in a normal course of business.

There are other differences. The potential for both management success and failure is much greater in projects. When the project contributes that new product or service and increases revenue for the company, the success is obvious. However, when companies don't provide the basic foundations or systems that can support project success or when the project manager is not well trained, the potential for failure is obvious. The potential for painful on-the-job training is also obvious.

There is often an innate conflict between the needs of a newly created project organization and the well-established functional organization which may never be resolved satisfactorily. Functional organizations provide a steady stream of revenue that keeps the company alive. Projects *might* lead to greater growth but they're not as certain as functions. Conflict can arise because of the contrasting requirements of survival versus growth. This conflict becomes certain when personnel, finances, and facilities have to be shared, causing predictable priority clashes within the organization. Occasionally, it's easier for top management to push off these problems rather than deal with them openly at the start. (That is what is happened to me.) The project foundation systems that are developed in the beginning should minimize those expected conflicts. The foundation solutions are then designed and presented in an implementation program. Since we live in an imperfect world, sometimes changes must be made if those systems are not accepted. This process is discussed in greater detail later in this book.

Even with the best of systems, I believe that the changing situation in most projects requires a greater flexibility and a greater talent as a professional management artist than any functional manager *usually* requires. (This may not be so if the total organization suffers a major change, but I'm referring here to normal kinds of operations.) The project-oriented management style of the successful project manager must change as the project matures. Starting a project is different from keeping it on track and very different from closing it down.

However, I have found that the problems that could lead to failure in managing projects are often matched by the opportunities for success. Starting and managing a project is an excellent way to learn general management. The same general management tasks are there, and it requires the same broad perspective on forecasting, measuring, worker evaluation, and operating skills. The project manager who masters project processes has less difficulty mastering total organizational processes. The scale is different but the experiences are similar.

HOW DO WE BEGIN?

Obviously parts of every project and organization are somewhat unique. It's also obvious that each project manager is also unique. But, fortunately in both cases, there are parts that are predictable. Every project and every organization have some kind of structure just as every person has a skeleton. That's predictable and understandable. We therefore start with things that one can predict and try to control. Those are the pragmatic basic project systems to help start and manage the project. There are also the potential behaviors that the project manager can adopt to improve personal management style. After that, every idea, concept, theory, and system has to be evaluated and modified to fit. My suggestions are only raw materials to adapt to suit the situation. (The management "needle" can be used with many different kinds of thread.)

CONTENTS OF THE BOOK

The book is divided into three parts. Each part deals with a different yet interrelated series of techniques. These techniques are then assembled into systems. Part I is concerned with overall organizational systems that support project operations. I call those systems *Forecasting* and *Measuring*. They describe the basic organizational environment needed for projects to succeed. I've used them, with minor modifications, in many different kinds of companies. They are practical working systems. They are predictable general solutions to the classical problems of most projects. These problems typically include lack of clearly defined authority, poor forecasting methods, and inadequate control procedures. As an example of the techniques used to resolve these problems, we discuss several types of cost-reporting methods and pick out those that seem to apply best to project operations.

Part II develops systems that refer more closely to the projects themselves. These are *Project Operations* and *People Management* and are intended to be transitions from the basics in forecasting and measuring to the creativity in the last system, personal management style. They describe how the project itself should be managed. Because of the differences among projects within the organization, these systems require more modifications. For example, motivation and individual reward systems are more variable and therefore apply differently to each project depending upon who the project team members are. The greater flexibility of these systems requires their development after the more predictable organizational systems are in place.

Part III describes the most variable and creative system of them all—*Personal Management Style*. This system deals with the project manager as an individual. Much broader systems design guidelines are used here because there always must be modifications due to individual differences. Now we are concerned with improving the management skills of one person, you, as the project manager. That means more variability, and the system really becomes a series of general behavioral recommendations. For example, we deal in this section with applications of various kinds of ideas about leadership and about ethics in decision making.

I've tried to base each chapter on what I consider to be the state of the art. Others may disagree with my selections. I have included few footnotes to impress the reader with my great learning and academic rigor. This book is intended to be a daily desk book, not a reference book that is rarely referred to. Quotations are used to allow a particular author to speak directly to you. Sometimes it's better not to interpret. There are no standard "forms" in this book that you can complete for your company. Instead, there are ideas, concepts, and descriptions of how they can be designed to fit your ideas and how they can be implemented. If the ideas are sound and practical, they can take any form in a standard document or computer program.

In addition to drawing on my experiences as a project manager, the chapters also draw on the suggestions and recommendations of hundreds of project managers who have tested and evaluated my ideas during my years of consulting and teaching in the field. Each chapter starts with an applicable case study. The case study is intended to illustrate the material covered in that chapter. Try to determine the answers to the case study before reading the chapter. Then answer them again after you finish the chapter. Finally, compare your answers with my suggestions. There are no answers that are absolutely right or wrong.

ACKNOWLEDGMENTS

During the rather extensive time I spent gathering material for this book, I had the opportunity to discuss these ideas with many successful project managers in all kinds of technical industries. They provided an intense, but helpful, test forum. Since listing all those hundreds of individuals would be impossible, I thank them all collectively here.

Of course, they are not responsible for any errors and omissions. I take sole responsibility for those.

BIBLIOGRAPHY

MARCH, JAMES G. and HERBERT A. SIMON, *Organization.* New York: John Wiley & Sons Inc., 1958.

MELVIN SILVERMAN
Cliffside Park, N.J.

Part I   Introduction:
The Work To Be Done

PERSONAL MANAGEMENT STYLE: THE INDIVIDUAL SYSTEM

PEOPLE MANAGEMENT: THE PROJECT SYSTEM

PROJECT OPERATIONS: THE PROJECT SYSTEM

MEASURING: THE ORGANIZATIONAL SYSTEM

FORECASTING: THE ORGANIZATIONAL SYSTEM

FOUNDATION: PROJECT MANAGER'S AUTHORITY

1.0 A LITTLE HISTORY FIRST

The term *project management* is relatively new but the idea is not. Organizing a temporary group of people for a specific purpose, with a manager or leader in charge, has been going on since time began. The purpose for forming such a group was usually to construct large public works such as a cathedral, a pyramid, or a road. People were brought together into a temporary organization, the job was outlined by the manager, and the work began. When the job was finished, the project was ended and the people left. It was simple, straightforward, and it worked. It was the only way to produce anything of great size or complexity because there were no large industrial, economically based organizations.

Before the mid-1700s, there were very few permanent organizations of large groups of people that produced goods and services. The army, the church, and the guilds were typical exceptions, but they performed different functions. The army provided protection, the Church provided guidance, and the guilds provided training. Small-scale, ongoing production from homes or very small informal factories

produced a trickle of consumer goods, and that was it as far as potential consumers were concerned. In the eighteenth century, things began to change. The Industrial Revolution harnessed steam power more cheaply than muscle power, and the development of precision measuring and production tools supported the effective employment of large numbers of people on a *continuing* basis. This inexpensive steam power and the increased productivity resulting from accurate tools were coupled with an emphasis on labor specialization. Products became cheaper as costs dropped. Demand rose. A new permanent kind of organization called a *factory* was created to meet this demand. The city became an important center of production and commerce. As urban populations increased with people moving off the farms into factories, life changed. Now people who were not related and did not live together had to appear at a given place and time and work together in a standard way for extensive periods of time. These new factories were not like projects, intended to solve one problem or build one building and then be disbanded. They were a new form of permanent production organization.

These new organizations carried some old management ideas. At first factories were managed like a very large extended family. This type of management was called *patriarchal management*. They were managed by the owner-manager. As the factories grew in size, they became physically distributed over large areas. Since the owner-patriarch-manager couldn't be everywhere, new methods of management had to be found.

1.1 THE EVOLUTION BEGINS WITH SCIENTIFIC MANAGEMENT AND HUMAN RELATIONS

As the factories grew in size, the patriarchal management methods began to be modified. By the end of the nineteenth century, these modifications were presented as a new theory called *Scientific Management*. This theory presented an evolutionary change from the owner-patriarch-manager to the employee-patriarch-manager. It still had many roots in the older familial system (the manager was the recruiter, trainer, paymaster, and disciplinarian), but managers were now employees and were *trained* to supervise. They no longer owned the factory, nor did they inherit their jobs from their owner-fathers. They did, however, retain some patriarchal responsibilities. They were, for example, assumed to be primarily responsible for the success of the factory. Therefore, they established the standards for production. The workers (i.e., the "family") were supposed to take orders and meet those standards. Since Scientific Management implicitly assumed that people worked best for money and that they were basically interchangeable, incentives were important. There was "one best way" to manage, and if the manager did not succeed, it was because that one best way wasn't followed. No modifications to that theory or the resulting systems were encouraged.

As long as the market was hungry for the goods produced by the factory, production and profits flowed. When savings were achieved in production, the com-

pany prospered. Part of those savings were disbursed as incentives, part as additional profits, and part helped to reduce the selling price. A reduced selling price was expected to increase sales volume and thereby increase profits again because overhead and incentives didn't rise as fast as volume. There were great economies of scale.

In summary, Scientific Management consisted of a nice, tight, closed set of ideas. The manager was now a semi-patriarch who gave the orders, and the workers were the "family" who followed them. Production was the main goal. It was logical and consistent. Costs were lowered and markets were expanded. And since products didn't change much, the theory worked well. But there were hidden flaws. Some of the basic assumptions were incorrect. One was the assumption about people: They were interchangeable and worked just for the money. This assumption was destroyed as a result of some experiments at the Hawthorne Works of Western Electric in the late 1920s and early 1930s.

These experiments were intended to determine the relationship between physical working conditions and a worker's output. Scientific Management held that workers would respond to financial incentives. The experimenters wanted to test this assumption, and the results were quite unusual. The experiments varied the physical conditions of work, such as the amount of light, the number of rest periods, and so forth. They also varied the amount of pay. The expectations were that the dependent variable, work output, would be affected as these conditions changed. They were not! Work output varied according to how the workers viewed the experiments. In other words, when the workers saw the experimenters (i.e., management) as "interested" and supportive, production went up. Otherwise, production was relatively unaffected by changes in physical inputs. The experimenters found that the factory's output could be affected by the *feelings* of the workers. Thus the experiment showed that physical conditions and pay were not the only variables that affected production; individual and group psychology played a major role as well.

The experimenters also discovered that there was an informal (or uncontrolled) authority hierarchy or organization that the workers followed. This hierarchy was reflected in the workers' culture, which was different from that expected by the company managers. This culture could also determine the volume of production. Therefore, management had to consider individual psychology as well as group culture as new and important variables in the work place. The workers were no longer passive resources who only responded to the pay they received and the physical conditions in the work environment. They responded to what they saw as their own self-interest. Sometimes the workers' self-interest was the same as management's, but sometimes it wasn't. When it wasn't, production was adversely affected. Management decided that something had to be done about this. As a result of these efforts, various types of *human relations* theories of management began to appear in publications. Worker's attitudes were now important and began to be measured through opinion polls and surveys. Work place conditions were somewhat improved in attempts to change individual attitudes (which did increase output somewhat), but management now had to consider how the factory culture affected overall production. When the culture or the informal organization supported management's goals, it was because of the supposedly improved human relations.

But when that informal organization became an impediment, it had to yield to management's dictates. Changing the plant culture was a one-way street. According to management theory after these Hawthorne experiments in the 1930s, the "irrational or emotional" results flowing from the informal organization could be disruptive and were therefore treated as a pathology. However the Human Relations School did soften some of the spirit of authoritarianism that Scientific Management attributed to the manager.

Even today we can find factory systems that use the discipline of Scientific Management modified, where required, by the human relations approach. The ongoing, functional organizations in many production companies typically adopt a hybrid scientific management-human relations form. Those organizations tend to produce uniform goods at predetermined levels of quality. They expect to optimize the use of the physical, financial, and human resources in order to maximize some revenue or profit return to the whole company. In this kind of an organization, following company standards and procedures means that things are working well and the company is operating effectively. If things aren't working well, then, according to the Scientific Management and human relations hybrid theories, it is because the manager has failed to modify the worker culture correctly.

These kinds of organizations are typical of many of today's factory systems. But these kinds of factory systems have a major problem: They don't produce new products or create improved services. They often can produce standard products very well. The workers are largely considered as resources, similar to money or machines. Creativity can be a problem when it disturbs production.

1.2 REENTER PROJECTS

The view of management described above began to be modified about the middle of the twentieth century. Consumer needs began to change very rapidly. The companies that responded with new and creative products and services succeeded. Those that couldn't failed. The factory that manufactured a cornucopia of standard products organized according to the scientific management–human relations theories couldn't change quickly enough. In spite of all its advantages as a production machine, it was too rigid. There had to be modifications to respond to rapid changes in the economic (and even political) environment. A sudden change in the market that made the standard "widget" obsolete or a new government regulation that declared the widget unsafe to operate had the same negative effect. Effective response and even anticipation of rapid change became all-important for economic survival, and the factory organization just couldn't do it fast enough. A new organization (and supporting theory) had to be developed to deal with these rapid changes. The new organization had to solve problems very quickly and then be dissolved, since there was no place for it in the ongoing company once its original purpose was accomplished.

Therefore, the project organization was reinvented. There is a major difference ·in its present use. Modern industrial and commercial organizations employ people who work using their heads, not their backs. These "information workers" (i.e., engineers, physicists, chemists, etc.) take a long time to train, are paid well, and have skills that are very difficult to replace quickly by the company. The expenses of hiring and firing these workers are very great. So projects are often organized from the ranks of the permanent information workers. When the project goals are achieved, those workers are reassigned back to their production or functional groups. In some cases, such as in *matrix management,* the workers remain under the authority of their functional managers and are temporarily assigned to projects.

Thus the older systems of management that worked well for the factory were supplanted, but they did provide legacies. The major legacy from Scientific Management was the idea of the trained and responsible employee-manager, the professional. The major legacy from human relations management was the realization that people control the quantity and the quality of work that they do. Both of these legacies are important to modern project management.

In many cases, project management worked well and was hailed as the new management system. The successes were well publicized as "universal" answers in the management journals of the fast-changing industries, such as those of aerospace and electronics. But not every company or every organization changes that quickly. And aerospace companies are not the same as insurance companies. The "universal" project manager or the "universal" project simply does not exist. Functional or production operations may be similar in many industries, but no two projects are entirely alike even within the same company. In earlier times when managers used the patriarchal model, the assumption was that it would fit everywhere. And in many companies that mass-produced products and services, this assumption was reasonable. The patriarchal model worked. But projects are, by definition, unique, and there is no "universal" model for the project manager. Creating something new is not like managing continuing factory operations. The project manager is concerned with building the factory. It's a one-time thing, not at all like managing the factory after it is built.

The project manager usually requires a team of creative, more or less independent project team members who, as information workers, solve problems and make nonrepetitive decisions within their own area of competency. On the other hand, the functional manager requires consistent, objective workers who follow predetermined organizational guidelines. The jobs of the functional manager and the project manager are different, and the necessary thinking process of the workers used in each type of organization are different.

2.0 THE ART, NOT THE SCIENCE (A Little Philosophy Here)

A project deals with an overall novel, unusual, and one-time project goal, and the tasks to achieve that end goal are not constant. Designing the foundations of a building is different from supervising the cement subcontractors who actually build it.

Designing is also different from managing the work of the plumbers, the heating-ventilating engineers, and the painters. And this difference applies to management. The project manager's tasks and roles change as the project matures. The relatively straight and repetitive tasks of the functional manager vary within fairly well defined limits. In projects, there is less science, less objective problem solving and more "art," or more subjective decisions, since the project manager (i.e., the "artist") and the situation (i.e., achievement of the project) are always interacting to create a new work of art (i.e., the new factory).

The ability to make appropriate decisions in unique situations is similar to the kind of adaptive thinking that an artist uses. Although the particular scene to be painted, the colors to be used, and the materials at hand may all be basic, the way they are used determines whether or not the work is a masterpiece. Management requires the same artistry. The raw materials of the project manager may be people, time, and capital, but the way they are combined, the way they are used, is new each time.

2.1 USING THIS ART

In my opinion, success in one kind of project management, for example, technical projects, includes all the problems and opportunities of *all other kinds* of projects. Technical projects contain all the problems of administration, design, development, procurement, construction, production, subcontracting, assembly, installation, field service, and customer contact. Therefore, while other industries may use project management in varying formats, if someone can successfully handle the situations in technical projects, they are probably very well equipped to begin to handle less complex situations.

A familiar design approach to learning technical project management is to start with general concepts and tools, then modify them to fit your perceptions of the situation. Which techniques will work together in a system to produce the best project goal? This provides a basic set of management systems. In other words, the "systems" are general groups of techniques or standard solutions to the more obvious or well-known problems of technical project management. These are similar to the basic systems of perspectives, colors, and drafting of an artist. But standards don't create a masterpiece. They are a necessary but incomplete set of resources. They have to be adapted to fit the situation, which is, of course, new.

Therefore, each of the systems described here should be redesigned to fit the unique situation in which you work.

2.2 DESIGN SEQUENCE

We will *not* follow the typical sequence of most projects. But just in case you wanted to know what that sequence is, it is

A. Evaluation or analysis

B. Design

C. Development

D. Procurement

E. Production or contracting

F. Testing

G. Implementation or shipment or installation

H. Field service

I. Project close down

Our goals are not limited by this one-time-through process. They require a continuing effort. They are to provide useful concepts to use in guiding any project sequence and to develop a supportive organizational environment in which any project can be successfully managed. These are moving, not stationary, goals.

Therefore, the projects themselves are not our major concern, although we will use the project sequence to illustrate various ideas. Our concern is centered on the project's management *processes*. In other words, it is determining and adapting project-oriented techniques and assembling them into systems. Then it is coordinating those systems into supportive frameworks and finally using leadership concepts that make those frameworks move. This effort never really ends. In one sense, we will be developing a more or less ongoing organizational environment whose purpose is to support the successful accomplishment of projects and whose systems and procedures are constantly under review and improvement. That's a tall order! Projects may end but the process of improving the supporting systems and leadership never does.

Therefore, we will be process-oriented. We'll analyze, develop, and adapt appropriate techniques as building blocks for systems that help to solve the typical problems of most projects. (While each project is unique, many of the problems that face them are not.) Then we'll explore the leadership concepts that help to solve the nonrepetitive project problems in a repetitive way. (By definition, nonrepetitive problems *may be* unique, but there are practical and repetitive ways to handle them, such as the scientific method. We'll deal with that a bit later.)

Our sequence will be to move from general techniques and systems to specifics. My reasons for this are:

1. *Every organization is a political system, not just a collection of individuals.* The number of variables impinging on it is so great that predicting the precise outcome of all of them is impossible. Over time, however, we can observe those variables that seem to affect it more. For example, one company may temporarily be primarily concerned with sales, irrespective of the consequent effect on profits. Another may be concerned with advancing the state of the art.

Understanding which variables are important and responding to them is a primary consideration in any organizational systems design. A sales-driven company will not easily respond to any system that recommends increasing profits by cutting off the lower volume sales distributors and concentrating on increasing profits (not sales) for those that remain. Similarly, a scientifically oriented company would not easily be able to accept a recommendation to cut down on product diversity and emphasize product standardization. In either case, the solutions may be logical but the politics may not be.

Therefore, any new system must reflect the dominant variables of the organization.

2. *Most of the important variables affecting the organization remain relatively constant over time.* The political system changes very slowly unless the organization suffers a sudden unexpected change. Then it will be more likely to respond to the new variables in that change. The change may be negative, for example, a sudden drop in sales of an important product. It could also be positive, for example, selling off a losing plant at a profit. Sudden changes may be opportunities in disguise.

Therefore we should determine if there is any pressure for improving project operations. If there has been a relatively sudden change in the economic or organizational climate of the organization, the potential for these systems to be successfully implemented might be greater. Where is the organizational source for the requirement to change? The higher in the organizational hierarchy for the source, the more likely that power will be exerted in support of those variables. Is the need for improved project operations constant, or has it occurred as a result of a change? Who "feels the pain"? The source of the need for the change and the organizational positions of those behind it or opposing it must be defined at the beginning. This means analysis before development and acceptance of general systems and concepts.

Most individuals in the organization are aware of the important variables that affect it and accept them most of the time. Every organization has an informal, but effective, hidden hierarchy of power. For example, in a marketing-oriented organization, it is reasonable that cost-reduction improvements to company products are not as important as the development of an expanded sales staff. Company budgets will reflect this, and everyone in the organization will generally respond appropriately.

Therefore, when change comes, it is evaluated by everyone. The change can either be disturbing to people or it can be invigorating. Will people involved in these new project systems be rewarded for "success" or possibly punished for "failure"? What will be the definition of success and failure? Change provides opportunities to develop and use new systems.

Most of these new project systems will not follow the usual organizational hierarchy because the projects themselves often cross organizational boundaries. There must also be provisions to adjust the existing functional organization's procedures when they interact with project systems. Project systems will impact functions. The major variables to which the organization responds and the positions of people affected by new systems must be factored into any design.

3. *The effect on the organization of the variability of individual behavior is increased as the organization decreases in size.* As the size of the group decreases, the contribution of each member becomes more visible and it can be more important. Consequently, when a project team is organized, the general procedures of the parent organization are usually modified slightly to fit the particular project goals and those of the project team members. The situation changes and the systems must be adapted accordingly. Individual creativity and commitment are necessary to solve nonrepetitive project problems. Systems are basically intended to help solve repetitive ones.

Taking the three reasons listed above into account, we'll design the overall or basic organizational project-supporting systems using very broad, noncontroversial ideas that minimize potential trauma. For example, some of these ideas would include using responsibility accounting; clear delegation of responsibilities; and specific internal agreements on technical, time, and cost project goals. These ideas can't be faulted in any organization that wants to use project management. If they cannot be accepted, it's better to gracefully withdraw from the field of combat right away. Organizational politics won't allow the battle to be won.

When the ideas described above are accepted, the project manager can design the more variable operating practices and systems that apply more closely to individual projects. Although these practices must be in line with the overall organizational environment, they will be more modifiable since each project situation is somewhat unique. These deal first with project teams in small groups and then with individuals.

For example, the engineering, quality assurance, contracting, and production teams might each have different project-oriented goals. Engineering is to design the state-of-the-art product. Quality assurance is to see that the product meets the customer's specifications. Production is to manufacture or contract that product as inexpensively as possible. The procedures for project team processes typically include methods for improving coordination between teams, for instance, standard reporting and meeting formats.

Then we consider the individuals themselves, the project engineer, quality manager, subcontracts administrator, and the manufacturing supervisor. They are different people motivated differently. The procedures now might include personal goal setting and specific reward techniques when relevant project goals are achieved. These goals are intended to optimize coordination among team members and improve individual commitment.

Finally, we'll move on to the most variable system of all, that which applies to the leadership, ethics, and management behavior of the project manager. I call it the personal management "style" of the project manager. This system will typically include suggestions for appropriate leadership techniques and making those difficult nonrepetitive decisions that cannot always be forecasted.

3.0 IMPLEMENTING SYSTEMS

Although this sequence of working from the general to the specific helps general ideas to be accepted easily by large groups, implementing the systems using these general ideas is best done in reverse, using small groups. For example, the general idea may be that "... everyone is entitled to life, liberty and the pursuit of happiness," but how does this apply to the person who steals a loaf of bread to feed his or her family? Is it the same as the engineer who overlooks a product redesign intended to eliminate a potential danger to the user because top management is so set on getting it to market next week? Whose life and liberty are we discussing? I feel more comfortable discussing theirs than yours, and yours than mine. When it gets to specifics, it's not as easy.

But specifics can also present opportunities. When implementing systems in small groups, it's easier to understand the politics. They're more obvious, and the ease of perception can lead to successful strategies. Then when the systems are successfully implemented on a small scale, that project can be a model for other projects within the organization. Of course, the systems for those other projects might be slightly modified, but the intrinsic procedures and ideas remain unchanged.

But even success can eventually fail when there is no corresponding redesign and reimplementation of improved project systems as the organizational politics, the economic and the legal environments, or the internal hierarchy changes. That change may be slow or fast. When it is slow, there is a subtle, continual updating or evolution of older systems. When it is fast, it results in almost a project type of major change through which older, less effective systems are redesigned and reimplemented. For example, a major change in project cost systems can occur when the project manager or top management suddenly realizes that the project has overexpended the original funds. This can happen without warning when the existing cost system is ineffective. In any event, the improvement process cannot stop, just the rate of change.

3.1 THE STANDARD SYSTEMS AND HOW TO DEVELOP THEM

In my opinion, five major systems are sufficient bases for successful technical operations: (1)Forecasting, (2)Measuring, (3)Project Operations, (4)People Management, and (5)Personal Style. Two of them are applied generally across the organization, and two are specific for the project. The two that are organization-wide can be used with little modification throughout the company for almost every project. They are

1. *Forecasting:* The methods used to solve and document today's answers to the potential problems of tomorrow.

2. *Measuring:* Reporting how well the answers that appear later match those problems.

The two that are project-specific must be more modifiable. The closer we get to the project itself, just as with the individual, the less predictable it becomes. As groups of people or groups of projects within an organization become larger, the group behavior is more predictable. Individuals do not behave the same but groups usually behave as an average of all the participants.

The two project-specific systems are

3. *Project Operations:* Organizing the project team and providing communication links that assist prompt, corrective management actions. This deals with people in small groups.

4. *People Management:* Developing the individual's commitment to project goals and dealing with people as individuals.

All of these systems interact and affect one another. They are not self-contained and independent.

The fifth system is the most important of all. It's also the most variable since it's concerned with only one individual, the project manager. It is

5. *Personal Style:* The project manager must develop his or her own appropriate management behaviors to fit the ever-changing situation of the project environment.

This last system is usually the most difficult to design since it is expected to determine the best behaviors to use in handling those nonpredictable, nonrepetitive decisions that every project manager faces. In other words, it's a repetitive systems approach to nonrepetitive decision making.

We'll start our design with the most important "basic" of them all: the project manager's formal authority. It's the *foundation* for all systems. We'll then cover these five interacting systems in sequence. We'll move from the more objective to the subjective. We'll improve systems design (or adaptation) skills as we progress from the more solidly based forecasting-estimating methods through measurement-accounting methods into the more variable project operations and motivation theories of people. Then we'll handle recommendations for the "artistic" personal styles of the project manager and how to use them. But as we follow this design process, there should be a parallel adaptation process being completed by the project manager to fit these recommendations to his or her specific situation.

3.2 ADAPTING SYSTEMS (Details and more details)

The adaptation process should be iterative. *Iteration* is the review, reforecasting, and redirection of a management plan. As different unforeseen problems arise and have to be solved, iteration provides for reevaluating and rerunning a test or a program. It helps to correct small discrepancies before they grow into major errors.

It is an important technique of project management. If we knew exactly what the end goal was, how much time it would actually take, and the final cost, we would never have to iterate. The wise project manager allows for many interim reviews and iterations since it is impossible to really know *everything* before beginning. Iteration, therefore, is almost a "given" in projects. To start to build that individualized framework, we'll use the familiar Scientific model of learning. That model is

1. Define your terms
2. Develop the theory (or the general patterns)
3. Set up the hypothesis (and how you will test it)
4. Test it (and if the results are satisfactory, your theory is supported; if not, go back and start all over again)

A variation of this model is as follows:

Define.

"*Exploration-* The individual begins to ask is there a problem? What kind of a problem? What do I need to know in order to be able to deal with it? Without some form of exploration the individual will be reduced to taking it on trust that the dogma of the teacher is both apposite and useful. . . . "

Develop.

"*Conceptualization-* The nature of the problem having been explored, the individual must conceptualize it. He must learn how to set this one experience of the problem in a more general context or framework. If this is done, he will be able not only to explain the first problem but all others like it. Conceptualization elevates the particular to the universal. . . . "

Test.

"*Experimentation-* . . . better understanding leads to better prediction, which could lead to more effective action. It is this action consequent on conceptualization that is here called experimentation. Here the individual as scientist is testing his hypothesis. . . . "

Satisfactory?

"*Consolidation-* This is the final stage. The concepts are internalized and begin to mesh in one's mind. The experimental phase is past, and the new hypotheses become the basis for future action. They have become part of one, so that habitual behavior is altered and affected. The lesson has been learned . . . " (Handy, 1981, p.21)

Therefore, we'll start with "definition." One of the first definitions is that of a *system*. It is usually defined as an assembly of parts that has an overall purpose. There is another, perhaps more appropriate definition: Systems are the basic, repetitive ways of problem solving that every organization establishes to handle problems similar to those that were handled in the past. In other words, systems are assemblies of tasks or procedures whose purpose is to solve *repetitive* problems. The systems themselves may be documented or not. When documented, they may appear in an operations sheet or a project manual. But even if they are not documented, they're still available when they are part of the organizational culture. In that case, they are transmitted orally. In all cases however, those systems are usually available for use by every project manager in the organization. Usually, one quickly learns "how things are done around here."

4.0 THE UNUSUAL SYSTEM: PERSONAL STYLE

Personal Style is also a system. It is the behavioral or visible output of some parts of your thinking processes. It is the *predictable* way that you, as the project manager, handle day-to-day *nonpredictable* problems. We can't see the working of the inner system of our own "thinking," so we'll deal with the outward result of behavioral "style" because this is observable. But a word of caution here. When observing behavior in others or else demonstrating it ourselves, we may not always clearly understand the thinking process behind it. Therefore, the other person's "thinking" is really our interpretation of the observation of the other person's behavior. Management Style is a most complex management system since it's always interpretable and much less repetitive.

5.0 THE MANAGER AS DECISION MAKER

Human beings have the unique ability to make decisions in novel and unusual situations. Those decisions can be based on inadequate, incorrect, or missing data inputs. (It's amazing that sometimes those decisions are even correct.) Therefore, they can manage organizations and situations that they have not experienced before. Consequently, managers must be decision makers. The decisions they make should be nonrepetitive. Repetitive decisions can come from computers. Managers, moreover, are decision makers who are expected to make the "right" decisions *most* of the time in *every* situation. We'll be concerned with systems intended to improve that decision-making process, to minimize the unknowns.

6.0 SUMMARY

In this very brief introduction, we've covered some of the background and history of management. Throughout most of history, projects were the only way to coordinate large groups of people to accomplish a task. When continuous operations became economically justifiable, the factory and functional managers took over. Within recent times, projects have again become important as short-term solutions to product and process changes and to creating new factories and equipment. Because designing and building a factory is a one-time task, it requires a high level of management "artistry." There has to be more commitment, creativity, and conversely, tighter controls on expenditures than in functional operations. You only get one chance to do it right in projects. In functions, there is always the possibility of doing it again next year.

To briefly review, there is a foundation of management authority upon which five systems that are basic to any technical project management are built.

1. Forecasting
2. Measuring
3. Project Operations
4. People Management
5. Personal Style

Management authority is the amount of freedom to act independently that the manager has. Systems provide repetitive processes to support problem solving. Managers are decision makers who use systems to classify and categorize problems before making nonrepetitive decisions that solve them. The important variables in decision making change as the project progresses. The familiar Scientific method provides us with a base to develop the next iteration. Revising or providing a project iteration is not the same as a "mistake"; it is more like an "improvement" that must be done because the situation has changed. In later chapters, we'll even be "iterating" or reviewing ideas and designs that were discussed in prior sections of this book. Iteration will become a familiar process.

Part I includes the foundation of management authority and the first two major systems: forecasting and measuring. The chapters in Part I are Chapter 1 (Definitions: The Important Beginning), Chapter 2 (Forecasting: Starting with the Charter), Chapter 3 (Forecasting: Techniques), Chapter 4 (Forecasting: Tying It All Together), and Chapter 5 (Measuring: The Second System).

Part II covers the second two major systems: Project Operations and People Management. The chapters in this part are Chapter 6 (Project Operations: The Third System—Structures and Teams), Chapter 7 (Project Operations: Building the Team), and Chapter 8 (People Management: The Fourth System).

Part III deals with the project manager as part of the least definable fifth system, Personal Management Style. It includes Chapter 9 (Personal Management Style: The Fifth System) and Chapter 10 (Getting Things Changed—Handling Stress and Ethics). The appendices at the end cover some of the basic management techniques that are useful in almost all projects:

Appendix 1:   One Way to Evaluate Projects
Appendix 2:   Project Management: A Goal Directed Approach
Appendix 3:   The Technique of Design Review
Appendix 4:   Program Evaluation Review Technique
Appendix 5:   Line of Balance

Chapter 1   Definitions: The Important Beginning

PERSONAL MANAGEMENT STYLE: THE INDIVIDUAL SYSTEM

PEOPLE MANAGEMENT: THE PROJECT SYSTEM

PROJECT OPERATIONS: THE PROJECT SYSTEM

MEASURING: THE ORGANIZATIONAL SYSTEM

FORECASTING: THE ORGANIZATIONAL SYSTEM

FOUNDATION: PROJECT MANAGER'S AUTHORITY

THE CASE OF THE DYING PHOENIX PROJECT
OR HOW *NOT* TO START A PROJECT

Cast

Charles Johnson, Vice President, Engineering, Acme Industries
George Winslow, Project Manager, Phoenix Project
Marge Bester, Manager, Quality Assurance
Peter Slifer, Design Engineer

The Acme company was a manufacturer of special machinery for the food-processing industry. A few years ago, a decision was made to expand, but something seemed to go wrong, especially in the Engineering department. As the work force expanded, more errors began to appear. Many were caught before the designs got out of the Engineering department. But some weren't discovered until manufacturing began. That was expensive. Then errors began to appear after some equipment had been sold and installed. Something had to be done. The company increased the number of field service personnel. That wasn't enough. There were some improvements but not all the errors were eliminated.

The final blow came early on a Friday morning. One of their latest model food-processing machines had stopped functioning three days ago at one of their most important customer's plants. Acme's service engineers discovered that a crucial electrical drive motor had to be replaced. They also found that it was not one of Acme's standard motors. It had been supplied for this one machine as a nonstandard component by a vendor who had since gone out of business. It had taken some creative work over a long weekend to temporarily install one of Acme's old motors that had been found in the company's inventory.

When Charles Johnson thought about it, he realized that this was not an isolated instance. Something had gone wrong in the engineering department. The field installation people had told him that shipments were often made with some installation diagrams missing and parts that were nonstandard. Engineering didn't keep their delivery promises to other departments as before, and designs weren't thoroughly debugged before they were put into manufacturing. The whole company could be in big trouble.

Charles had an inspiration. George Winslow had been with the company for over four years. George was an excellent engineering team leader and was due for a promotion.

Charles called George into his office, that afternoon.

Charles: George, our problems are growing. Our present management systems can't seem to handle the load anymore, and I've been reading a lot about this new way called *project management*. Maybe this is a good time to start project management ourselves.

I'd like you to be our first project manager. Take charge of redesigning the way Engineering operates. There are problems in design, standardization, and quality. It's a big job. I've named it the "Phoenix" project because it will give our company new life. Find out what we have to do to change and improve things, and then do it. I've written this memo to all the department heads informing them that you're in charge of the "Phoenix" Project. I've put together a file showing the types of problems that we've had and their frequency. I've also noted which equipment designs are giving us the most trouble. That should define things. Just go out there and get this thing resolved. Oh, by the way, try not to let your regular work down because we'll need those conveyor designs within the next six weeks.

George: Charlie, I really appreciate this chance. I know there's a lot of improvement that can be accomplished but I can't do it alone. How can I get some of those other groups that are not in Engineering to cooperate with me?

Charles: This memo appointing you as project manager should solve that. It will be circulated throughout the company. I'm sure all the other vice presidents will agree with it.

George: O.K. By the way, what should I do if I have a conflict between my regular engineering job and being project manager?

Charles: No problem, I'm sure that you'll handle it.

Charles stood up and mentioned something about an important meeting down the hall. George quickly stood up and walked back to his office. What a promotion! He was really pleased.

Six months later, George's feelings had changed. The Phoenix Project had become unsolvable. Everyone was willing to tell him what was wrong but no one wanted to cooperate when the time came to do any work. His old job as Design team leader was falling behind. Disaster was ahead. It got very bad after he met with Charles and all the department heads at the monthly management meeting. He gave each department head a copy of his report at the meeting.

The meeting almost broke down when the report was discussed. The Drafting Department manager objected to suggestions for handling design changes because the paperwork was not the same as they had been using for the past 10 years. The new procedures were going to take too much time to implement and the drafting department was overloaded right now. How could they handle that increased load?

Then the food industry machine design manager made it very clear that she didn't like George's idea for a central testing and acceptance laboratory for the whole company. She preferred the present system. Every design group did their own testing, and she liked that. She pointed out that processing apples wasn't like processing tomatoes. They were too different processes. In her opinion, standardization across the company wouldn't work.

She also didn't like the suggested computer-assisted design system either. She had heard that Jones company down the road had tried computer-assisted design and it had failed. She also said that she couldn't spare any people to set up the necessary design standards.

The last straw came when they read George's conclusions. The project was estimated to take two years. It would have a budget about 25 percent of the total Engineering department budget for one year. And it would probably require more people to run the new systems for the first two years after implementation.

After the meeting, Charles asked George to stay after the others had left.

Charles: George, this report and the meeting were both disasters. I was so sure that you could handle this project, and it looks like you're going to fail me. Maybe setting up a project manager to get this thing done wasn't a good idea after all. This Phoenix Project is a serious test of our future ability to manage. Our present situation is completely unsatisfactory. More time and money aren't enough. What are we going to do that's different next time? I need some recommendations from you, and I need them within a week. This meeting is over.

George walked back to his office, sat down in his chair, and stared at the report. The answers didn't leap off the page. Peter Slifer, one of the design engineers walked by, noticed George just staring at the report, walked into the office and sat down.

Peter: Hi, what's going on? I noticed that the management meeting didn't last as long as it usually does, and my boss mentioned that Charley Johnson said something about getting rid of this whole project business when he came out. I sure hope not. Our engineering group has been operating in fragments for years, and it's about time that we all got together. Each design group puts new components into our products because they're the latest state of the art without telling anyone else. When the Purchasing department finds out about it, they complain. They say they have qualified vendors supplying equal components that are listed in the standard parts manuals.

And that's only one example of what goes on around here. Our drafting group is a bunch of prima donnas, who won't make a mark on a print until you go through a big requisition and scheduling process with them. Even Quality has dropped because. . . .

Just then Marge Bester walked by, happened to hear Peter's last statement, and walked into the office.

Marge: Yes, Pete, why has quality dropped? (There was an ominous look on her face.)

Peter: Probably because your group doesn't cooperate enough with the Engineering department, they just tell us what to do and never help us do it, and another thing. . . .

George interrupted them.

George: This sounds like the management meeting. Can't anybody agree to work with anybody else?

Marge: Look, George, telling people to cooperate doesn't mean that they will. I used to work in a company where they had quality "control". There we would tell people when they made "bad" products and when they made "good" ones. It hasn't worked and it never will. They still make "bad" ones and "good" ones. We're now going to something different called quality "assurance." That means we're going to help people to develop and follow a *process* of doing things right. We've given up on the idea of trying to inspect quality after the fact. Why not try the same idea with your project?

George: That's a great idea. I'll develop a systematic *process* approach to this and all company projects. Perhaps I can outline a project manual that will be a kind of plan or process to follow when completing my project. It will outline the project manager's responsibilities, the controls, the forecasting, and measuring systems, and all the interactions that are needed to complete this project. I'll send it to Charley Johnson as the report he asked for. If he approves it, I'll call another meeting of the major team members. This time, I'll ask them to review the manual in specific detail. Let them separately modify it where they think it should be modified. Then we'll get together and revise the manual to fit these modifications. It won't be perfect, of course, but the general ground rules and the processes will be there. At least we'll be able to understand who is really responsible for parts of projects, what they are supposed to do, how they are to interact, and what the outcomes should be.

It's like learning to write a contract or a technical product specification. You have to know the clauses and descriptive systems before you can do it right. Why not do the same thing in projects? We have to define the processes first. We might even be able to use it for other projects.

Marge: That's fine except what happens when another project is started that's entirely different from this one? And what happens when we have projects of different sizes? And finally, what happens when another project manager or project team gets going? These are all differences that no standard project manual can handle. What then?

Questions

1. Do you think that this project can be saved now? How? If so, who should do it? If not, why not? What should be done?

2. Can your answer to Question 1 be implemented now? At any time in a project? Are there any special times? Why?

3. What would be your answer to Marge's point about not being able to use a standard project manual for a company?

4. What should George do about the requirement to keep up with his present work in addition to taking on the project management job?

5. Is this a familiar situation? What happened?

Note

A case study really only applies to a particular situation, but look for the underlying ideas that you can use, not the story itself (unless, of course, the unusual has happened and the situation has really been repeated). Those underlying ideas come from applicable research findings.

Try to answer the questions before you read my suggested answers. Sometimes we learn by seeing differences between our perceptions and those of others.

1.0 WHAT WE COVER

We start with definitions, which are the basics of management. Then we'll distinguish among varying organizational levels and between functional and project management. We'll learn the effects of being promoted and changing from technical expert to manager. Management involves decisions and some of them are repetitive. Those repetitive decisions are designed into organizational systems. Project systems design follows organizational design. The foundation of those systems is adequate project management authority. There are potential risks and benefits that accompany this design process. These are outlined.

2.0 DEFINITIONS

Most definitions of management begin well but finish poorly. They begin by referring to a process, not an end goal. They are often inadequate because they don't differentiate among different levels of management. As an example, consider these two common definitions for management: (1.) "coordinates the work of others," and (2.) "plans, organizes, directs, and controls." These definitions seem to assume that the general manager is doing exactly the same thing as the assistant production manager. That's not true. There are many differences between the tasks of those two managers, but these and most definitions that I have seen don't address them.

However, if we define management as a decision-making process within an organization, we can then begin to classify the different levels according to the decisions made. I believe that the major classification device is a subjective concept called *uncertainty*.

2.1 MANAGEMENT CLASSIFIED
BY UNCERTAINTY—IN THE ORGANIZATION

We'll use the following definition: "Management is an inferred process of making nonrepetitive decisions by *absorbing uncertainty*. The decisions made usually involve people." Decision making is inferred because we can't observe it happening. All we can do is observe the results. We then evaluate them according to our own interpretations. The decisions are nonrepetitive because all other decisions are supposedly in the company's standard operating procedures and its culture. The standard operating procedures include documented, organizationally accepted answers for similar situations. One definition of the culture is the nondocumented, but equally known and organizationally accepted answers for these similar situations. The standard operating procedures are often explicitly defined in some company manual. The culture may not be explicitly defined, but it's just as real. When those systems are adequate, the manager only has to deal with unique problems. There is minimal waste of valuable management time and effort. But in dealing with these nonrepetitive problems, the decision maker often receives input data that includes inadequate or incorrect information. Which data will be suspect is not known. Some data may even be missing. What's missing is also not known. But decisions must be made anyhow, and they always involve other people because those people provide most of that input data.

The lone researcher seeking the elixir to change garbage into gold is not a manager, according to this definition. The amount and correctness of the raw data are fixed by Mother Nature and the perceptions of the researcher. It's a very private arrangement. If the researcher is clever enough to ask the right questions and understand the answers, the results are invariably logical and consistent. There are no other people involved. This changes when the researcher becomes a team leader and has to deal with subordinates. Asking clever questions correctly, then, may or may not result in logical, consistent answers. The individual attitudes of team participants can color answers inconsistently. So can the thinking processes of the decision maker. Managerial decision making, therefore, involves both the inferred thinking of the decision maker and those of others, as interpreted by the decision maker. Purely scientific decision making does not.

Generally, decisions can be made under three conditions: certainty, risk, and uncertainty.

1. *Certainty* is complete and accurate knowledge of the consequences that follow on each alternative available to the decision maker. Those decisions don't even need humans after the first decision is made. A computer can handle them.
2. *Risk* is defined as correct knowledge of the probability distribution of the consequences of each decision. When the decision maker has that, logical decisions should be based upon an objective weighing of gain versus cost. If that doesn't happen, the decision may seem to be based in risk but it's really based

in uncertainty since the decision maker's emotions, not logic, are the important consideration now.

3. *Uncertainty* is a set of consequences for each alternative that the decision maker thinks exists. However, it is impossible for the decision maker to assign definite probabilities to the occurrences of the particular consequences. (Adapted from March and Simon, 1958, p.137)

If the decision maker feels that there really was no probability distribution, all decisions would be equally acceptable and the decision maker could make one at random. That's not what happens. The decision maker, who is a human being, "knows" (or thinks he or she knows) subjectively that a probability distribution exists but doesn't "know" exactly what that distribution is. Therefore, it's impossible to definitely predict the probability distribution of any decision made. But it's made anyhow. It's a fairly common process of everyday life, which, when we think about it, is really very unusual and wonderful.

A trivial but illustrative example of this is the housewife who buys vegetables that look and smell good but she's not sure of their quality until she gets them home and starts to prepare them for dinner. The probability distribution for their "goodness" could vary anywhere from completely acceptable to not acceptable at all. However, if she feels that there's a very close correlation between outward appearances and usefulness, she acts accordingly. If there were no correlation, her probability distribution fails. If there really was a verifiable probability distribution between outward appearance and usefulness, that would be risk, not uncertainty. But even though she's not sure of the probability distribution, she thinks that it exists and acts accordingly. Tomatoes should be large, red, and slightly soft. Carrots should be stiff, not too large or small, and so forth. As we all know, these criteria don't always apply, so she's not using "risk" but "uncertainty."

Decisions made under certainty don't require human intervention. Programming a computer will do it. Making decisions under conditions of risk is basically a subjective evaluation of the objective reward-cost ratio of that particular decision, since we always have the correct knowledge of the consequences. For example, we know objectively that the number 4 will show up one-sixth of the time if we throw one die of a set of honest dice long enough. We then have to decide subjectively if the potential reward of a particular throw is worth our evaluation of the risk of a lost bet. Risk is usually related to an individual. Some of us are risk takers and others risk avoiders. (Most engineers are risk avoiders. I like to believe that when I'm riding in a commercial airliner designed by conservative engineers.) But although we, as individuals, may objectively know the final reward-cost ratio, it's how we personally value it that's different for each of us. It's probably based in our upbringing and, in general, the kind of person we are.

Uncertainty is very different. I've proposed that the amount of uncertainty that the manager has to absorb seems to vary directly with the position in the organization. The higher the position, the less objective the available data is and the greater degree of uncertainty the particular manager must absorb, either cognitively or emotionally.

2.2 UNCERTAINTY—OTHER FACTORS

In addition to organizational position, there are other factors affecting uncertainty that are associated with the individual managers themselves and with the different kinds of organizations. For example, it's obvious that *no two decision makers are alike*. An entirely novel problem may be solved differently according to different people's mental maps. This has little to do with the actual number of variables but with those that we *select* as important. Emotional probability distributions are based on who we are and what has happened to us in the past. Then we have to consider that in every managerial situation *we have many situational variables and few controls*. The number of variables that affect managerial decision making is much greater than those in any reasonable scientific experiment. Many of these variables are never known by the manager. The best that can be done is to subjectively assume that the relevant variables are known and develop repetitive systems or answers as standard operating procedures. There are few absolute controls. And even when those systems do work, they become modified with time. Things change.

Obviously, uncertainty is associated with both a particular position and with the person in that job. We can deal more easily with the details of a management position here in this book because the goals (and the variables for decision making) are relatively limited when compared to those of a person's heredity and experience. The latter are beyond the scope of this book. Therefore, we'll confine most of our discussion to the organizational uncertainty level. How far we progress as individuals in benefiting from this discussion depends, of course, upon us. One of the purposes of this book is to increase one's knowledge and consequent capacity to make better decisions. Using this knowledge is a personal task.

3.0 UNCERTAINTY CLASSIFYING THE JOB

Using the idea of uncertainty, we can now begin to compare one managerial job with another and differentiate among them. As noted before, as one goes higher in the organizational hierarchy, I feel that there is an increase in uncertainty. This seems reasonable because there is usually less hard, objective data as one climbs the organizational pyramid. It is also reasonable that the consequences of those decisions will affect larger parts of the company. Uncertainty and responsibility must travel together in any organization. In other words, decisions and the organizational effects of those decisions appear to be directly related to the manager's organizational position.

This relation of increased managerial responsibilities with increased "uncertainty" happens in many conventional organizations that are *not* project-oriented. At lower organizational levels, the policies, procedures, and practices are relatively clear and the level of uncertainty under which decisions are made is fairly low. As one rises in the hierarchy, there are fewer procedural guidelines and less factual or objective data. There's more judgmental or subjective decision making. There may be many more inputs and reports provided to the company president, but the recommendations in those reports invariably contain uncertainty. If they were based on unequivocable, hard, objective test data, there wouldn't be any unknown probability distribution and no real need for management decision making. The answer would be digital, not analog.

Some uncertainty may be displaced into risk as experience is gained, but the effect of those uncertainty-based decisions will still be great. Let's look at a classical, pyramid-shaped organization chart to see how it works.

Figure 1-1 shows a conventional pyramidal-shaped organizational structure with Tom as the chief and two subordinate managers, George and Peter. I have assumed that there is a decreasing curvilinear relationship between "uncertainty" and "frequency of decisions," because it seems to be empirically correct, in my experience. The higher the position, the fewer the decisions. These decisions are made under conditions of greater uncertainty and they can have a greater impact upon the organization. For example, the CEO (chief executive officer, Tom in Fig. 1-1) might decide if and when the company should expand into large highway-type earthmoving equipment as a future growth pattern, building upon the company's present experience in building smaller equipment for building contractors. Similarly, the chief engineer (Peter in Fig. 1 1) might be concerned with many more decisions during the same time period, but these decisions would not have the same overall impact as Tom's. The chief engineer's decisions might include approving a relatively new vendor as a potential supplier of hydraulic pumps. Peter would also be able to confer with other managers such as the purchasing manager or the quality assurance manager in order to reduce his uncertainty even more. Tom has no equivalent managers to help reduce his uncertainty.

In a reasonably well-managed organization, fewer decisions will be made at

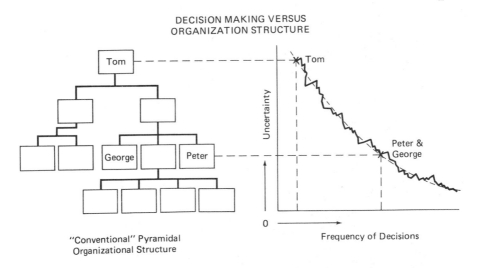

DECISION MAKING VERSUS
ORGANIZATION STRUCTURE

"Conventional" Pyramidal
Organizational Structure

Figure 1-1. Decision making versus organization structure

upper levels, those that can't be handled at lower levels. The uncertainty entailed in decision making and the size of those problems should be greater than the uncertainty-absorbing capacities of everybody below those upper levels. But, as noted before, decision making is appraised by the results, or lack of them, because *how* it happens is almost an unknown process. Our own thinking processes are somewhat like a "black box" to us. We can define some of the inputs and observe the outputs, but we're not quite sure what's happening inside. Therefore, this inferred uncertainty-absorption process is our interpretation of the adequacy of the outputs, given the inputs.

And unless someone else knows more or has better experience than the manager who makes the decisions, no one is better equipped at that *time and place* to make a better decision. Assuming that a decision is made at an appropriate organizational level, if someone later on asks, "Why did you do that?" The answer, of course, is "It was the best decision that could have been made then in that place." The proper question, of course, is "How can we improve future decisions, given what we've learned?" not "Why did you do that?"

3.1 PROJECTS DEFINED

Project structures are not the same as those of conventional functional organizations. The uncertainty at the beginning of a project is usually very high. At the end, it's very low. Initial data are usually sparse, and it's difficult to completely determine needed resources. As the project proceeds, the data input becomes clearer, uncertainty begins to decrease, and nonrepetitive decisions decrease. But they never completely end. In other words, the beginning is highly uncertain, and this uncertainty decreases at the end of a successful project, but it never completely disappears. No project closes with all the potential problems answered. We know that all the variables can't be considered for the obvious reason that they're not always known. If the inventor of the pesticide DDT knew all the implications of his invention before marketing the product, he could have minimized the potential damage caused to the environment. But he didn't know at that time, so the best decision was made at *that* time.

Therefore there are always unknowns in projects and everyone implicitly understands this. But if this is so, why use them since there's always so much uncertainty? One reason is that it's the best way we have to handle very unusual problems and organize the efforts of specific groups of people. Projects can do things that functional organizations can't. Companies turn to project-oriented structures

1. When it is absolutely essential that they be highly responsive to two sectors simultaneously, such as markets and technology.
2. When they face uncertainties that generate very high information processing requirements.

3. When they must deal with strong constraints on financial and/or human resources.
(Davis & Lawrence, 1978, p.134)

But these project-oriented structures must be designed to minimize unforeseen problems. They should have few restrictions placed on both the project manager's freedom to define problems and pursue solutions and those of the project team members to contribute. They should have the following criteria:

1. It has a budget of its own.
2. It has a clear purpose tied to larger purposes of the organization.
3. A formal mechanism exists for making the output of the project (or matrix) group an input to the rest of the organization.
4. The project manager has as much authority as functional managers do.
5. People are given released time from functional tasks to work on project tasks.
6. Successful project work does figure in raises, bonuses, and/or promotion.
(Weisbord, 1978, p.29)

In other words, the project structure must be designed to promote investigations and solutions in a nonpunitive environment. There are only three basic, simply defined "golden limits":

a. technical achievement
b. elapsed time
c. cost.

They are called *golden*, because they circumscribe or outline the project's boundaries. Any solutions that don't affect these limits could be acceptable, but exceeding those boundaries requires corrective action at a higher (i.e., "golden") organizational level.

Projects have a specific organizational structure that is created with the understanding that it will be dissolved at completion. (Therefore, a "project" with just one person in it is not really a project as I have defined it.) The design of the project structure within the overall organization probably has two different axes: (1) the regular organizational or functional axis and (2) the project-oriented axis. According to this definition, a "task" with several people within a department wouldn't be a project either. The levels of uncertainty would be limited enough to be adequately handled by the functional manager of that department. Drawing a project structure onto a company chart would show a vertical axis (the functional) and a horizontal axis (the project). (Figure 6-1 is a representation of these two different axes.) Not only the design of the project structure, but the building and dissolving processes for these structures are complex. With that complexity, there is often confusion when information must be passed along.

The essence of project management is that it cuts across, and in some cases conflicts with, the normal organizational structure. . . . Because a project usually requires decisions and actions from a number of functional areas at once, the main interdependencies and the main flow of information in a project are not vertical but lateral. Up-and-down information flow is relatively light in a well run project; indeed any attempt to consistently send information from one functional area up to a common authority and down to another area through conventional channels is apt to cripple the project and wreck the time schedule. (Stewart, 1965, p.274)

3.2 FUNCTIONS VERSUS PROJECTS: TASKS, STRUCTURES, AND MANAGERS

Functional managers don't have the same limits on their tasks as project managers do. Since their overall jobs go on "forever" (even though day-to-day tasks may change), they have almost no time limit on the life of their group. There is an ongoing stream of specific tasks to be accomplished, but these are within a limited range of uncertainty and within the capability of the group. Some typical functional groups or departments are Engineering, Purchasing, Shipping, Sales, Manufacturing, Quality, and so forth. These groups are almost "eternal" and continue as long as the organization exists. These functional managers have a primary responsibility to *train* their personnel. They are important as developers of people.

Coordination among functional groups is intended to be done formally by the next upper level of management; e.g., functional sections within departments are coordinated by the department manager, and functional departments are coordinated by the general manager. However, it's not the same with projects. Project managers lead a group that is intended to have a limited life as a group. It is not "eternal." Projects are organized to solve a problem that is considered to be beyond the uncertainty limits of any one functional group. Not only do projects have a limited life, they usually have very limited budgets and defined goals to be accomplished within that life span.

Although both types of managers can be responsible for coordinating across groups (functional managers across groups within departments and project managers across team managers), there are two major differences that distinguish between these two general types: familiarity with the work being done and training subordinates. These differences contribute to the levels of uncertainty that these managers must absorb.

3.3 FAMILIARITY WITH THE WORK BEING DONE

Functional managers can be quite familiar with the work done in their group because that work is relatively limited in its complexity. For example, the chief engineer may not be immersed in the latest design of the newest cement pump, but if cement pumps are part of the usual product line, the chief engineer could probably do a creditable job of design after minimal familiarization. The chief engineer absorbs an average level of uncertainty over time. That level, of course, is higher than those of his or her subordinates. It may temporarily rise with some minor emergency and then go down when work slacks off, but the band width is fairly constant.

Conversely, project managers are usually less familiar with all the work done by the technical people assigned to the project since the project itself has been set up to solve a "new" and highly unusual problem beyond the normal scope of the equivalent functional manager. Project managers *rarely* can do the detailed work themselves if the project is really complex without a great deal of training. Their decision making, therefore, depends to a great extent upon evaluating inputs from and accepting the *expertise of others*.

In technical projects, project managers manage interfaces among the various kinds of work being done, since the assumption must be that "experts" have to do the work. This doesn't mean that the work is ignored by the project manager, just that it's often done by others who are more technically competent than the project manager. Managing the interfaces, however, is vital when work of one kind must be related to work of other kinds. Occasionally, the project may be so unusual that the project manager is familiar with only one aspect of it, the aspect from which his or her initial expertise was drawn, for example, from the fields of engineering, purchasing, quality, and so forth.

In other words, project managers are concerned primarily with the places where the various technical specialties come together. His or her decisions are very similar to those of a general manager who coordinates functional team members as they work on various phases of the project. Like general managers with diverse groups and less specific technical knowledge, project managers must depend more on the competency and motivation of people working on the project. Therefore, a project manager should have an increased sensitivity to and understanding of interactions among personnel. And there's a limited overall time period to get to know them.

3.4 TRAINING SUBORDINATES

In a very simplistic sense, functional managers either train or replace people, and they have more time to do it since functions are "eternal." Conversely, most project managers are concerned with relatively shorter range goals and therefore usually don't have the time to emphasize "training." The project team members are expected to be well trained when they arrive. While some training may be done on projects, it's not a primary goal. Project managers are primarily "users," not "trainers," of people. In effect, functional managers provide the "trained" personnel, and the project manager "uses" them. Training subordinates in the functional environment helps the functional manager to understand his or her people, rewarding progress through raises, promotions, and so forth. The project manager doesn't have this primary responsibility or the advantages that come with it. The control that functional managers have over subordinates is, therefore, much greater than that of project managers. Later, in Chapter 8, we see how the project manager can support the functional manager in the issuance of raises, promotions, demotions, and so forth for the team member.

4.0 DEFINING UNCERTAINTY LEVELS

Now that we have reviewed some specific differences between these two kinds of managers, we can compare the totality of uncertainty as one variable, and time as the other variable. As noted before, the more uncertainty, the more responsible the position. While it is impossible to plot actual uncertainty versus time, it's still possible to visualize it. By definition, we don't know what the actual probability distribution of uncertainty is, but we can assume some measurements. Let us consider a scale from 1 to 10, where 1 is low uncertainty and 10 is maximum uncertainty

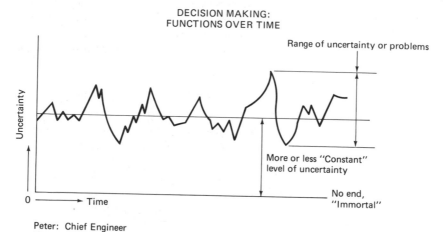

Peter: Chief Engineer

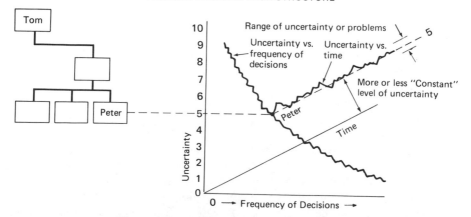

Figure 1-2.   Decision making: Functions over time

Consider the amount of uncertainty that a typical functional manager and project manager are supposed to "absorb" when making decisions. Taking "functions" first, we assume a conventional pyramidal organization chart (see Fig. 1-2). The uncertainty curve over time for a functional manager is usually a random distribution around some average, or acceptable, level of uncertainty. This can be interpreted for the typical Engineering department as follows:

> The Engineering department always has work to do, but the tasks are neither very large nor very small in relation to its capabilities. And while the tasks themselves are not entirely predictable, most of the uncertainty is. And it goes on year after year.

As you can see, this functional department may have a random frequency distribution, but the variations in uncertainty occupy a relatively predictable band. It's possible to categorize the kinds of problems that this department faces. Peter, the chief engineer, can concentrate on nonrepetitive problems by developing "standard" heuristic (experientially based common sense) solutions and then possibly using lower level personnel for minor changes or even computers for no change in decision making. He'll also assign the tough problems to his best, higher level people. Therefore, the department, as a whole, maintains an average level of uncertainty by shifting problems to appropriate people and by using heuristic solutions to more familiar problems.

The situation is different for project managers. I have found that the uncertainty level for most technical projects starts rather high, then decreases, and eventually levels off at a low level as the project closes (see Fig. 1-3). It starts high because the goals are usually more complex than those of equivalent functions, they

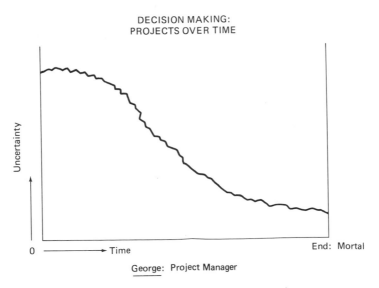

Figure 1-3.   Decision making: Projects over time

are more poorly defined, and they deal with more complex tasks. When the project crosses organizational boundaries, uncertainty increases further, because each department has different goals, personnel, and systems. These differences must be either incorporated into or superceded by the project systems. Since the project will end at some point, we know that eventually the goals will be reached with a lower level of uncertainty. The degree of uncertainty changes a lot over the course of a project.

Uncertainty never reaches zero because we never really know if we've solved all the project problems or just those we know about. Some uncertainty is always left over when the project ends.

5.0 THE CHANGE FROM TECHNICAL EXPERT TO MANAGER

While there are differences between these two types of management, it's interesting to note that successful functional and project managers both seem to come from the same general population. Many managers achieved their position because they were very good in an entirely different set of circumstances. When we were above average engineers, scientists, chemists, systems analysts, physicists, or whatever, it was often assumed that we would be just as effective as managers if promoted into those positions. That doesn't always follow. There is less predictability in dealing with individuals than there is in purely technical problems. The big change is in dealing with people.

> "It came as quite a shock to me to realize that my success as a manager was so dependent on what other people could and would do and how effective I was in working with them. It wasn't this way when I was an engineer; then my success depended on how effective I was as an individual working alone. Being a manager is so much different from being a technical person; one must not only be able to see the big picture but see how the parts and the people relate and fit into the big picture.
>
> My training as an engineer taught me to look closely at the technical details of the work. This was a form of myopia; my particular engineering discipline helped me to develop a form of tunnel vision. This presented no problem as long as I remained in engineering work.
>
> I was an outstanding engineer—in fact so outstanding that I was promoted to be a manager of engineers. Now my future is in someone else's hands. Now if I don't produce by working through my staff I will be a failure. This is not a comfortable feeling." (Cleland and Kocaoglu, 1981, p.45)

Our previous successful technical educational processes might even be a hindrance now since they were based on an orderly and predictable view of the world. It was pragmatic. By understanding and accepting its rules, it was possible to provide an answer to the particular problem being faced. That doesn't equally apply to management. Rules are less obvious since they are based in the emotions of human beings. There are many more of them and they are less predictable. It requires a different kind of thinking to deal with our fellow humans well. Like the sciences, management is concerned with predicting and controlling change, but that change is almost wholly dependent upon the creativity of individuals, who are much less predictable than nature. There are many more variables. This applies to both functions and projects.

The process of becoming a successful manager after being a technical specialist can be difficult. At least, the technician can use the successes of others without much modification. For example, although someone else discovered the law of gravity, it always works the same way for everybody. But managers and artists, by definition, are more limited in their ability to use the work of others since each follows a somewhat different method from every other one. Management situations differ a lot, more than technical jobs do. There are fewer guidelines than those we had in previous jobs.

> "You don't become an artist by painting someone else's dots. There are some complex human activities and management is one of them, in which it is the aesthetic fit (i.e., the form and harmony among things) that matters most." (Nichols, 1971, p.135)

Learning the best fit requires participation. Management is not an observer sport.

One way to learn is by evaluating progress and reforecasting or redirecting the next steps of the project. This process is called *iteration*. All projects are iterative, which means they proceed in steps. After each step, they cycle back for review and restructuring and then proceed again. We start our progress with something familiar, the past solutions in the organization's "memory" (standard procedures and culture) and in our own experience.

5.1 LEARNING TO BE A MANAGER

Procedures and past experience help us to start. Then we can use a familiar tool to continue learning. That is the Scientific method of definition, theory building, and testing. Knowledge is gained using this process. For example, "Did it work?" is one major question. In more scientific terms, was the hypothesis supported? Controlled tests are impossible in an ongoing organizational situation. Therefore, the learning process requires some open-ended, on-the-job testing of various hypotheses. Consider the following modification of the Scientific Method:

1. Start with objectivity (it's easier) or the successes and failures of others, but understand that even "objectivity" or success was defined by the people who achieved them, and that wasn't you.
2. Set up personal management theories; use them "objectively" in the search for a basis for the hypothesis; try out those hypotheses that test the theory.
3. Use inferential thinking to change theories as the data come in from the tests

of the hypotheses. Optimize where there is a higher chance of success or where commonalties with past or present practices and solutions can be used.

4. Use "incremental" thinking; implement slowly and in small stages. Organizational change takes place very slowly, even when it gets going at all.

5. Take successive limited comparisons. Did the last solution work? Why? What can be used from it to help solve the next problem?

In a more or less summarized and slightly more formal process,

1. Devise several alternative hypotheses within the theory(ies) selected.

2. Develop a crucial experiment (or several of them) that are intended to *exclude* one or more of those hypotheses.

3. Do the experiment to get an understandable and definite result.

4. Recycle to eliminate other hypotheses or else sharpen up those we have, etc.

5.2 ANOTHER VARIABLE: THE SITUATION

There are other problems to consider. Management techniques are not universally applicable. For example, building a bridge is not at all like building a computer-based information system. The project techniques that apply must be modified as required. For example, bridge building requires specific, directive control techniques to control many subcontractors. Here PERT would probably be excellent, (see the Chapters 2 through 4 on Forecasting for details and Appendix 4), providing clearly outlined technical tasks. The bridge blueprints must be exact. The materials list must be complete and the work contracts drawn by legal experts. By comparison, the computer-based information systems would use less well defined, iterative control techniques and less directive, more collegial project coordination (here periodic design reviews and sign-offs might be more appropriate). There are similarities, of course, but building a bridge is not like the iterative process required in programming software.

Therefore all project systems and techniques are not entirely transferrable as a linear, total work of art. Don't try wholesale adoption of project systems and tactics without consideration of the need for changes due to differences among industries, companies, or even the people working on these projects. Some project systems or bundles of techniques are basic, and we start with those in our design. But as our design develops, there will be more flexibility.

6.0 QUICK REVIEW

We have already defined projects and management. We differentiated among various management levels according to the amount of organizational uncertainty using a graph of uncertainty versus frequency of decisions. We then defined the differences between the two major kinds of organizational managers—functional and project managers—in terms of uncertainty versus time. Functional managers absorb relatively the same amount of uncertainty as time goes on. They may change the distribution of work to make decision making more repetitive, but the management job is "eternal." Project managers have a changing amount of uncertainty to absorb. There is more in the project's initial stages, probably requiring the project manager to operate at an organizational level *above* that of his or her functional counterpart during the very critical beginning phases of the project. Later, when the project is ending, the project manager seems to operate at a level *below* that of the comparable functional manager. Projects have a life cycle. When they are completed, the project is dissolved.

We should begin to design project systems that have the possibility of being useful in many project situations by studying those that have been developed by others. Why reinvent the wheel? In the prior definition of management (making nonrepetitive decisions involving people by absorbing uncertainty), there is no reference to the freedom of action that the manager has. Freedom of action defines the limits within which others allow that authority to be exercised, and authority is the ability to cause someone else to behave as the manager wishes. Uncertainty can also be defined as the totality of responsibility and authority inherent in both the organizational position and in the particular manager. Up to this point, we've considered mostly the organizational position aspects. Now we should deal with a totality. It's not a static variable or measurement. For example, any manager can move some daily situations out of uncertainty into risk with increased experience. Typically, it can be done by outlining the authority and responsibilities of subordinates as part of the delegation process. Then with successful delegation, as situations are moved from uncertainty into risk, with repetition they can possibly be moved later from risk into certainty by documenting standard procedures. The company might even connect parts of uncertainty to particular management jobs through job descriptions or other administrative channels. But as a totality, uncertainty is modified almost daily to some extent by the decisions of the manager and others. After this brief review, we're ready to deal with the foundation of every project—the project manager's authority.

7.0 PROJECT AUTHORITY: THE FOUNDATION

The degree of authority is another aspect of uncertainty, and it is different for functional and project managers. The difference lies in the basic aims of project and functional managers, and there is an endemic conflict between those aims. Functional managers are primarily concerned with optimum use of human and economic resources to solve a more or less "eternal" stream of problems. These problems fall almost randomly within a range of uncertainty for a particular department. In contrast, the project manager is primarily concerned with using the optimum resources to solve a particular and very unusual problem. The functional manager represents continuity, and the project manager represents change and discontinuity. Most organizations favor the status quo, and understandably so, since, the functions provide the ongoing stream of revenue. Unfortunately, functional management is less able to handle the major tasks that projects can. Completing these tasks promotes growth.

When a project must cross organizational boundaries in handling a large and novel problem (which the project itself was developed to resolve), it can be very disruptive to the smooth operations of most functions. The epitome of disruption is the matrix, where people are assigned part time to various projects but remain under the administrative control of their functional managers. That's when conflict can be serious for both project participants and project managers. (Chapter 6 deals with project participants.) We now deal primarily with the project manager.

The project manager is intent on directing team members in accomplishing project goals. This means using authority. If the functional manager issues conflicting directions that move team members away from project goals or assigns them to other projects that are supposedly more important, authority conflicts will appear. It is better to resolve these conflicts at the beginning of a project with a clear-cut distribution of formal authority.

The functional organization is usually quite vertical with authority and responsibility, as part of uncertainty assigned to fit each organizational level. Functional authority can be delegated downward, to the appropriate people charged with getting the particular tasks done. It's a relatively understandable and workable technique. Authority can be delegated and it's gone but, of course, responsibility cannot be entirely given to others. It only expands. A task, the required authority, and the responsibility can be delegated, but the delegator still has the original amount of responsibility.

Projects, however, are usually horizontal. They cross functional boundaries, and occasionally authority and responsibility for accomplishing project-related tasks are sometimes unclear. This leads to conflict. This conflict can also be minimized by defining the project manager's authority correctly at the outset of the project. The source of the project manager's authority is not fixed in a particular organizational location like the functional manager's is. It varies depending upon the size, complexity or importance of the project. Now we come to the foundation of the project manager's authority, which is intended to minimize most endemic conflicts: the definition of the project manager's reporting level.

The project manager *must report to a level in the organization at least one step above that from which his or her personnel are drawn.* If this does not occur and the project manager reports to the same (or worse yet, a lower) level, there is a built-in organizational trap causing eventual project delays and possible failure. Most conflicts such as those involving assignment of personnel and resources will usually be resolved *against* the project manager when there is a clash between the immediate needs of the project to meet limited targets and the longer range needs of the particular function to optimize the use of the departmental resources. This has been my experience, probably because some upper-level executives who mediate these conflicts feel that *now* is better than growth in the *future*.

For example, the correct authority-reporting level for project manager in Fig. 1-4 is the division manager. Since the project draws people from Engineering, Man-

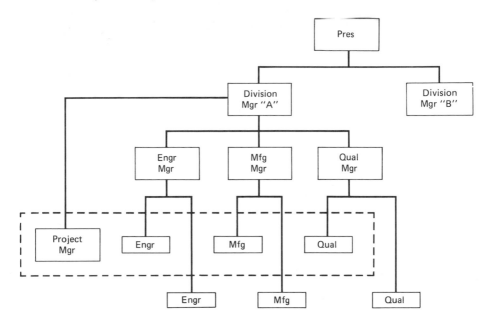

Figure 1-4. Reporting: Small projects

ufacturing, and Quality, the project manager's reporting level must be above these functional managers. With a larger project (see Fig.1-5), the project manager should report to the company president. Project managers who report to the wrong organizational level have no authority. Their freedom to act is limited. They should *not* try to be project managers at all. They are actually *project coordinators,* who must refer most organizational problems upward to the real decision maker. The difference between a project manager and a project coordinator is, therefore, very important—almost like life and death!

Project coordinators can't absorb uncertainty by making nonrepetitive decisions involving people assigned to the project. They only report the particular as-

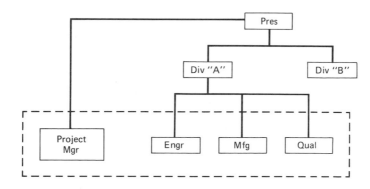

Figure 1-5. Reporting: Large projects

signment problem that needs a decision to someone else who is the real project manager. Coordinators don't decide. They communicate.

Therefore, the project manager's authority and reporting level depend on the organizational structure. The freedom of action must be defined. A notice of the appointment by itself is insufficient, especially when it's the usual memo, such as

> Mary Smith has been appointed to be project manager of our new anti-gravity program. Your company expects that this program will certainly put us ahead of all other anti-gravity companies, and I'm sure that you'll all give Mary your support in this new venture.

This is only the beginning of the definition of the formal authority or the amount of freedom that Mary has. It's like getting a legally issued hunting license! All that means is that the holder is now *legally* entitled to go out and try to shoot tigers; of course, the next step is to get the tigers to cooperate. There's no indication of the size of the tigers, the type of weapons available, or of the time limits for the hunting season. In like manner, this memo merely begins the definition. There has to be a more detailed definition because the authority or freedom granted a project manager automatically violate the classical concept of the unity of command. (Unity of command means that each person has only one immediate superior.) In other words, authority must be defined adequately, otherwise the inevitable conflicts between the various project and functional managers over work assignments for various people within a particular function will usually be resolved against the project.

Some criteria must be considered in defining the project manager's authority. Information was obtained by Goodman, who asked each manager the following question:

> In your company does the project manager have the *final* authority to make the crucial project decisions regarding the areas listed below?

1. Initiate work in support areas?
2. Assign priority of work in support areas?
3. Relax performance requirements, i.e., omit tests?
4. Authorize total overtime budget?
5. Authorize subcontractors to exceed cost, schedule, or scope?
6. Contract change in schedule, cost, or scope?
7. Make or buy?
8. Hire additional people?
9. Exceed personnel ceilings when a crash effort is indicated?
10. Cancel subcontracts and bring work in-house?
11. Select subcontractors?
12. Authorize exceeding of company funds allocated to the project?

13. Determine content of original proposal?

14. Decide initial price of proposal?
(Goodman, 1976, p.303)

Since no two projects, or project managers are alike, it might be helpful to use the list given above, or any other list, to outline the appropriate authority that should apply to a particular situation. The following describes how to make such an outline:

1. List all the factors that are considered to be vital for this project.

2. On an imagined linear scale from 1 to 10 (1 means unacceptable and 10 means perfect), subjectively assign a number to the *present* situation: What can the project manager do now?

3. On the same scale, assign a number to what *should* exist. Take the numerical difference between "what is" and "what should be."

4. Taking the largest mathematical difference first, because that item is most important to you, then determine what you can do now with that item, and how long it will take to change it to where it should be?

As an example (see Fig. 1-6 for some examples), if the present situation allows the project manager to only "influence" item 13 (determine content of original proposal), it might be assigned a *present* score of 2. Since the project manager *should be* part of a final review team to determine the content of the proposal, the *desired* score might be assigned a 9. The difference is 7. Since 7 is a larger difference than 1 for item 14, it should be part of the project manager's authority to do as noted in the job description. This assumes that the project manager is writing his or her job description and is determining the relative importance of various kinds of decisions.

It is, therefore, necessary to outline the formal authority in some kind of a job description. This must include the correct reporting level. An adequate job description is a vital foundation block for any project system. One way to design the project manager's authority is noted in Fig. 1-7.

	Should	Present	Difference
12. Authorize exceeding of company funds allocated to the project?	5	5	0
13. Determine content of original proposal?	9	2	7
14. Decide initial price of proposal?	4	3	1

Figure 1-6. Project manager's authority

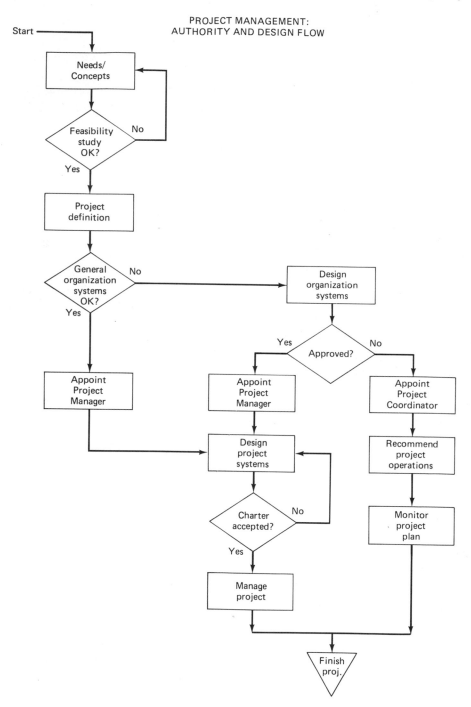

Figure 1-7. Project management: Authority and design flow

8.0 THE SYSTEMS

With the authority foundation block defined, we move on to general organizational systems. The first two general interrelated systems are Forecasting and Measuring. The *Forecasting system* is concerned with thinking *today* about solving problems we think will occur *tomorrow*. For example,

- What are the solutions that we have to provide today to solve tomorrow's problems? For example, what should the technical specification sheet include?
- What management tools shall we use to solve them? For example, shall we use Gantt charts, PERT diagrams, or what?
- When will it be possible to define our technical goals and their measurement well enough to forecast the entire project?
- How will we know it?

The *Measurement system* reports how the forecasted goals are being met. This system applies to the following considerations:

- What measurement procedures shall we use in order to determine if we're making satisfactory or unsatisfactory progress against our forecasts? For example, what kind of accounting information is useful for project managers, and how can it be obtained?
- Do we want to develop a general or specific information system? For example, how will we be sure that they are responsibility-based and fast enough to respond to the changing conditions in the project?
- What will be processed through the information and cost systems, and to whom will the data be reported?

As general systems, the Forecasting and the Measuring systems are expected to be applicable across every project in most organizations with very little modification. Both the bridge building and the computer design company should be able to use them in almost the same way.

The next two project systems are used in the actual operations of the project. The *Project Operations system* specifies how the project will be organized and how it should proceed on a regular basis. It includes the following:

- What kind of structure shall we have? How much autonomy will team members have?
- What kinds of meetings should we have, including procedures to be used, and who is supposed to attend them? For example, what are the ground rules to make them effective?

- How should internal reports be completed?
- How should project changes be handled? For example, which ones require notifying upper management and which ones don't?

The most important project system, the *People Management system*, deals directly with people and includes recruiting, evaluating, and motivating.

- How shall we recruit project personnel? For example, are there different procedures to follow when recruiting within the company and from the outside?
- How should agreement be reached concerning the scheduling of work, reporting on progress, reviewing accomplishments, developing organization charts, and fixing responsibilities? For example, what records are necessary?
- How should performance be evaluated and competence rewarded? For example, who shall do it and when?

The Project Operations and People Management systems are less standardized than Forecasting and Measuring. They apply more closely to the projects themselves. They have to be modified as required. That's where there will be a greater difference between the bridge building project and the software development project. These four systems are usually enough to provide a stable basis for a successful project. This assumes that the most variable system of them all—the *project manager's Personal Management Style*—is appropriate. We deal with that later.

8.1 A PLAN AND THE SYSTEMS

Every plan has three interrelated parts, such as,

1. Forecast
2. Measurement against that forecast
3. Management strategy to guide action when there is a difference between the forecast and the measurement

The first two sections of the plan—the forecast and the measurement—are in the first two general systems, Forecasting and Measurement systems, which are the common language or communications channels of most technical organizations. The second two systems—Project Operations and People Management—are more flexible, since they deal only with a particular project. They define the project management strategy. They have to be adapted depending, typically, upon the size of the project, the cost, the importance of the problems, and the people involved.

The last system—Personal Management Style of the project manager—may also be part of the strategy. This system fits the idea of a repetitive "system" least of all, since it applies to a unique individual, the project manager. It covers individual decision-making (and problem-solving) behaviors that might appear to be repetitive (for example, delegation, leadership, etc.), but that are only repetitive for the *individual* manager. Each project manager selects different sets of management behaviors as he or she sees fit. The last system is most complex. No one else but the project manager can decide

- How to properly delegate tasks.
- What the appropriate leadership behaviors are.
- How to resolve conflict.
- How to handle change.
- How to provide an ethical base for decisions.

The five systems are all interrelated. They are intended to have a certain logical flow from the simple to the more complex. Most organizations have less difficulty defining the forecasting and measuring-costing systems to be used in all projects than in defining project operations and people management systems. They have most difficulty with personal style because it is dependent upon an individual and therefore least predictable.

Even though all these five systems are interrelated, we have to consider them separately merely to simplify the design process. In effect, our systems design process will be our own project. We'll design and then iterate as we learn and improve.

8.2 POTENTIAL RISKS

There is a word of caution. Modifying or "tailoring" systems to fit a situation, especially the Personal Management Style system, requires the ability to evaluate and possibly change one's "mental filters" or one's own perspectives. This is not easy, since those filters have probably worked well in the past. But it's a new situation now. It's helpful to consider the tools (and theories) of others. They have proven successful, and even if they can't be applied totally without modification, they can be used to begin one's own designs. Management, in general, and project management, specifically, is situationally dependent, but some things are basic. Consider, for example, the definition of the correct project authority given previously, as reporting to at least one level higher than the sources of resources. That level has to be dependent on the particular situation.

We will start with overall strategy before tackling tactics. It's rarely practical to reverse that order and start with tactics. Overall agreed-upon goals have to be defined before any actions on the project manager's part. It's a safer path, but no path is risk-free. There could be problems ahead.

"Even the best project manager, moreover, can hardly accomplish his project objectives without antagonizing some members of management, quite possibly, the very executive who will decide his future. In one instance, a project manager who had brought a major project from the brink of chaos to unqualified success was let go at the end of the project because, in accomplishing the feat, he had been unable to avoid antagonizing one division manager. Such difficulties often lead a project manager to look for a better job at this time, in or out of the company." (Stewart, 1965, p.285)

Starting with general ideas that are accepted at the beginning can help to uncover many potential problems and to minimize or eliminate them. When these systems are successfully designed and implemented, the management skills you learn often include a healthy amount of diplomacy. That's not bad either.

8.3 POTENTIAL OPPORTUNITIES

On the other hand, as mentioned before, project management can offer as many, if not more, personal opportunities as there are risks. Successful project managers are usually exposed to a greater variety of experiences than their functional equals.

"Successful experience in operating under a matrix constitutes better preparation for an individual to run a huge diversified institution like General Electric—where so many complex, conflicting interests must be balanced—than the product and functional modes which have been our hallmark over the past twenty years." (Davis & Lawrence, 1978, p.132)

This book is intended to help you develop your own designs. It is, therefore, not a spectator manual. Unless one tries to apply these ideas to one's own projects, neither this book nor any other will be of much help.

Projects are real. Their problems are real. And the pragmatic systems that help you to manage them must be real too, not just written about, read about, and forgotten.

9.0 SUMMARY

We started by outlining basic definitions and foundations for project systems design. Project managers are able to develop a surprising degree of personal authority (or freedom), since there are fewer limits imposed by the job than equivalent functional managers have. The right place and the right time to develop general systems that support the project manager's freedom to manage are at the beginning of the project. Higher levels of uncertainty make this necessary. Consider the flow diagram on Fig. 1-7 as a possible way to start. It's a possible "map" to follow when a project manager has been assigned a project.

My Suggested Answers to the Case Study

1. I believe that the project can be saved, but it requires prompt and effective action by George. George should first outline his own description of his duties and reporting responsibilities as the project manager. He should then prepare a forecast showing the technical goals, the amount of time needed, and the costs to be incurred. After the forecast is outlined, he should prepare an organization chart showing the major project team managers that he needs and a brief outline of their project-related responsibilities. George should then present his "process" ideas to each of the functional managers concerned, pointing out the advantages to them. After discussions and acceptable minor modifications, Charles should approve it. A project that doesn't have top management approval is like a cannon with no explosives; it's utterly useless. If you can't get approval at the beginning, you'll probably never be able to get it.

   If this project doesn't succeed, Acme will not be able to compete with companies that use a project-oriented format, and present problems will continue to increase. The rate of change in Acme's markets and in its internal organization seems to be too great for the regular functional organization to handle.

2. Yes. A reorganization into a project-oriented format can be done at any time. It doesn't need any special time because a project can be started at any point.

3. At this time, there's no need for a standard project manual for the whole company, just those projects intended to solve present problems. If successful in a limited fashion, the project systems that apply can be extended later, when there is some history behind them. A company-wide project manual is almost a contradiction in terms. Perhaps a better idea would be to develop the "basics" for the whole company and then let each project manager modify them to fit. The "basics" could include the first two systems—Forecasting and Measuring.

4. Sometimes logic cools down "impossible" demands. For example, George could lay out a quick time schedule showing his estimates of the amount of time needed by his present jobs and then superimpose the time needed by the Phoenix project. Then he can provide potential solutions and schedule alternatives to Charles. If Charles wants to, he can determine the sequence that he wants George to follow.

BIBLIOGRAPHY

CLELAND, DAVID I. and DUNDAR F. KOCAOGLU, *Engineering Management*. New York: McGraw Hill, 1981.

DAVIS, STANLEY M. and PAUL R. LAWRENCE, "Problems of Matrix Organizations," *Harvard Business Review*, May-June 1978,131-42. Copyright 1978 by the President and Fellows of Harvard College; all rights reserved.

GOODMAN, RICHARD ALLEN, "Ambiguous Authority Definition in Project Management," *Academy of Management Journal*, December 1976, 301-15.

HANDY, CHARLES B., *Understanding Organizations*, New York and London: Penguin Education, 1976. Second Edition 1981.

MARCH, JAMES G., and HERBERT A. SIMON, *Organizations*. New York: John Wiley, 1958. Copyright 1958 by John Wiley & Sons. Reprinted by permission.

NICHOLS, RALPH G., "Listening Is a Ten-Part Skill," in *Readings in Interpersonal and Organizational Communication* (2nd ed.), eds. Richard C. Haseman, Cal M. Logue, and Dwight L. Freshly. Boston: Holbrook Press, 1971.

STEWART, JOHN M., "Making Project Management Work," *Business Horizons*, Fall 1965, 266-85.

WEISBORD, MARVIN R., *Organizational Diagnosis: A Workbook of Theory and Practice*. Reading, Mass.: Addison-Wesley, 1978.

OTHER READINGS

AVOTS, IVARS, "Why Does Project Management Fail?" in *Matrix Organization and Project Management*, (Michigan Business Papers, 64,) eds. Raymond E. Hill, and Bernard J. White. Ann Arbor, Mich.: University of Michigan, Division of Research, Graduate School of Business Administration, 1979.

BRUNER, JEROME S., "The Conditions of Creativity," in *On Knowing: Essays For The Left Hand*. ed., Cambridge, Mass.: Belknap Press of Harvard University,: 1962.

COTTLE, THOMAS J., "The Mosaic of Creativity," in *Confrontation: Psychology and The Problems of Today*, ed., Michael Ethheimer. Glenview Ill.: Scott, Foresman, 1970.

DAFT, RICHARD L., "Learning the Craft of Organizational Research," *Academy of Management Review*, 8, no. 4 (1983), 539-46.

HOLLANDER, EDWIN P., and RAYMOND G. HUNT, *Current Perspectives in Social Psychology*, (4th ed.) New York: Oxford University Press, 1976.

KRUPP, SHERMAN, *Patterns in Organization Analysis*, New York: Holt, Reinhart & Winston, 1961.

LASDEN, MARTIN, "Overcoming Obstacles to Project Success," *Computer Decisions*, December 1981, 114-77.

MACIARELLO, JOSEPH A., *Program Management Control Systems*. New York: John Wiley, 1978.

MORTON, DAVID H., "Project Manager, Catalyst to Constant Change," *6th Annual International Meeting of the Project Management Institute*, 9, 1974.

ORNSTEIN, ROBERT E., *The Psychology of Conciousness*. New York: Penguin, 1975.

PASCALE, RICHARD T., and ANTHONY G. ATHOS, *The Art of Japanese Management*. New York: Penguin, 1982.

PIRSIG, ROBERT M., *Zen and the Art of Motorcycle Maintenance*. New York: Morrow, 1974.

RANDALL, CLARENCE B., *The Folklore of Management*. New York: Mentor Executive Library Books, New American Library, 1961.

SAYLES, LEONARD, and MARGARET CHANDLER, *Managing Large Systems*. New York: Harper & Row, Pub., 1971.

SCANLON, BURT K., "Philosophy and Climate of the Organization," *Vectors*, 4, no. 5, (September-October 1969).

SILVERMAN, MELVIN, *The Technical Manager's Survival Book*, New York: McGraw-Hill, 1984.

Chapter 2   Forecasting: Starting with the Charter

PERSONAL
MANAGEMENT
STYLE:
THE INDIVIDUAL
SYSTEM

PEOPLE MANAGEMENT:
THE PROJECT SYSTEM

PROJECT OPERATIONS:
THE PROJECT SYSTEM

MEASURING:
THE ORGANIZATIONAL SYSTEM

FORECASTING:
THE ORGANIZATIONAL SYSTEM

FOUNDATION: PROJECT MANAGER'S AUTHORITY

THE CASE OF THE ETERNAL PROJECTS

Cast

Milt Rector, Vice President, Alpha Paper Company
Sam Albanese, Chief, Design Group
Bob Snowden, Chief, Project Management
Mary Murdy, Chief, Field Operations

Alpha Paper Company is a company that prides itself on being a producer of low-cost, high-quality, fine writing papers. When the company was first started about 50 years ago, top management decided that they would develop their own paper-making machinery in order to achieve and maintain a technological leadership position in the industry. The project engineering group was responsible for the design and installation of most of the special machinery that contributed so heavily to the company's economic success. For most of the company's history, that group had consisted of a few long-term, highly qualified engineers and scientists, but within the past few years this situation has changed. The staff increased. The present level was about 30 people. When the company suffered a minor decrease in sales, all managers were requested to re-evaluate their budgets to determine where costs could be cut. The Vice President of Engineering, Milt Rector, had asked his section managers to meet with him on Friday morning to review the group's worker requirements for the next fiscal year. As usual, the meeting started promptly at 9:00 a.m.

Milt: You all have gotten a copy of the memo from the president requesting us to take another look at our labor costs for the next year. I've reviewed your requests, and they all look in line with our operations. I noticed that there has been small increases in both Project Management and in Field Operations, while Design has remained almost constant. Why is that?

Bob: Milt, we all know that one of the reasons the company has done so well in the past is because we've all been able to get the latest technological improvements into the plants very quickly. Our project teams design, build, and install that stuff faster and better than any of our competition. And Sales continually demands more products at a lower cost, but at the same quality levels. We have been trying to meet those demands. I've had to hire four additional engineers just within the last year to keep up with the demand for engineers to supervise installation, and even this sales downturn hasn't dropped our load very much. The plants always want their equipment renovated and updated, and we're always very busy because of it.

Mary: Bob's right. My people are spending a lot of time out in those plants making sure that the equipment continues to run. Remember that while Bob is putting new machinery into the plant, the plant often doesn't discard the older machinery. Sometimes they modify it in their shops and use it for other purposes. Some of that stuff is over 20 years old, and we're still making changes on it because a plant manager may want to try out a faster motor drive, speed up the dryers, or other ideas that they have.

Milt: Sam, you haven't said a word yet. Maybe that's because you haven't increased your requests for labor for the next year. Why is that?

Sam: I really think it's because we installed those computer-assisted design stations last year. They help a lot. My people have always been able to support the Project Management group in the design and drafting work. Of course, there may be an increase in the "build or install" part of it, because a lot of that is done at the plant sites, but I have been able to handle the increase in requirements for Design.

Milt: Something is wrong here. Project Management and Field Operations are increasing, but Design is staying constant. What's going on? I need some answers. We'll take this up again at the next staff meeting next week.

At lunch that day, Mary, Sam, and Bob were reviewing the meeting.

Mary: I've gone over my budget again since this morning, and I still find that I need more people to service the plants. In fact, one of the old line dryers that we developed about 15 years ago just went down again at our Perth Amboy plant. I have to get out there this weekend, and you know how I hate that town. Sometimes I wish that we could get rid of those maintenance jobs. Why not let the plant handle them?

Sam: Well, doesn't the plant manager have a maintenance and repair crew to do that kind of work?

Bob: Sure, but they always claim that the original project wasn't done right and it's our responsibility to fix it when it goes down.

Sam: Bob, I think that you just gave me an answer to the problem. I can even tell you what's gone wrong. It's so obvious, and we have all overlooked it for too many years.

Questions

1. What do you think the problem is?
2. How can it be solved?
3. Who is responsible for solving it?
4. Outline the specific process to be followed in implementing your proposed solution?

1.0 INTRODUCTION

Everything starts from a plan, and plans usually start with forecasts. In many ways, forecasting is not a prediction of the future. It's a decision *today* concerning expected future problems. A forecast can never be "wrong" if it was made with the best data available at the time. When and if that future problem appears and the decision proposed is no longer appropriate, a new decision is required "today." Not all forecasts are equal, nor are all forecasters. The fewer the unknowns in the forecast, the higher the coincidence of forecast and the eventual future. The more experienced the forecaster, the more likely the forecast and the measurement will agree. "Experience" is what the forecaster learns when the measurements occur; differences are analyzed and the causes for these differences are understood. Better forecasts are made when that learning process teaches the forecaster how to determine the sensitivity of the forecasted variables. Sensitivity analysis is using experience to select those sensitive criteria or variables that can most affect the present decisions about future problems. It can also point out which of those criteria are relatively insensitive and can be disregarded. We learn from the comparison of actual history to prior forecasts. We learn what is variable in the project (and therefore important to consider carefully) and what is not. We learn which criteria or "filters" to select and which to ignore. It comes from experience.

2.0 THE FORECAST BEGINNING: SHOULD WE DO IT?

From an overall viewpoint, any forecasting system may be visualized as a progressively finer set of filters through which the project must pass. The development of a proper "filtering" process is an important systems design task. It is somewhat repetitive when the particular project being evaluated is similar to others that the organization has completed. Conversely, there are always some unknowns. The filtering process is supposed to separate the knowns from the unknowns. It is supposed to emphasize what *is not known* about a project, rather than what *is known*.

The first of these filters defines who the "customer" or eventual "user" is: Who is the decision maker for the project's eventual acceptance? When there is a contract with an outsider, that outsider is obviously the customer. If the project is to build a new factory, the plant manager is the customer, not the facilities manager. If the project is to develop and install a new data processing system, the department manager who is to use it is the customer, not the data processing manager.

Therefore, clarity in defining the customer and his or her needs is part of the first filter. The complete filtering process may be neat and rational. The questions may be standardized. Using a numerical rating, it is relatively straightforward to determine the project's acceptability. (See Appendix 1 for a typical type of questionnaire that covers the initial evaluation process for a data processing project. Appendix 2 covers the entire project life cycle, not only the project's initial evaluation process.) The outcome of this filtering process for the potential customer is a Request for Proposal (RFP) that outlines the task as defined by the customer (see Fig. 2-1). Outside customers provide this willingly, in most cases. Insiders rarely do. The prospective project manager may have to patiently and diplomatically extract these data.

REQUEST FOR PROPOSAL	**COMPANY "FILTER"**
A. Scope of work	1. Is this our kind of project?
B. Financial or payment criteria	2. Do we want to enter or stay in the business?
C. Progress reporting types and schedules	3. How does this RFP fit in with our present operations?
D. How response will be evaluated	
E. Delivery schedule	
F. Proposed management systems	

Figure 2-1. Company "filter" questions

2.1 SHOULD WE BID?

A series of standard questions should be developed before forecasting begins (see Fig 2-1):

a. Is this our kind of project?

b. Do we want this business if the project is unfamiliar?

c. How does the Request for Proposal fit in with our present operation?

d. What are the possibilities of getting the job if we bid?

Other questions that are pertinent could include the following:

a. Can we do it technically?

b. Do we have the talent and the capacity to do the job in accordance with the customer's request?

c. Do we have any similar solutions that can be applied to the RFP?

d. What modifications do we have to make in our standard project management operations in order to satisfy the potential customer's administrative and management needs?

If the project passes these questions, or any other set devised by management (see Appendix 1), the project plan is ready to start with a forecast.

2.2 EXTERNAL AND INTERNAL PROJECTS

How things will be done is often as important to the knowledgeable customer or client as what is to be done. A standard response format outlining the how as well as the what is, therefore, often developed. The typical customer RFP might include a description of the scope of work (SOW), financial or payment criteria, progress reporting schedules, and indications of how your response is to be evaluated. In addition, the RFP usually requests a delivery schedule and a description of proposed management systems to be used in controlling the project.

Projects are sometimes started for reasons that may not be as rational as we would like them to be. For example, we might have

The Sacred Cow: The project is suggested by a senior and powerful official in the organization. Often the project is initiated with the simple comment such as, "If you have the chance, why don't you look into? . . ."

The Operating Necessity: If a flood is threatening one's plant, a project to build a protective dike does not require much formal evaluation.

organizational systems supporters and the functional managers at the beginning of a project, there are fewer hurt feelings, confused messages, and entrenched positions to deal with later on. There's a well-known aphorism that says, "We never have the time to do it right at the beginning, but we always have time to do it over." The responsibility of the project manager is clear at this point: define what needs to be accomplished and the systems that are needed to do it.

An experienced project manager in one of my seminars responded to this idea of issuing a charter as being comparable to a declaration of war on the functional managers. He felt that even if the boss did get the functional managers to agree to the charter, they could adopt a "We'll just see if this hot shot will be able to ride herd over us" attitude and subversion would be quite high.

My response was to ensure that everyone understood that cooperation was less expensive than subversion by including a general statement in the "Administrative Procedures" section of the charter (see item 7 of Fig. 2-2). That statement would require the project manager to issue a "cost impact" statement if the project was affected negatively *for any reason*. If this intent is clearly documented beforehand as part of the charter, it is a clear signal to everyone that delays cost time and money, and the additional resources requested will show that cost. It is quite interesting that when that statement is in the charter with the approval of management, it rarely has to be used. The combination of the project manager's reporting level to upper management and this documented *requirement* to issue a "cost impact" statement if trouble occurs produces a two-edged sword that affects both the project manager and the functional managers.

The project manager must issue a "cost impact" statement if the project is delayed, otherwise he or she has to absorb the time and cost overrun and probably suffer for it later. The alternative is to ask for additional resources immediately. That will bring the problem to the surface. If the functional manager does not

---

PROJECT CHARTER

---

1. The description
2. Project priority
3. Standards
4. Change procedures
5. Personnel and physical resources
6. Price/cost and how measured
7. Administration procedures

Figure 2-2.   Project charter

cooperate, the additional project costs should be his or her responsibility. Conversely, if the functional managers cooperate extensively that should be noted as well by the project manager in the project summary when the project is finally ended.

I believe in diplomacy, tact, and cooperation. These will be rewarded with more than subversion or uncooperation will in both the short and long run. To provide for those good things, we have laws. Those laws generally *don't* impact on most of us as average citizens, but it's nice to know they're around if we run into trouble. And it's better to pass the laws before something bad happens rather than try to fix it later.

3.1 PROCESSING THE CHARTER

The charter can be an internal document when dealing with internal customers. It can also be the basis of a formal contract used in satisfying the needs of an outside customer. (In the latter case, I suggest that it be much more complex requiring legal support.) But whether inside or outside, the charter is intended to initially define how the project will be managed.

In my opinion, completing the charter for the first time requires only the project manager. It is the first definition or top-down iteration. The next iteration that will be done by the project team is the second or bottom-up phase. There might even be others as the project progresses. The charter is first completed by the project manager. The total process should look something like the following:

   a. "Top down" (the project manager completes the charter, providing the initial input for management approval)
   b. "Bottom up" (proposed project team managers revise the charter and attach their justifications)
   c. "Final review" (the optimum answer is presented to management for acceptance and consequent transmission to the potential customer)

Any changes that occur to the answer to the RFP when it goes through item c above as a result of "final review" directions should be treated as any other new direction to the project. That is, if the changes are expected to result in a modification to the original budget or technical scope of work (as defined in item b above), a charter bottom-up review should be repeated.

The project charter becomes part of the standard operating procedure (SOP) that simplifies and clarifies the way your projects operate. It can even be used to satisfy any external or internal customer's request for a proposal when project management systems are requested. The *charter* primarily describes *how* the project will be managed. The *project flow diagram* primarily describes *what* the project will accomplish. They complement each other.

4.0 THE CHARTER CONTENT

The major elements of the charter are as noted in Fig. 2-2.

4.1 THE DESCRIPTION

The description deals with the technical achievement or end result. This may seem to be redundant because the customer has already outlined what is wanted in the RFP. We all see the world differently, and I've seen some amazing differences between what the customer seemed to have said (in one way or another) and what appeared in the actual project description. This is the description or the scope of work. It must include the three familiar "golden limits" of the project:

1. What will be delivered or accomplished?
2. What will be the time limit?
3. What resources (costs) will be needed?

There may or may not be a functional specification, but even in the best of all possible worlds where the functional specification is clarity itself, this part of the charter should be completed to avoid a difference of interpretation.

When a technical end result cannot be adequately defined or else it's very far out in the future (e.g., a general cure for cancer), the charter should be completed for smaller subprojects that *can* be defined. These are often called *feasibility studies*. For example, a subproject could be limited to determining why some cancer cells remain on the original site and some travel to other parts of the body.

Some of the major flaws that can occur in the development of general specifications are

A. *Omissions*: Whatever is required should be fully described. . . .

B. *Nebulous requirements*: A specification requirement that can be interpreted in more than one way could result in having the contractor provide for the more simple and less costly requirement, generally to the detriment of the customer.

C. *Unclear accuracies*: Accuracies are generally expressed in percentages. Disputes can arise if the base for the percentages is not clearly cited. . . .

D. *Inconsistencies*: The details of a specification must be consistent throughout. . . .

E. *Impossible requirements*: A requirement may have been expressed in the specification or the schedule which, as the project progressed, proved to be impossible to satisfy . . . the one dominant question raised would relate to who was the more knowledgeable party at the time of the contract. In the absence of any other overriding circumstances, the customer is usually considered to be more knowledgeable. . . .

F. *Deficient Compliance*: The courts have taken the position that if contractors are in substantial compliance with the specification or delivery requirements, the contract cannot be defaulted. However, . . . they are not relieved from submitting to an adjustment in the contract price. . . . (Hajek, Victor G., *Management of Engineering Projects*. New York: McGraw-Hill, 1977, p.117. Reproduced by permission.)

This list covers only major flaws. The writing of descriptions or specifications is therefore a prime consideration in the development of any charter.

4.2 PROJECT PRIORITY

There is rarely a single project that exists in splendid isolation from all other work to be done at that time. Establishing priorities for this project with respect to all others is therefore important, and it can't be done some time in the future. Priorities apply only to the here and now or the very predictable future. They cannot be established with respect to some absolute external standard. They are existential, referring to the present situation.

We must assume that there are no "absolute" priorities (except for death and taxes), just relative ones. The project manager must assign a priority to this project after considering all the other work that has to be done. An existential priority means the priority "today." If the existential priority assigned is not satisfactory to upper management, it can always be changed. When that happens, it is clearly understood that there will be consequences. These include reprioritizing all the other work affected by the change. There are no elastic clocks. Everything can't be a number-one priority. If one priority changes, it follows that probably all of them will change.

4.3 STANDARDS

What are the acceptance criteria that are to be used? How will the final test criteria apply? I remember an old joke, "If you don't have an objective acceptance test, you don't have a project. You have a career." What are the measurements and how will they be taken? In effect, what does acceptance mean to the customer or user?

4.4 CHANGE PROCEDURES

I've heard that it's a rare and beautiful thing when a project is completed without any kind of a change. I've never seen it happen. Project changes follow from the definition of a project as the solution to some previously unsolved problem. Obviously, if the answer were completely known at the project beginning, there wouldn't be a need for a project. Even if the initial specifications are defined in crystal clear terms, changes can still happen because of the following two major factors:

a. Inability to meet agreed-upon or predetermined goals and standards due to unforeseen factors usually based on the laws of nature. For example,
  - Our design for the space ship life support has to be reworked, since the last two times we tested it, the mice didn't survive.
  - The specs for this large control valve allow a maximum of two drops of leakage per hour. We've already spent six months and $50,000 in design time and our analysis of the state of the art now indicates that the best that can be done under the limitations of the specifications is three drops.
  - The preliminary flight tests indicate the weight distribution required by the specifications for this small missile causes inherent instability and self-destruction due to high-frequency flutter at certain speeds.

b. Changes imposed by management or the customer that were equally unforeseen. For example,
  - Management: I'm sorry about this, but I'll have to "borrow" three of your designers for two months beginning next Monday. But I'm sure you'll still be able to finish your project on time.
  - Customer: Wouldn't it be nice if we could change the design a bit to add the ability to submerge and become a submarine to that supersonic jet fighter you're developing? After all, how much more would it cost? I'm sure that it wouldn't stretch your budget, or even if it does, you must have a few dollars set up for emergencies that you can use.
  - Take Your Pick: I know that you can handle this within the original contract terms, but we've decided to supercharge the engines on half of those battle tanks that were ordered a year ago without changing anything else in the design.

Developing procedures for handling changes is probably the most difficult part of any charter. With an "internal project," a minor change (i.e., one that will not affect the three "golden limits"—technical achievement, time, and cost—doesn't require a change order (or an impact statement, as defined next). However, if any of these three limits will be affected, it requires a notification and a formal change notice, because additional resources are needed. Of course, with an external customer, a change order or a change in scope might be issued with *any* change that affects cost.

Change notices should be in the form of an *impact statement*. The impact statement is written as soon as the problem or change is discovered. Don't wait. Waiting usually exacerbates the situation. Impact statements should follow the general design of any written technical report, to include (a) The address and definition, (b) The recommendations, and (c) Everything else (see Fig. 2-3).

a. *The Address and the Definition*: In addition to the subject, the address also includes the usual data such as the report distribution, the writer, the date, and the identifying number. The identifying number ties the impact statement to the project and to the status at the time. It may also identify the project work package and the contract. It is the number that is used to keep track of all internal documentation. The definition describes what the original plan was and the deviation from that plan.

b. *The Recommendations*: This section entails the solution being recommended, including people, time, and money; how they will be applied; and when they are needed. These are the major elements of the impact statement. The recommendations also include an automatic release date that determines when the answer to the recommendations must be acted upon by the impact statement receiver.

c. *Everything Else*: This is where all the background, test, or contractual backup data are shown. If the receiver of the impact statement wants to analyze the data, it's there to be analyzed. If the receiver feels comfortable with the recommendations, this section may not even be read. In either event, it is required as a historical record of what happened on this project. This record is one of the first sets of documents to be reviewed when other similar projects are to be done at a later date.

4.4.1 Automatic Release Date

There are reasons for impact statements to include an *automatic decision* or *release date*. Projects have limited life cycles, and time is a major resource that shouldn't be wasted while management or the customer cogitates. The automatic release date prevents non–decision making. This means that the requested resources will either be automatically given to the project or the contract will be appropriately modified as recommended if there is no negative response within a predetermined time period. Impact statements, therefore, should be stopped only with a negative. If the receivers of the impact statement can't decide on time, it's not the project manager's problem. Of course, there must be a reasonable time limit set on impact statements for either the rejection or the silent approval to be received. If a receiver rejects an impact statement, the rejection must include either an alternative proposed solution or a change in the scope of work (SOW). While there is always a possibility that the work could still be completed without the issuance of an impact

---

THE iMPACT STATEMENT

---

1. The address and definition
2. The recommendations (with an automatic release date)
3. Everything else

Figure 2-3. Impact statement

statement—if we can assume that the project manager is the expert on this change—that possibility is relatively slim.

Conversely, management may not like the idea of "If I don't hear from you within so many days, we'll assume acceptance." The alternative can, of course, be offered: "If I don't hear from you in so many days, we'll assume rejection and, of course, stop the project." Finally, if neither of these is acceptable, the only other alternative is: "If I don't hear from you in so many days, we'll just keep spending time and money until we run out of funds and time." I don't recommend this last alternative. I believe that every change *must* have a time limit. The responsibilities for a time-based management (or customer) response to changes must be clearly outlined at this point in the charter.

Project managers sometimes feel that issuing impact statements might take up an inordinate amount of time or that they might be construed as being "disloyal" by upper management. If the concept of the impact statement is well defined as the beginning (such as in this charter), they will come as no surprise to any receiver. When many impact statements are issued during the project, it could possibly be an indication of poor planning or many unforeseen changes. In either case, recognize problems as quickly as possible and ask for help if resources are inadequate.

4.5 PERSONNEL AND PHYSICAL RESOURCES

This section lists who, what, and when. It describes the personnel and the physical equipment needed and when. The success of any project depends to a great deal upon the people who participate in it. Human beings are not interchangeable. Either list actual names of personnel required or the categories of skills, such as three engineers of at least a grade 2B, two quality assurance technicians from the electronics group, and so forth.

The project might require wind tunnels, computers, test stands, or other machinery on a full- or part-time basis. If the equipment is not available when it is needed, project deliveries, costs, or achievements will fall behind schedule.

4.6 PRICE/COST AND HOW MEASURED

What will be the cost for the entire project, either to the company if it's an internal project or to a customer as a basis for the eventual price quotation? The usual process for costing a project includes a familiar pattern:

a. Defining the scope of work (in the initial description)
b. Forecasting the task relationships including a work breakdown of some sort
c. Determining who will be responsible for each of these tasks or the parts of the work breakdown
d. Estimating the costs considering the resources available
e. Adding it all together including cost adjustments for contingencies

f. Modifying the aggregate if there are savings because of the aggregation process or adding to it if there are problems due to that aggregation process

If we're dealing with an outside customer, this process might determine the price, assuming that price is related to cost. In some cases, of course, price is not related to cost. In a purely competitive situation, it's related to the ability of the customer to pay.

This section is only the cost or price included in the first iteration of the project charter; it's not the usual formalized procedure.

4.6.1 Project Cost Envelope

The first forecast of project costs is often accurate enough for top management to make a preliminary determination as to whether they wish to commit the time and money for a more complete, team-coordinated effort. Why even get a team together if the first forecast shows a completely unsatisfactory economic solution? Naturally, this approach assumes that the project manager has sufficient overall technical and business knowledge about the subject of the project. It's a reasonable assumption. One doesn't usually become a project manager without some applicable experience and background.

This "first pass" will not be as accurate as a complete estimating effort, but it is rare that it will be off by more than an acceptable error, say about ± 20 percent. Usually, this is close enough for a preliminary evaluation. Of course with experience, the project manager improves in accuracy. This is a "guesstimate," but it's satisfactory for a preliminary decision. The result of this top-down "guesstimating" process is called the *project cost envelope*. It's an overall number that is subjectively arrived at by the project manager or the project estimator.

If this "envelope" looks acceptable to top management, the next step would be a more complete iteration of the forecasting and estimating process using a bottom-up approach, which requires the team members to estimate each of their own tasks and to coordinate the integration of these tasks into a total project task cost. It refines the cost numbers, provides a more accurate "envelope," and also commits the technical personnel providing the inputs since they have contributed their thinking to the eventual forecast.

4.6.2 Responsibility for Costs

One of the primary ground rules for a project cost envelope is that the "guesstimate" should be limited to those costs that can be controlled by the project team. There is a concept in accounting terminology called *responsibility accounting*. It means that costs are only assigned to the manager who can change them. Using this concept, there should be no assignment of manufacturing overheads, corporate administrative costs, or other charges that are not the responsibility of the project manager. They do not belong in any project cost reports.

For example, if the project manager is required to review project personnel and is responsible for recommending salary changes, those salaries should be legitimate charges for the project. However, if salaries are not controlled by the project manager but hours are, the project cost reports should be in those hours, not in dollars. Remember that costs can mean time as well as money.

Responsibility accounting should be a mainstay of all projects. Otherwise there's no objective way to measure financial success.

4.7 ADMINISTRATIVE PROCEDURES

Administrative procedures are special systems and techniques intended to standardize and simplify project operations. They should be outlined here as part of the charter. They include the techniques described later in the Project Operations and People Management systems. Because these systems are extensive, we cover them separately later in more detail. We'll review just two typical techniques used in both of these systems: (1) project organization and (2) project design review.

4.7.1 The Project Organization

An *organization* is defined as an expected set of the behaviors that people exhibit in a work environment. When these behaviors are determined by the company, the subset is called the *formal organization*. When they are determined by the participants themselves, the subset is called the *informal organization*. Both the formal and informal organizations are quickly learned by newly recruited or transferred people. Companies can recruit all kinds of people but only those who fit stay. Expected behaviors are either quickly learned or the individual becomes uncomfortable enough to leave most functional organizations. Projects, however, have a limited time to indoctrinate newcomers. Therefore, recruitment of the right people is even more important in projects than it is in functions. Consequently, the charter should state that if recruitment is done inside the company, it will initially be handled through direct negotiations between the project manager and the functional manager from whose group those team members are to be recruited. If it is done outside the company, the initial requirements are defined by the project manager before being placed in the personnel department. Personnel may handle the initial recruiting and screening activities; however, the project manager will make the final selections. In order to expedite integration, the project will always have an organization chart and project-oriented task descriptions.

As part of the "guesstimate" of the price or cost, there is consideration given to the type of skills needed and the amount of time to be spent by each skill. This is also needed for recruitment. For example, let's assume that a project engineer is needed to handle all the engineering tasks and that recruiting is to be inside. The project manager should then contact the chief engineer and outline the project needs. There should either be a request for a particular engineer or a recommendation from the chief engineer for someone to fill the position. The nature of the tasks to be completed and their duration should be part of the outline. When the project manager and the functional manager agree on an individual, the request is offered to that person. If there is a refusal or a scheduling problem that prevents the project manager from obtaining the originally requested person and a substitute is required, the project cost envelope should be re-evaluated. If there is a major change, management should then reapprove the project charter. (This process is covered in greater detail in Chapter 6.)

When the selection of the project engineer is agreed to by all three—the functional manager, the project manager, and the particular engineer—the engineer then participates (as do other members of the "team" selected similarly) in the next iteration of the project-planning process, the bottom-up forecast. This second forecast refines the definition of the technical result, the cost, and the time estimates. The recruiting process is basically the same when the particular engineer is recruited from outside the organization.

4.7.2 Project or Design Review

The other technique we examine here briefly is *project design review*. Project planning includes review of achievements against forecasts. It should be clearly understood that there are three possible outputs of a design review:

1. Continue on with the project.
2. Rework project design to some acceptable standard and then continue.
3. Stop the project.

Predetermined dates must be noted for total project design review meetings. (See Appendix 3 for a complete description of a typical design review procedure.)

5.0 THE CHARTER: IN BRIEF SUMMARY

Development of this type of internal charter as part of the Forecasting system is probably one of the most important tasks in effective project management. The problems that are solved are classical and quite repetitive. If they can't be resolved before a project begins, the conflict, confusion, and inability to reach well-known project goals (meet specifications on time within the budget) will quickly follow. The charter helps in defining plans for:

a. People
b. Achievements
c. Resources

6.0 IMPLEMENTING THE CHARTER

The charter is first completed by the project manager. If the resulting data look acceptable, then a full-scale forecasting sequence and proposal documentation should be completed. In summary, there is a

   a. Top-down forecast (the project manager completes the charter), then a

   b. Bottom-up forecast (the personnel who would be primarily responsible for work to be done on the project makes revisions), and then

   c. Top review (present the answer to the RFP (Request for Proposal) back to Marketing, Sales, and top management group for approval before submission)

Any changes that occur to the answer to the RFP when it goes through section c as a result of the top review should be treated as any other change to the project. If the changes are expected to result in an overexpenditure beyond the original budget or affect the technical scope of work (as defined in section b), an impact statement should be issued. This happens if the RFP cannot be satisfied within the budgets that were negotiated.

7.0 SUMMARY

Solving the repetitive organizational problems of projects before they become critical using the internal contract or charter follows another more or less familiar contract technical design and development sequence that many of us used when we began our technical careers:

   A. Define the scope

   B. Set up priorities

   C. Determine schedules and estimates

   D. Lay out test and change procedures

   E. Specify resources

   F. Outline the organization

This is a familiar general organizational systems design that outlines *how* any assignments are to be managed in a company. It often is part of the standard operating procedures. It can be an important part of the foundation upon which modifications can be built to satisfy any customer's RFP.

This Could Be What Happens Next in Our Case Study

The answers to the questions are noted in the following conversation. Do you agree with them?

Bob: Well, Sam, what's your answer?

Sam: We have been designing, building, and installing our machinery for customers who never have to be responsible for or accept our machines. Our "customers" are the plants. We've never gotten any of the plant managers to agree to a contract or a set of acceptance tests within that nonexistent contract. That means that we put all that machinery in, with their approval of course, and they never have had to "accept" it, because we've never defined what "acceptance" really is. It follows logically that we're becoming more involved in their maintenance engineering and modifications. As time goes on, the manpower requirements of the original Machinery-Design Engineering group stay about the same since the level of activity for "new" equipment has either remained constant or has even gone down a little. That's one of the reasons why my group can remain fixed but Project Management and Field Operations are increasing. How obvious!

Mary: But isn't maintaining and improving installed equipment what we're supposed to be doing?

Bob: No, Mary. Corporate Engineering is supposed to design and install new equipment, not maintain existing equipment. That's the plant's job, and I see a way to solve some of our problems. We can present a report to Milt offering him two alternatives:

   A. Engineering gets out of the maintenance business. This means that the plants have to agree to some sort of acceptance test of our machinery. After that, they have to either maintain the equipment themselves or else use intercompany charges to pay for our maintenance labor.

   B. Engineering stays in the maintenance business. This means that both Mary's and my staff will increase over the years as the installed machinery becomes older. This should stop the complaints about the increase in labor costs.

Mary: That's a great theory but how will we get it across? After all, they have been relying on us for years. Why should they stop now?

Sam: How about this? We ask each plant manager what he or she would consider to be an agreeable acceptance test. If we agree on a test, we then put our machines to it. Any machines that "fail," Engineering will set up a repair or alteration program to fix the equipment until it passes. It may cost us in the short run, but it will get us out of the long-term maintenance business.

Bob: I'll get a report ready for next week's meeting with Milt. Meanwhile, let's review all our old machine prints to get a rough estimate of what tests we would suggest and how much it would cost to meet them.

BIBLIOGRAPHY

Assad, Michael G., and G. P. J. Pelser, "Project Management: A Goal-Directed Approach," *Project Management Quarterly*, June 1983, pp. 49-58. Project Management Institute, P.O. Box 43, Drexel Hill, Pa. 19026.

Hajek, Victor G., *Management of Engineering Projects*. New York: McGraw-Hill, 1977.

Meredith, Jack R., and Samuel J. Mantel, Jr., *Project Management: A Managerial Approach*. New York: John Wiley, 1985.

OTHER READINGS

BAUMGARTNER, JOHN STANLEY, *Project Management*. Homewood, Ill.: Richard D. Irwin, 1963.

BEER, STAFFORD, "The Shorter Catechism of Stafford Beer," *Datamation*, February 1982, pp. 146-55.

GOODMAN, RICHARD ALAN, "Ambiguous Authority Definition in Project Management," *Academy of Management Journal*, December 1976, pp. 301-15.

HORNGREN, CHARLES T., *Cost Accounting: A Managerial Emphasis* (5th ed.). Englewood Cliffs, N.J.: Prentice-Hall, 1982.

MAKRIDAKOS, SPYROS, and STEVEN WHEELWRIGHT, *Forecasting Methods and Applications*. New York: John Wiley, 1978.

MARTIN, CHARLES C., *Project Management: How to Make It Work*. New York: Amacom, American Management Associations, 1976.

ROSENAU, JR., MILTON D., *Successful Project Management*. Belmont, Calif.: Lifetime Learning Publications, 1981.

Chapter 3 Forecasting: Techniques

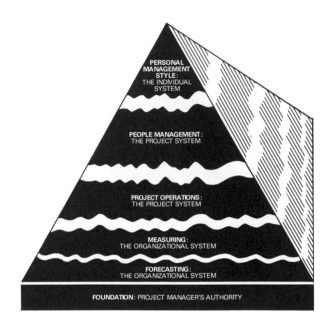

PERSONAL
MANAGEMENT
STYLE:
THE INDIVIDUAL
SYSTEM

PEOPLE MANAGEMENT:
THE PROJECT SYSTEM

PROJECT OPERATIONS:
THE PROJECT SYSTEM

MEASURING:
THE ORGANIZATIONAL SYSTEM

FORECASTING:
THE ORGANIZATIONAL SYSTEM

FOUNDATION: PROJECT MANAGER'S AUTHORITY

FIBRO MACHINE COMPANY

Cast:

Martin Wassal, Project Manager

Dianne Pascal, General Manager

Lew Golder, Project Engineer

Alex McAndrews, Subcontracts Administrator

Charles Dobbs, Chief Engineer

It was early Friday morning, and Martin had left home a little early to prepare for the weekly staff meeting that started later that day at 4:00 p.m. in Dianne's office. As he maneuvered his car through the early morning traffic, he reviewed the situation. It was a little more than three months ago that Dianne had called him into her office and given him the assignment as proposal manager for the turboencabulator project. This project was started in response to Mammoth Industries' request for a proposal for a novel automatic pipelaying machine. Fibro had been making mechanically controlled machines for years, but this one was one of the largest and most complex that they had ever considered. It had more power and more controls (since it was the first machine with on-board computers) than anything else that they had every built.

During the first month Martin completed his internal project charter and made his general presentation to Dianne and her staff. The charter was accepted so Martin recruited his project staff and proceeded to develop a detailed bottom-up response to the proposal. Martin met with the various functional managers in order to recruit the project team members. Most of these meetings went well. The people he asked for were available and assigned to his project. But the meeting with the chief engineer, Charles Dobbs, didn't go well. Charles said that he couldn't release Lew Golder, whom Martin had asked for as project engineer. Lew had to finish up a design that was behind schedule, and he wouldn't be available for two weeks. Martin delayed his bottom-up meeting for those two weeks.

Two weeks later at the meeting, Martin passed out copies of the request for proposal and his charter. Lew Golder proposed that they use a new optical disc computer that he had just read about in a technical magazine. It was built by Topnotch Computer Company. That company had an excellent technical reputation. Lew also said that he thought that the modifications to fit the computer into the Turboencabulator would be minor. Alex McAndrews promptly protested. He said that Fibro hadn't done any business with Topnotch before, and he wouldn't agree to buying a new component like this without a thorough investigation. Lew reminded Alex that he was the project engineer and his technical opinion should overrule anybody else's. Martin had to agree with Lew.

Then Lew became very hesitant about committing himself to meeting the design and development time targets but eventually agreed. The meeting finally ended when Martin asked them all to come back in two weeks with detailed estimates and their review of the charter.

The follow-up meeting was delayed for four weeks because Charles had suddenly assigned Lew to another job. But it finally came about and the project cost was established. The computer cost was 20 percent of the total project cost. The job would involve about 75 of Fibro's best technical personnel and take 14 months to complete. It was accepted at the staff meeting, and a selling price was set and then presented to Mammoth Industries. Mammoth rejected the quotation since it was several weeks after the due date.

Fibro's sales team then met with Mammoth representatives. They tried to convince Mammoth that Fibro had the best proposal. After several hours of discussion, Mammoth accepted the proposal but with a 5 percent reduction in price.

The next day, Dianne phoned Martin to tell him that Fibro had gotten the Turboencabulator job and that she would post the notice that day that he was appointed project manager. She also mentioned that because of the delay, the Fibro sales team had to reduce the price by 5 percent in order to satisfy the Mammoth negotiating team. The turboencabulator contract arrived in the interoffice mail later that day.

After he reviewed the contract, Martin called a meeting of the proposal team, told them that the Mammoth job was a project, and requested that they review their parts of the proposal to see if those parts had changed since the proposal was sent in. He told them about the reduced price too and asked them if it was possible to reduce costs to meet it. At the project "kickoff" meeting a week later, Lew flatly stated that he couldn't guarantee that the engineering tasks would be done within the original budget, and, of course, any reduction would be impossible. He said that he had updated and reviewed the job with his functional boss, Charles Dobbs, and Charles had told him that Engineering test equipment could not be made available to the project as originally scheduled. Engineering would need it for the next year. Lew also reported that Charles now felt that the time and cost for the engineering design tasks had been underestimated, and he was going to increase them as soon as he got around to it, within the next several weeks. Lew said he would tell Martin about the new numbers as soon as they were ready.

Alex had another bit of news. Topnotch had raised their computer prices by 20 percent within the last week, and they wanted at least 50 percent premium in addition to modify their standard Model A-32 for the turboencabulator. They also were very busy and couldn't deliver for 18 months. The turboencabulator job seemed to be disintegrating before Martin's eyes. What could he do?

Questions

1. How should Martin have set up the bottom-up meeting? What should he have done about the dispute between Lew and Alex at the meeting? What should he have done about the delay that Charles caused when he assigned Lew to another job?
2. What should be done about the reduction in the overall price by 5 percent? How and by whom?
3. Should anything be done about Lew's report? What?
4. How should the Topnotch computer problem be resolved? By whom?
5. What should Martin do now to prevent these things from happening again? What should Dianne do?
6. Did you ever have a similar situation? How did you handle it?

1.0 Forecasting: A Quick Overview

At this point, we have made a promising start in designing project systems. The project manager's authority has been correctly defined, and the initial top-down charter is approved. This is the core of the project manual or the project standard operating procedure (SOP). We're now ready for the bottom-up iteration, the first revision to the charter. This revision happens in the first of many project meetings that are called *organization meetings*, a series of problem-solving meetings with an agenda or a schedule for discussion. We will cover several alternative agendas. Then we begin discussing the actual forecasting techniques used in the meeting, as well as many others. Iterations begin even at this stage.

The bottom-up charter revision or organization meeting provides inputs to solve two important questions:

1. Does each of us really know what we have to do?
2. Are we committed to do it?

BRIEF SUMMARY   NOW

1. Define the scope.
2. Set up "existential" priorities.
3. Determine schedules (time) and estimates (cost).
4. Lay out test and change procedures.
5. Specify resources.
6. Outline the project organization.

A. Do we really know what we have to do?

B. Are we committed to it?

Figure 3-1.   Bottom-up questions

Iteration is almost a never-ending process. It occurs throughout the project life cycle. Figure 3-1 is a bottom-up iteration that is partially an expansion ands revision of some of the charter elements. After a review of those charter elements, preliminary forecasting can begin. Figure 3-2 illustrates a very simple flow model that

EXPANDING THE CHARTER ELEMENTS

A. Establish or review objectives.
B. Do task planning.
C. Do work breakdown schedule.
D. Layout schedule, performance versus time.
E. Estimate time and costs.
F. Clearly outline work assignments.
G. Develop procurement policies.
H. Define internal physical resources.
I. Design information system.

Figure 3-2.   Expanding the charter elements

could be used as a preliminary agenda for such a meeting. This agenda follows the usual design sequence: first technical objectives, then time, and finally cost. Figure 3-3 gives alternative preliminary agendas for such a meeting. A typical agenda could be as follows:

A. *Establishing the Objectives*
   Review the charter, customer request, and management inputs for changes and modifications.
B. *Do Task Planning, That Is, Forecasting Technical Tasks*
   The logical development of sequential and concurrent tasks is the intended result of task planning.

MORE QUESTIONS	FINAL PROCESS

a. Can we do it technically?

b. Do we have the people (talent) and capacity to meet the RFP?

c. Have we done anything like this before?

d. How should we modify our "standard" charter?

1. Top-down review
   Complete the charter

2. "Bottom-up review
   Team review and proposal iteration.

3. Top review

CHANGES

1. Identify the change and how it varies from previous forecast.
2. Forecast (a) technical, (b) time, (c) cost.
3. Identify interactions with other parts of the project.
4. Ensure that all other team members are informed.
5. Provide for implementation.

Figure 3-3.   Alternate meeting agendas

C. *Do a Work Breakdown, That Is, Relate Those Tasks to One Another.*
Separate the knowns from the unknowns and determine if multiple parallel paths or feasibility studies should be used. With time constraints, multiple design approaches or parallel paths might be appropriate. It's expensive but effective in getting a technical solution. With cost constraints, feasibility studies are better. They take a bit more time, but the costs are a lot less.

Determine the size of the various work packages based on the required technical complexity or organizational size. The more complex the work or the greater the unknowns, the smaller the work package. But when work packages are too small (i.e., less than a week), it costs almost as much to forecast and measure them as it does to do the work. As a rule of thumb, try not to have any work packages last less than three weeks.

D. *Lay Out the Schedule or Performance Versus Time: Time*
Start with the end date, forecast elapsed time back to the beginning date. Don't forget to consider the resources you have and when they will be available. For example, do we have the people, financing, and physical resources to complete the various work packages in time? Are there any other limitations, such as contractual obligations, that affect time?

The actual techniques chosen for forecasting can vary from the relatively straightforward techniques to complex mathematical models. This chapter and the next one cover Gantt or bar charts, PERT diagrams, line of balance, work breakdown schedules, and the coordinated forecasting process. (See Makridakis and Wheelright, 1978, or Georgoff and Murdick, 1986, for more techniques.)

E. *Estimate Expenditures: Costs*
Working or "chargeable" time is not the same as elapsed time. They are related because the working time may be dependent upon availability, but they are not the same thing.

F. *Clearly Outline Work Assignments*
Who is responsible? The project manager is supposed to manage interfaces among the various assigned tasks and work packages. Therefore an initial job is that of allocating major parts of the project to the team managers. That is, the project engineer handles technical work, the subcontracts manager does the outside contracting, and so forth. But the project manager is always very concerned with *who* will do it and *how* they intend to get the work done. Therefore, after the major parts are allocated, the team managers prepare their own forecasts. These forecasts are reviewed by the project manager and then integrated into a whole project. This sequence is needed because the team managers, as well as the project manager, need to know when project interfaces will be expected to happen.

The following task description (Hajek 1977, p.152) is typical for recruiting the project team:

Date: 15 January 19-

Project: Turboencabulator No. 21

Quantity: One (1) prototype

Assignee: Project Engineer M. Shoes

Type of task: Design of encabulator drives

Description of task: Design the encabulator drives to match customer's specification (Attachment A). Design approach shall be consistent with that proposed in Proposed Task Design, Project 234 (Attachment B).

Task schedule: The schedule is shown in Attachment C.

Budget: A preliminary estimate of engineering hours is shown in Attachment D.

Reports: Cost Prediction Report (nonlinear beginning in one month); Line of Balance Report (monthly); Manpower Loading Display (monthly).

Attachment A: Specification X 48.

Attachment B: Design approach, Encabulator, Project 234.

Attachment C: Task schedule, Project 234.

Attachment D. Budget: Charter estimates, Turboencabulator No. 21.

G. *Develop Procurement Policies, Make or Buy?*

What are the criteria for judging where to get the materials, goods, and services that the project will need? How will you "value" product quality, security, delivery, and all those other variables that occur in addition to price? These criteria and their applications involve all the managers on the project team.

H. *Define the Internal Resources Needed*

What physical equipment will be needed? The cost estimates usually have been done by now, but this refers to people, raw materials, test equipment, facilities, and machinery. Cost estimates often assume that required internal physical resources will be available when needed. This is where it's all spelled out.

I. *Design Your Information System*

It is rare that the functionally oriented internal administration or cost-reporting systems can be used without modification by the project. Functional managers' jobs have a generally consistent range of uncertainty over time. Project managers have a larger amount of uncertainty than equivalent functional managers do when the projects are just beginning. And they have much less time at the end. Therefore, in order to absorb the same amount of uncertainty (if plotted vertically) over time (plotted horizontally), project managers require more and faster feedback information at the beginning of projects and less at the ends. (see Fig. 3-4)

A correctly designed project information system, then, must report on a nonlinear basis over time. We need lots of information reported very frequently during the formative stages of the project when most of the technical

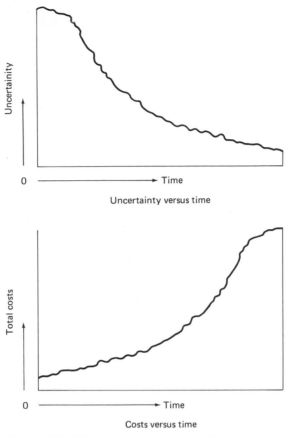

Uncertainty versus time

Costs versus time

Figure 3-4. Uncertainty and costs versus time for projects

FREQUENCY OF REPORTING IN PROJECTS

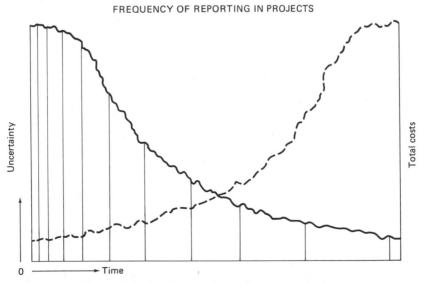

Figure 3-5. Nonlinear frequency of reporting in projects

decisions will be made, but not much time or money will be spent (comparatively speaking). Then, we need less information as the project matures and moves into manufacturing, even though that's when a lot of money is usually spent. In other words, there is more uncertainty and less funds expended in project beginnings. There is less uncertainty but higher costs incurred at the end. That's not the way most organizational reporting systems work. They usually report on a linear basis over time. For example, the cost reports for the previous month always come out by the second week of the following month. This is not acceptable for projects. (See Fig. 3-5.)

2.0 THE CHANGING ROLE OF THE PROJECT MANAGER

Forecasts are the selection of an answer *today* to some problem that is expected to occur in the future. The forecasting technique itself is also a communications process that helps coordinate the efforts of the team. It provides a common language for the team and promotes understanding. Therefore, there has to be an agreement that whatever technique or method is selected, everyone interacting with it understands it, even if they may not always agree with the output or the forecasting recommendations themselves. Every method has its advantages and disadvantages, and it's difficult to select those methods that will fit every situation. We will cover several appropriate forecasting techniques. Those that are actually selected by your project organization should be available from some general organizational systems manual. It should even be possible to select several methods and compare the results or the scenarios of each.

In the second iteration of the project, called the *bottom-up forecasting process*, the role of the project manager changes. Initially, during the completion of the project charter, the project manager is the major forecaster and planner. Now he or she becomes

a data tester
''Why do you think that the hydraulic system should be made of titanium rather than aluminum, and what is the basis for your opinion? You know that titanium costs more and it has a longer delivery time. Do we have any objective data supporting your recommendations? What methods can we use to compare your opinion against those of the purchasing people, who say they have a vendor who can make an aluminum system that is just as good and less expensive?''

a coordinator
''When I added all the elapsed time of these tasks, their total was more than twice the contracted project length. What can we do about consolidating them or running them concurrently?''

or an adjudicator
''As project engineer, you say that you'll be sure enough of your initial findings to eliminate the preliminary wind tunnel tests next March. But the

representative from Quality Assurance insists that every design should have preliminary wind tunnel tests. How will we resolve this difference?

Obviously these changing roles are not independent of one another. For example, it's possible (and may even be necessary) to be a data tester and a coordinator while in the role of an adjudicator. By definition, we've said that project managers may not be as technically qualified as the team participants, but they have to be able to test and coordinate the inputs that these people provide. The project manager must make multiple comparisons of seemingly objective data to detect discrepancies, uncertainties, or unsolved problems. It's much more difficult than dealing with regular and relatively predictable technical problems found in equivalent functional organizations because of the very complex, nonrepetitive human interactions that are often the basis for these project-oriented situations. Projects are new and different. Project forecasting techniques that are used in the Forecasting system are intended to define these unknowns and begin the process of solving them by changing them into knowns.

3.0 STARTING THE FORECAST—GROUND RULES

All forecasting techniques explicitly require that the team managers on the project assume the responsibility for their parts of the project. The people are more important than the particular technique chosen. With all the various scenarios or agendas designed to build a team, never forget that it's the individual project team members who have to do the actual work. The procedure is only used to help them. (See Hunt, 1979, for more of this idea.)

There are only parts of the future that are predictable, such as potential results that occur due to the actions of physical laws such as gravity, chemical reactions, and so forth. The rest, obviously, is not predictable. There will be unforeseen problems that can disrupt even the most detailed forecasts, and there are unexpected changes that can be imposed by the customer, your own organization, or technical considerations. It always happens. Therefore, although the following discussion of techniques may seem to suggest a kind of linear process, in practice the process is rarely straightforward. There are always iterations that occur because of a lack of knowledge in earlier phases of the project or because of current changes in the project direction. Even an obviously linear project such as reading this book is somewhat iterative, since some subjects are treated in later chapters again after being introduced earlier. Some are even discussed several times, but from different viewpoints. You might possibly want to review (iterate) by going back through these chapters and by matching the contents with your situation as you go along.

3.1 THE FIRST FORECASTING TASK:
DEFINITIONS AND OVERVIEW

Definitions are integral building blocks of any system. If there is little coincidence between the project's definitions of its own goals and those of the customer, there is always trouble. Similar troubles occur when there are differences of definitions of project subgoals among project team managers. Some overall definitions must be obtained from the customer. A *customer* is usually defined as an organizational outsider. A *user* is usually an organizational insider. While many definitions have been covered in the initial top-down charter, some may have slipped through the cracks. While partially repetitive, these might be considered as part of the bottom-up review.

A. What are the objectives and how are they measured?

B. What will the end product look like?

C. How will it be tested?

D. Are there any contractual or technical constraints?

E. What resources do we need?

F. What are our organizational constraints?

G. When does it have to be done?

H. Are there intermediate steps that require the customer's approval or coordination? What are they? What do we have to do to complete them?

I. Are there any management, financial, or reporting systems that have to be used as part of the project?

J. What is the cost? Price?

3.2 QUALITATIVE AND QUANTITATIVE COMPARISONS—MATRIX DECISION ANALYSIS

Since projects are always concerned with creating something new, they are sometimes defined qualitatively first. "This project should result in a new injector nozzle that will increase our share of the diesel engine market by 20 percent." This opinion is qualitative, even though it uses numbers. Qualitative measurements, therefore, are important and therefore we must learn how various team managers think about the assigned project tasks. We must be able to measure differences of opinion. Qualitative doesn't mean nonmeasureable; it's just a different kind of measurement. There are qualitative, but measurable, decisions made every day by consumers. Some people buy bicycles and some limousines. Both purchases provide transportation but there's a major qualitative difference.

An overall measurement system allows us to define all forecasts and compare them. It includes a sequential movement from qualitative into absolute quantitative measurements as follows:

1. *Nominal: Nominal measurements* are simply different from one another. An example is red and black checkers. Red is different from black, not better or worse, just different.

2. *Ordinal: Ordinal measurements* can be defined as nominal measurements plus

or minus an additional tangible factor. Picasso's paintings are different from Fragonard's; that's nominal. If a museum will trade several Fragonard paintings for one Picasso, we have an ordinal measurement. By that standard, an ordinal measurement is tangible. Another example is as follows: Charlie is a much better project manager than George. In fact, on a scale from 1 to 10, where 10 is tops, Charlie's about an 8 while George is a 5.

3. *Interval: Interval measurements* can be defined as ordinal measurements plus some scale relationships. For example, 60°F is 30° hotter than 30°F. It's a more exact (i.e., replicable) measurement than ordinal, but it's still not absolute since 60°F is *not* twice as hot as 30° F, since there's no "zero" point.

4. *Ratio: Ratio measurements* can be defined as interval measurements plus an absolute zero point. For example, 80°K is not only 40° hotter than 40°K, it is also twice as hot since the Kelvin temperature scale has an absolute zero. The Fahrenheit scale doesn't.

In effect, we pass from completely personal judgment to very objective measures. Ordinal measurements are very useful in project forecasting, since they help us to compare different *thinking* processes. (Remember the comparison of Charlie and George?) The least exact scale, the nominal scale, is not as useful because there's no numbering to compare amounts or sizes. It's just a bigger or smaller, better or worse kind of measurement. Nominal scales can differentiate between two measurements, but they don't help in determining opinions of *how much* better or worse. Jumping over ordinal to interval scales, we find that they are more exact because they have uniformly acceptable scales. But in management, we deal with opinions and mental processes, and they are far from linear. Everyone doesn't think in uniform increments. Finally, ratio scales are limited to scales with a zero value and therefore cannot be logically applied here, again because we are dealing with human decisions and those have no absolute zero, just relative ones. Even in the natural sciences, there are very few ratio measurements and they apply to basic electricity, temperature, and so forth.

In my opinion, ordinal scales are perhaps the most useful in measuring and comparing forecasting and management decisions. For example,

"How sure are you that we can develop that new pump to meet the customer's schedule of three months? On a scale of 1 to 10, where 10 is absolute confidence, what score would you give your forecast?"

"An 8."

"Why an 8 and not a 9 or a 7?"

The ordinal scale is helpful in communicating and evaluating the data inputs upon which we, as managers, can base decisions. It's possible to use ordinal numbers to resolve differences in viewpoints during the forecasting process when comparing alternative solutions. These are often used in a technique called the *decision matrix analysis*. For example, Fig. 3-6 shows three possible solutions to a

MATRIX DECISION ANALYSIS

CRITERIA	SOLUTION 1			SOLUTION 2			SOLUTION 3			TOTAL
	A	B	C	A	B	C	A	B	C	
Cost	4	6	5	2	7	6	6	8	8	52
Reliability	8	7	5	7	9	5	5	7	8	61
Size	3	4	3	4	4	5	1	4	6	34
Performance	7	6	5	6	6	6	7	8	9	60
TOTAL	22	23	18	19	26	22	19	27	31	

A likes Solution 1 on an
 overall basis
B likes Solution 3
C likes Solution 3 too

Solution 3 is the most costly
Solution 2 is the most reliable
Solution 2 has the best size
Solution 3 has the best performance

Figure 3-6. Matrix design analysis

problem. We ask individuals A, B, and C to give us their *expert opinion* of the various criteria associated with each solution on an ordinal scale of 1 to 10, where 10 is wonderful. By adding vertically, we can see which solution A, B, and C like the best. By adding horizontally, we can see which criteria seem to be most important. It does work. In this example, A likes solution 1, B likes solution 3, and C also likes solution 3, on an overall basis. Going a bit further, we can possibly discard "performance" and "reliability" as criteria, since everyone gives them high scores for all solutions and concentrate on the problem criteria of cost and size. The problem is becoming more easily defined, through ordinal (subjectively quantitative) measurements.

There are other ways of looking at measurements that might help us in our forecasting techniques. For example,

It is common for those who oppose a project, for whatever reasons, to complain that information supporting the project is "subjective." . . .

Subjective vs. Objective: . . . All too often the word *objective* is held to be synonymous with fact and subjective is taken to be a synonym for opinion; where fact = true and opinion = false. . . . A measurement taken with reference to an external standard is said to be "objective." Reference to a standard that is internal to the system is said to be "subjective." . . . A yardstick incorrectly divided into 100 divisions and labeled "meter," would be an objective but inaccurate measure. The eye of an experienced judge is a subjective measure that may be quiet accurate.

Quantitative vs. Qualitative: . . . The true distinction is that one may apply the law of addition to quantities but not to qualities. Water, for example, has a volumetric measure and a density measure. The former is quantitative and the latter is not. Two one-gallon containers of water poured into one container give us two gallons, but the density of water, before and after pouring is still 1.0.

Reliable vs. Unreliable: . . . said to be reliable if repetitions of a measurement produce results that vary from one another by less than a prespecified amount.

Valid vs. Invalid: Validity measures the extent to which a piece of information means what we believe it to mean. A measure may be reliable, but not valid. . . .

To be satisfactory when used in an evaluation/selection model, the measures may be either subjective or objective, quantitative or qualitative, but they must be numeric, reliable and valid. (Meredith & Mantel, 1985, pp. 54-55)

In my opinion, ordinal measurements provided by experienced team managers are very useful. They may be subjective but they are usually valid and most probably reliable.

4.0 NOW WE CAN BEGIN THE ACTUAL FORECAST

To begin the actual forecast, first determine the sequence of the technical objectives, since those objectives are usually the rationale for the project itself. It's been my experience over the years that there has been more money made or lost in forecasting these objectives than either time or cost. If the technical objectives are unsatisfactory or uneconomic, there's very little that can be done by Purchasing, Manufacturing, Sales, or any service function to improve the situation later. This seems to be confirmed by research findings.

For each project, we asked both the government administrators and the company management to rank their criteria of success in order of importance. Technical performance was by far the most important consideration, being ranked first by 97% of the government administrators and by 63% of the laboratory and project managers. Meeting delivery schedules was second, and achievement of cost targets was third in importance. . . . Performance to schedule and cost performance is more easily measurable than technical performance and is also more comparable across the full spectrum of research and development projects. (Marquis, 1969, p. 82)

It's interesting that the success of many projects is often defined by the secondary measurements of meeting schedule and cost targets, even though those targets are directly dependent upon the success in meeting technical targets. Some organizations seem to believe that schedule and cost targets are more objective than technical targets. They're not.

If we agree that the first objectives are to be the technical ones, we can start, as usual, with definitions that allow testing of results: How will they be measured? How will they be tested? What happens if the test results are not clear? Then, we can define the second set of objectives of time. Both the technical and time objectives are begun at the end of the project. ''The end result will be a revolutionary new widget that will have wonderful characteristics and do marvelous things for the widget owner.'' This is the overall technical objective. Now we can work backward defining the criteria for the widget bearings, the housing, the drive motors, and landing gear, almost as if we were disassembling the still-to-be-designed widget.

Next, let's consider the time objective. This widget has to pass final acceptance tests by next March. Working backward again and considering the backward-outlined technical breakdown, ''the housing has to be purchased by the previous September, machined by November, and tested by December.'' So, we start with the ''end'' and work backward. It should always be understood, however, that we are defining targets, not the actual completion date(s). A *target date* is a presently expected completion date that will occur in the future. It must include provisions for some subjective contingency.

Finally, we can set budgets or cost targets. As noted before, some project forecasts seem to concentrate on this last objective, but it really is the least important one in most projects. However, this may change if conservation of resources becomes a primary target. When this happens, both technical and time targets may temporarily become secondary, and the forecast revolves around the resources available. In other words, although there is an optimum sequence of forecasting that the project manager should follow, there can be occasional modifications to fit changes in the situation.

4.1 TECHNICAL OBJECTIVES—FORECASTING

Not all technical objectives are equal. It is wise to be concerned about those with the highest unknowns, because they usually have a major impact on the project. After all, what's the reason for the project to exist if it isn't to solve a novel or prior unsolved technical problem?

Technical objectives dealing with a greater number of unknowns obviously should include many more checkpoints than those with lower uncertainty levels. *Increasing* the number of checkpoints in the forecast decreases the number of problems, since each checkpoint measures a smaller increment in the project that must be satisfied before the next increment is attempted. When each step is completed, the next step (whatever it may be) is released by the project manager. The amount of uncertainty is limited to the amount allocated in the prior step. The team manager or participant in charge of that smaller step is then like a subproject manager handling the appropriate uncertainty level of that subproject.

4.2 TIME OBJECTIVES—FORECASTING

A forecasted date is not a fixed promise; it's a target. There's always some uncertainty involved. If the organization attempts to define forecasts as some kind of a rigid contract that must be met, the practical answer is often to increase the amount of time for contingencies. The time added for contingencies is related to both the task uncertainty and the organizational pressure that the forecaster perceives. It's a practical way to handle this kind of pressure.

Contingencies and uncertainties can be decreased by increasing the frequency of reviews. This should lead to more frequent forecasting iterations, but iterations don't have to happen at equal intervals. The amount of time between project reviews depends on the particular stage of the project. For example, "We ought to meet twice a week while we're working on this tricky nozzle design." It also depends on the amount of time being forecasted. For example, "We're reasonably sure that we can use the test stand for one week next March. We'll review the situation again in February to be sure. Someone may have a higher priority then, and we might have to subcontract it out." Remember that every forecast has a built-in *decrease* in usefulness with a longer end date. The further out in the future the forecast extends, the less useful it is, and that's usually on a geometrically decreasing basis with age.

Therefore, if an organization expects a time forecast to be met absolutely, it will probably be either disappointed or else the forecasts will be padded to make sure that there is enough time to cover almost any eventuality. Since forecast accuracy is important, let me repeat that probably the most important factor supporting accuracy is an adequate definition of the project scope. The second factor is the subjective evaluations and testing of forecasts by the team managers, the project manager, and then finally even by upper management. The more exact the definition, the more apparent become the differences between the knowns and the unknowns. The unknowns can cause most of the trouble. The project manager, therefore, should concentrate heavily on minimizing the unknowns through

A. Following up more closely

B. Creating smaller sections of the forecasts for the unknowns

C. Allowing some contingency time in the overall forecast summaries for those unknowns to be used if necessary

If project managers are to be held responsible for accuracy irrespective of the technical unknowns, the obvious answer is to create sufficient slack to compensate for those unknowns.

4.3 COSTING OBJECTIVES—FORECASTING AND RESPONSIBILITY

Cost forecasts should be made by the same people who provide the technical and time forecasts, since they are responsible for the particular tasks to be accomplished. When the funds are not spent by the same people as those who create the cost forecasts, I have found that something unforeseen always happens, and no one is willing to take the responsibility for it.

5.0 CONTROLLING THE VARIOUS FORECASTING INPUTS—THE RESPONSIBILITY MATRIX

Controlling the forecasting process itself is somewhat like a production control process. Even relatively straightforward projects require some minimal management to schedule and control the forecasting process. A *responsibility matrix* can be very useful. List the particular forecasts across the top of a sheet of paper, the names of the responsible managers down on the left side, and specified code letters at the intersections denoting the results. The code letters could signify when a particular forecast is due, the input needed to complete it, and the expected output. A minimal set of code letters would involve the initial scope and schedule for each forecast and preliminary assignment of action items. Every project forecast is iterative, including task rescheduling and assignment restructuring.

Unless the project involves a low degree of complexity, it probably is a mistake to do very much detailed forecasting at the first project meeting. There are still too many large unknowns that have to be analyzed and major design decisions made before details can be forecasted. When these are completed, subsequent meetings can be used for detailed forecasting. The responsibility matrix can help in controlling this process (see Fig. 3-7). A typical task relates Peter to the Quality assurance input from Sam.

RESPONSIBILITY MATRIX

	Configuration control	Quality assurance	Contracts	Etc.
A. George (Engineer)	D. 3/1 —	E. 3/1 —	B. 2/15 —	—
B. Mary (Finance)	— —	E. 3/1 —	E. 1/15 2/1	—
C. Sam (Manufacturing)	A. 4/1 —	E. 3/15 —	B. 2/15 —	—
D. Laura (Quality)	B. 2/1 In Work	E. 2/15 In Work	B. 2/15 —	—
E. Peter (Project manager)	C. 4/15 —	C. 4/1 —	All 3/1 —	—

Figure 3-7. Responsibility matrix

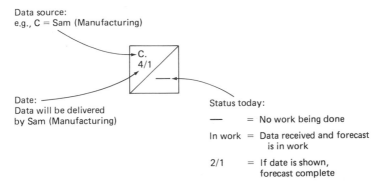

Data source:
e.g., C = Sam (Manufacturing)

C.
4/1

Date:
Data will be delivered
by Sam (Manufacturing)

Status today:

—   = No work being done

In work = Data received and forecast
           is in work

2/1   = If date is shown,
           forecast complete

Figure 3-7. (cont.)

6.0 TASK OR EVENT FORECASTING: INTRODUCTION TO THE WORK BREAKDOWN SCHEDULE

The project has now been broken down into major tasks or events. These are not final and unchangeable. They are probably not even the tasks that the project will actually use without modification; rather, they are a starting point to guide project thinking. The simplest of forecasting tools, such as a logical hierarchy of tasks, is therefore probably a good beginning point.

The development of an appropriate hierarchy usually follows a decision-making process called a *means-end analysis* (MEA). An MEA uses successive approximations starting with the general goal to be accomplished (that is, we begin at the end), discovering a set of very general means for achieving that goal, then breaking down those means into more detailed means. This is continued until a level is reached that can provide specific known results. We work from the "end" back to "today."

> When the new goal lies in a relatively novel area, this process may have to go quite far before it comes into contact with that which is already known and programmed; when the goal is of a familiar kind (e.g., a Red Cross disaster program in a particular area), only a few levels need to be constructed of the hierarchy before it can be fitted into available programmed parts. (Metaphorically, we imagine a whole warehouse full of parts in various stages of fabrication. The plan for the new structure must be carried to the point where it can be specified in terms of those stocked parts.) (March & Simon, 1958, p. 191)

At each step in the hierarchy development, there is a judgment about the need for greater detail. Do we need more or fewer work packages? The number of steps or levels that are chosen should depend upon the complexity of the project (i.e., the number of unknowns) and the size of the organization. If we're not sure what the machine will look like, those metaphoric parts in the warehouse will only consist of standard screws, nuts, bolts, and possibly sheet steel. Conversely, if the project is to build a minor adaptation of a standard turboencabulator, for instance, those metaphoric parts will probably be standard components such as an entire engine, a motor frame, the usual wheel assembly, and an air conditioning unit.

In other words, with increased project complexity and unknowns, there is an *increase* in the number of levels of the hierarchy and a *decrease* in the size of the work packages. This costs money to control. Nothing is free. This means that with greater unknowns, it is sometimes possible to provide more simultaneous activities and consequently increase the speed of problem solving. However, increased speed of problem solving also increases the need for control over interfaces among various work packages, since there will be more of them. Therefore, there is an increased requirement for the project manager to communicate and interact with project personnel when there are many unknowns.

As a straightforward example, assume that Jim is selected as the project manager appointed to build a swimming pool for his local high school. The tasks could be related to one another in a simple, but logical, hierarchy (see Fig. 3-8).

1.0 Build the pool
    1.1 Review designs
        1.1.1 Concrete in-ground structures
        1.1.2 Steel in-ground structures
        1.1.3 Above-ground structures
    1.2 Check local building codes
        1.2.1 Plumbing codes
        1.2.2 Electrical systems
        1.2.3 Construction codes
    1.3 Evaluate resources
        1.3.1 Financial assets
        1.3.2 Time assets and limitations
        1.3.3 Size requirements and limitations

Each of the entries preceded by a three-digit number (e.g., 1.2.1 Plumbing codes) defines a *work package*. The total assembly of these various work packages, or how they relate to one another, is a *work breakdown schedule* (WBS). There are other tasks that Jim probably also thought of, such as determining the ground load capacity, checking construction, operations problems, and so forth. However, the items listed will serve as an example for now. Each task, such as Work Package 1.2, includes the summary of all the tasks below it (that is, 1.2.1 + 1.2.2 + 1.2.3). It also includes the work that must be done within the task itself (that is, within 1.2) for integration of those tasks below it and to move it up to the next higher task, which is 1.0. This simple forecasting WBS assumes that the project manager has a minimal, yet sufficient, knowledge of pool building to develop the initial hierarchy for this project.

POOL: W.B.S.

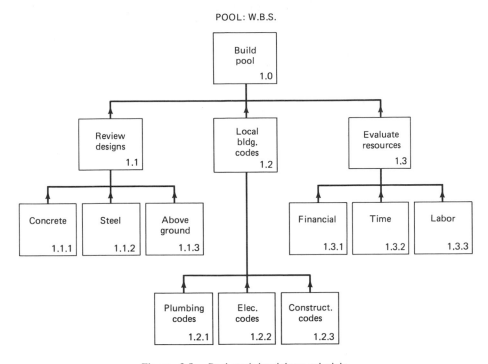

Figure 3-8.   Pool work breakdown schedule

There are relatively few work packages in this example, since this project is not really breaking any new technical ground. With fewer work packages, each of the submanagers (construction manager, contracts manager, and so forth) can handle a larger proportion of the project. Thus, the project manager's task is minimized.

In addition to being affected by the complexity of the tasks to be completed, the size of the tasks is also affected by the size of the organization. Larger project organizations sometimes require increases in the amount and complexity of reporting to provide project teams with effective communication links. The formal (functionally based) monthly reports are inadequate, since projects have to deal with higher uncertainty in the beginning than do equivalent functional organizations. There has to be more frequent reports, meetings, and reviews.

It follows that simpler projects and smaller project organizations should use less complex and less detailed forecasts. Less complex forecasts can be more easily modified, since there are informal meetings, reviews, and interoffice memos to provide prompt, effective feedback data. Also smaller organizations have fewer requirements for formal reviews, since there are fewer people with whom to communicate, since almost everyone knows what's going on anyhow. This means that the size of the work packages is sometimes inversely related to the size of the organiza-

tion: the larger the organization, the more frequent the number of reviews and therefore the smaller the size of the work packages. This is not a hard and fast rule, of
course. There are large organizations that allow substantial delegation, which
means larger work packages but when this occurs, we are again dealing with small
groups. The contents of the various work packages can also be affected by the resources available and the project time limitations.

The ideal way to use resources would probably be to produce a smooth requirements curve over time that would minimize the resource need and still meet the
project's delivery time (see Fig. 3-9). Therefore, there could be two constraints: (1)
the availability of resources during a specific time and (2) the overall project time.

1. There can't be more resources required during a specific time period than are
   available to the project. For example, "I'm sorry but all we have is three concrete mixers and one of them is in constant use. I guess you'll just have to
   schedule your tests to fit the availability of the remaining two mixers."
2. The project must meet the overall elapsed time targets. For example, "If you
   really need three mixers to meet your project deliveries, maybe we'll have to
   go outside the company and subcontract some mixing outside."

Sometimes a hierarchy may not be adequate to show more complex relation-

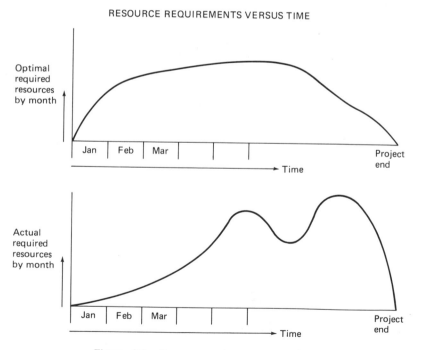

Figure 3-9.   Resource requirements versus time

ships among various tasks. Gantt or bar charts may then be used; however, the initial step is the same: Define the general tasks first.

6.1 GANTT OR BAR CHARTS

Gantt charts (see Fig. 3-10) show tasks versus elapsed time. They are named for their inventor, Henry Gantt, who was a pioneer in management theory in the early part of the twentieth century. Gantt charts usually don't show quantities of time within tasks, just how long the particular tasks will take in elapsed time. They don't show the amount of work done but when the tasks are expected to occur and how they're related to one another. For example, if we assume that a typical task, say, that of "digging a foundation" for some building, will require about three employee weeks of labor, but those three employee weeks might occur over two months, the Gantt chart will only show the two months, not the three employee weeks.

Advantages of Gantt Charts
  a. Easy to assemble and update.
  b. Activity- and time-oriented.
  c. Least costly to use.
  d. Can comprehensively cover a limited project life cycle.

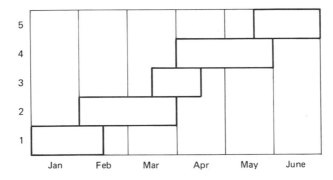

GANTT or bar chart

1. Preliminary design
2. Development and test
3. Procurement
4. Manufacturing
5. Field service and training

Figure 3-10.   Gantt or bar chart

e. Easy to make breakdowns of smaller parts from more complex projects.
Disadvantages of Gantt Charts

a. Doesn't clearly show the interrelationships among activities that do not affect one another directly.

b. Cannot be used for long-term or very complex projects.

c. Frequently unreliable because the judgment of the forecaster may change over time.

d. Inadequate to handle significant project changes that require an entirely new forecast.

e. Cannot handle development phases when time standards are not well known.

6.2 PERT CHARTS

PERT (Program Evaluation Review Technique) and CPM (Critical Path Method) are basically the same. (For a more detailed explanation of PERT, see Appendix 4.) The major difference between them is the number of estimates of time required for a given activity. These are once-through techniques specifically designed for forecasting projects. The technique consists of two major components: (1) a network of interrelated events and (2) estimates of time for each activity within those events (see Fig. 3-11). The critical path is the *longest* time through the network; in Fig. 3-11, it is Start—A—D—G—END at 14. All other paths are shorter.

Advantages of Pert and CPM

a. Very good for simulating alternate plans and resource allocations. Provides a means to obtain a sensitivity analysis of the proposed effects for various changes in the plan.

b. Can schedule work where the tasks are not well defined, such as development.

c. Clearly identifies activities and events.

d. Points out the critical path and directs management attention to it.
Disadvantages of PERT and CPM

a. Quite complex and difficult to implement. With less than 30 events (in my opinion), a PERT chart could be updated manually without a computer. If there are more than 100 events, a computer will probably be necessary, but the cost of updating the chart will probably be greater than the value of the data that you receive by the time that you get it.

b. Although very useful to control outside vendors and coordinate their efforts, it is not responsive to internal company projects where functional supporting managers may be quite independent and nonresponsive. Reports that these managers have suddenly become part of the critical path may not result in appropriate correction from those managers.

PERT CHART

Job Label	Job Description	Immediately Preceding Jobs	Normal Time
A	Procure materials	Start	3
B	Prepare site	Start	6
C	Prepare request for engineering approval	Start	2
D	Prefabricate building and deliver to site	A	5
E	Obtain engineering approval	C	2
F	Install connecting lines to main system	A	7
G	Erect building and equipment on site	B, D, E	4
End	Final inspection	G, F	2

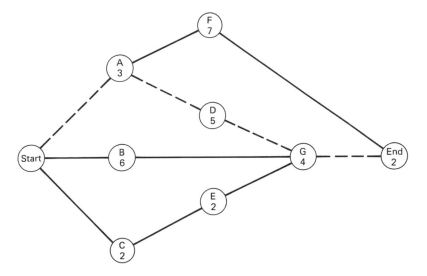

Figure 3-11.   PERT chart

According to one of the few detailed research reports on the effectiveness of PERT,

> Our analysis reveals no significant difference in technical performance between those projects that used PERT and those which did not. There was, however, a significantly lower probability of cost/schedule overrun if PERT was used. PERT has this result because it requires extremely careful planning from the outset of the project. (Marquis, 1969, p.85)

The basic conclusion here is that PERT/CPM helps because it requires thinking through the relationships of tasks to one another and their duration. It assists in allocating resources but has no influence on the excellence of the end product.

However, the networking technique itself is useful since it requires the forecaster to show how events and activities are expected to be related. Those expectations and the actuality that follows when the project is in work may not be the same. These are excellent forecasting tools, but they are not very useful as a measurement tool because it is very costly to obtain measurements or feedback of actual achievements to compare against those predicted to occur on the critical path. It's a disciplined, expensive proposition to require people to report actual achievement against a PERT diagram. Since few projects today have a network with more than a hundred events, it's also easier to get feedback through meetings or a few phone calls. There is research that indicates the futility of even updating the PERT diagrams as the project moves along. For example,

There is an apocryphal story going around . . . about an . . . operations researcher who visited one of the major aerospace companies in the United States to learn more about PERT. He was initiated to the subject by an extensive and elaborate briefing, explaining what PERT was and how the contractor used it. There were millions of dollars in savings and major milestones were being completed years ahead. PERT charts were being drawn by automatic drafting machines run by digital computers. To provide the instant creation of PERT charts, CRT tube–equipped terminals were provided. Actual control of the project was accomplished in a two-story-high lecture hall where one of the walls was covered with an immense PERT chart. Employees of the firm were regularly trained in courses provided by highly successful outside consultants.

After a briefing, the . . . visitor . . . began his own investigation. As he left the control room, he questioned the lady who kept a log book of all people entering the control room. He observed the principal users of the room were employees of the firm who updated the PERT chart. The project manager entered the room only at irregular intervals, and the log book showed that he was invariably showing around important visitors. The chief engineer's name did not appear in the log book.

Further detective work uncovered that the voluminous computer-produced PERT reports were delivered in numerous offices and filed in rows of filing cabinets. However, our . . . visitor found it difficult to find people who were actually reading the reports and observed that recipients of PERT printouts were often equipped with the powerful "Shredmaster Conveyor 400." The project manager solemnly declared that he could not possibly run the project without PERT. Under the pressure of questioning, he finally triumphantly showed a cardboard with his PERT chart displaying six major milestones. (Vazonyi, 1970, p.89)

Finally, if the concept of uncertainty is useful, then most feedback is needed in the beginning of a project when uncertainty is highest. But the feedback received at the beginning of the project will show very few expenditures of time and money.

6.3 LINE OF BALANCE

The transition from the development stage to the production stage requires a different control technique. Now the need is to forecast production quantities of re-

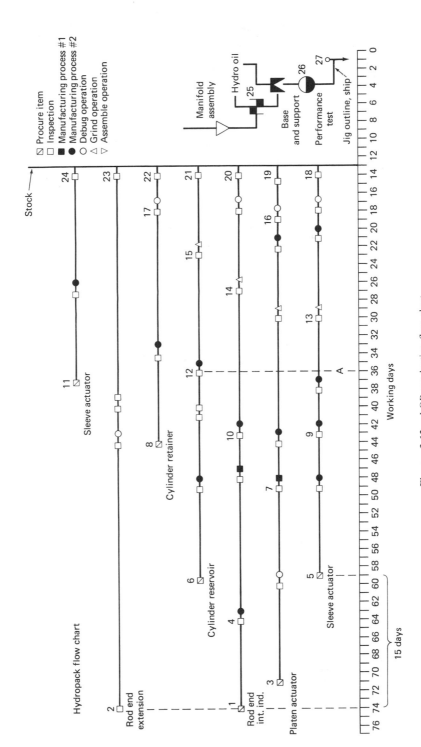

Figure 3-12. LOB production flow chart

91

quired components to support shipment of the required number of assemblies at the right time in the future. A technique called *line of balance* (LOB) helps to do this. It has four related parts, as follows:

a. *The production flow chart*, a time-based description of how the proposed product is to be produced. The number of *working* (not calendar) days is the horizontal axis. The length of time a particular component is expected to remain in production (not the actual production time) is shown on the body of the chart. This means how many days the shop or the particular installation will have the product in work. For example, with a regular five-day week, the horizontal axis will show about 20 working days per month, not the usual 30 or 31 days in the month (see Fig. 3-12).

b. *The cumulative schedule of shipments*, a schedule showing the cumulative number of deliveries on the vertical axis and working days on the horizontal axis (see Fig. 3-13).

c. *The line of balance chart*, a chart showing checkpoints taken from the production flow chart on the horizontal axis and the cumulative number of components that have passed that checkpoint on the vertical axis. There are two plots for each time period: the expected number that passed the checkpoint and the actual number that passed the checkpoint (see Fig. 3-14).

d. *The production report*, a periodic report showing the differences between requirements and actuals (see Item c above) and the explanation for these differences (see Fig. 3-15).

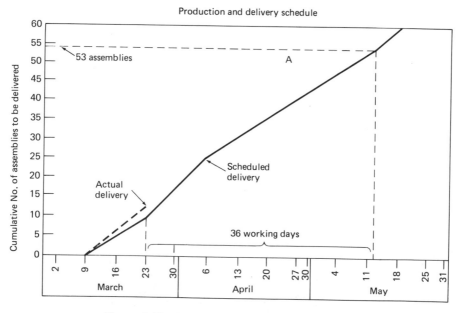

Figure 3-13. LOB cumulative schedule of shipments

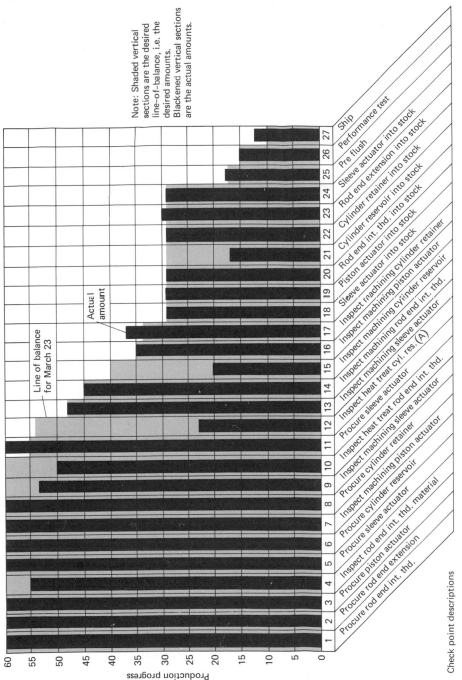

Figure 3-14. Line of balance chart

93

PRODUCTION REPORT − 23 March				
Check point	Description	LOB Req'd	LOB Actual	Corrective action
2	Procure rod end extension	60	58	2 pieces due in tomorrow
4	Inspect rod end internal	60	55	5 pieces in process
9	Inspect machined sleeve actuator	60	53	7 pieces in process
10	Inspect H. T. rod end	60	50	Vendor delay − 15 pieces tomorrow
12	Inspect H. T. cylinder reservoir (A)	53	23	Vendor delay − 40 pieces tomorrow
15	Inspect machined cylinder reservoir	40	20	Holdup − checkpoint 12
21	Cylinder reservoir into stock	28	17	Holdup − checkpoint 12

Figure 3-15.   LOB production report

The details behind this brief overview of the technique of line of balance may be found in Appendix 5. The same figures are used in the more lengthy explanation.

Advantages of LOB

a. The planning of the production flow chart and the cumulative schedule of shipments often points out problems before they occur in production. This is a typical advantage of all networks.

b. Both underachievements (production is behind schedule) and overachievements (excess inventories) can be used for corrective actions.

c. Can be used to forecast funds needed to meet production goals.

d. Is easily updated; data gathering is easy.

Disadvantages of LOB

a. Inflexible if major changes modify the project plan. The network has to be replanned.

b. Provides little allowance for uncertainties.

7.0 ESTABLISHING CHECKPOINTS OR MILESTONES

Every task has at least two milestones attached to it. A *milestone* is not a point in time; it is (1) tangible, (2) deliverable, and (3) the acceptability of it is defined by the receiver. The effective use of the three criteria makes milestones one of the best forecasting and measuring tools available. Don't be misled by its simplicity; it's very useful. Let me explain.

When a milestone is used, the definition of acceptability is the joint responsibility of the deliverer of that milestone and the manager who is to receive it. The receiving manager, of course, has the last word because he or she is actually the "customer" and is accepting (or rejecting) the product or service that the milestone is supposed to represent. I have found that having the receiver determine the adequacy of the deliverable "milestone" eliminates a lot of conflict later on. It can prevent the receiver from not acknowledging receipt or else rejecting the milestone because of previously unspecified requirements. It is therefore a vital forecasting task to effectively define not only *when* that tangible milestone will be delivered but also *what* acceptance really means and *how* that acceptance is defined. At a minimum, every work package or task has at least two milestones. The first one begins the work package, because the particular work package manager (for instance, manager A) has accepted the delivered milestone from the manager of the prior work package. The last milestone shows the end of the work package when the results are delivered by manager A to the manager of the next work package, manager B. This signifies that A's work package or task is completed. Manager B determines if the delivered milestone meets predefined criteria. If it does, the milestone is acceptable. Any problems with deliveries, such as milestone timing or acceptability, quickly rise to the attention of the project manager. When this happens, the work package interfaces are in trouble. Naturally, the very last determiner of project milestone timing and adequacy is either the eventual customer, the user, or (in the case of internal projects) Quality Control. In summary, if it's not tangible, not a predefined deliverable, and not accepted by the receiver, it's not a milestone and doesn't exist.

8.0 DEVELOPING COSTS: HOW MUCH?

Developing costs is the last forecasting assignment. It follows technical tasks and time. The process is relatively straightforward to understand but, like many kinds of management tasks, it is sometimes difficult to follow. Generally, we estimate costs from the knowns to the unknowns. Have there been any projects in the past that were similar to this project? What can we learn from them? If raw material is relatively stable when compared with labor, start with that.

Management uncertainty about a work package or in the total estimate determines the number of contingencies that are forecasted. This part of the total amount of uncertainty is subjectively related to a specific project manager, not to the relative organizational level. Using the example of material versus labor, it is apparent that there would be a higher contingency in the labor portion of the project if the cost of raw materials is more stable or predictable. The labor portion may even have several levels of contingency. Consider research and development labor versus production labor. Contingency factors need to be adjusted accordingly.

The contingencies in the forecast should also generally follow the project uncertainties. The first tasks have the greatest uncertainty and, therefore, should have greater contingencies than tasks done later in the project, when uncertainty is lower. This is not subjective or related to a particular project manager. It's organizationally based because of the specific nature of project. Therefore, to minimize contingencies, it might be reasonable to make the size of the first few project tasks fairly small and allow sufficient contingency in them to provide for reanalysis, retesting, or whatever. Later, when the project design is stabilized, the work packages can be larger and the contingencies in them can be decreased.

Conversely, if one feels that it might be appropriate to increase contingencies in the work packages that are a year away, as opposed to those that will be done next month, that's moving into subjectivity again, since it doesn't follow the assumed uncertainty curve for our typical project.

9.0 SUMMARY

Forecasting is the first of the five major project supporting systems. It's defined as the process of selecting the best solution *today* to some future problem. The organizationally acceptable forecasting techniques, such as bar charts, PERT/CPM, line of balance, and others should be part of an overall organizational system because they can be basic for all projects in a company. As a project proceeds, variances or differences between the forecast and the actual measurements will require some kind of management strategy, which is really limited to three options. Managers can decide to (1) modify present actions to fit the forecast, (2) change the forecast, or (3) do nothing. The first choice—modify present actions—is usually selected. If the forecast has been handled as well as possible, it could still be realizable. The manager has to take corrective action now to fix the variance. The next choice—change the forecast—is chosen when new data indicate that a decision made *today* about future problems no longer matches those made *yesterday* and when there's nothing that can be done today to diminish the variance, the forecast must be changed. In other words, deciding to change the forecast is reasonable and necessary when it becomes apparent that the existing forecast doesn't provide the optimum answers today or doesn't include future problems that we now see *today*. Finally, if the manager chooses alternative 3 and does nothing, it's probably because a present evaluation indicates that the variance is minor.

The forecasting process is not only a decision-making tool, it's also a communications device that supports data transfer among team managers. Those processes, therefore, should be chosen with care and should be documented as part of the project charter as a standard operating procedure.

To reiterate, the sequence of the forecasting system is (1) technical objectives, (2) time, and finally (3) cost. *Technical objectives* are the main reasons for the project itself. *Time* often is very difficult to extend, so it's next in importance. Finally, we have *cost,* which can often be extended. It's usually easier to get more funds than to change technical goals or to get additional time. Milestones are tangible deliverables and are used in defining specific technical objectives at a given time with a given cost.

In the next chapter we put the total forecast together after we've chosen and used the various forecasting processes, such as Gantt charts, PERT, LOB, and so forth. We'll emphasize the various kinds of work breakdown schedules to assemble the various forecasts and show how to integrate the team's forecasting efforts using the coordinated forecasting technique.

My Suggested Answers to the Case Study

1. The meeting should have had an agenda. The sequence for discussion should have been, first, technical problems and proposed solution, second, time required and what should be done sequentially and what in parallel, and, third, a rough idea of cost.

    Lew and Alex should have stated what the knowns and the unknowns were about the Topnotch Computer. Since part of the argument was based in personal opinion, a decision matrix could have separated the areas of agreement. Then, the unknowns could have been attacked through feasibility studies, vendor evaluations, or any other miniproject.

    Martin should have immediately written an impact statement. Then he should have shown it to Charles and informed him that the statement was going to Dianne the next day. He should have sent it if the situation remained unchanged. The decision to be late is then out of his hands.

2. There are two major alternatives:
    a. Reduce the profit allowance and allow the original full cost to the project. The reasoning is that the project can be responsible only for those costs to which it has committed itself. When sales reduced the price, their decision had nothing to do with cost.
    b. Reestimate the project to determine where costs can be cut. Every part of the project is not equally fixed. Perhaps if materials are a large part of the cost, a reduction can be negotiated with appropriate vendors or a new make-or-buy analysis can be made.

3. The availability of the engineering test equipment should be the subject for an immediate impact statement. The additional costs to have the testing done outside or to displace the Engineering department from its own equipment are unexpected and will definitely affect the project completion date.

    Charles Dobbs' feelings are primarily Lew's problem. Some of the questions that Martin should suggest that Lew should get answered are:
    a. What are the differences between the original tasks "as quoted" and the tasks "as presently evaluated" in cost, time, and technical solutions.
    b. What should be done by Engineering to minimize this change?
    c. What do the decisions look like now as compared with those before?

    Lew, as the project engineer, is the main contact for all technical decisions, and he, in effect, is now a miniproject manager within his own function. He is the technical manager on the project and should be required to provide potential answers when potential problems arise in his area. It's up to him to resolve Charlie's feelings, if he can. Or if he can't, the next step is for Martin to take over. If Martin can't get anywhere with Charles, another impact statement is sent to Dianne. She can then resolve the problem.

4. The first step is to determine the alternatives available and place a value on each of them. Then, a new plan dealing with the alternatives must be presented to Martin by Alex. Some alternatives to be evaluated include
    a. Negotiating an overtime price with Topnotch to meet Mammoth's delivery schedule.

b. Doing a vendor evaluation to determine if there are other vendors who can supply.

c. Determining if the computer can be built in-house.

5. Martin's major concern is with managing project interfaces. Therefore, when Lew and Alex presented potential solutions to these problems, his main tasks were to

a. Determine how these solutions will impact the project time, cost, and technical solutions.

b. "Test" the solutions presented to him. For example, by asking the following questions: How did you come to these conclusions? What are the relative pluses and minuses of each of your solutions? What percentage of risk can be assigned to these answers?

Martin should not initially get involved in the solution process itself. The expert (in this case, Alex) is closest to the vendor situation and knows most about it. Therefore, he should be the first to provide alternative solutions.

Dianne's major concern is with the overall operations of the division. Assuming that she has satisfied herself that the data presented by Martin are the best available at that time, she has only two alternatives: provide the additional funds to the project from company funds or renegotiate an increase with Mammoth. Both of these decisions are beyond Martin's scope of responsibility as project manager.

Do you agree?

BIBLIOGRAPHY

MARCH, JAMES G., and HERBERT A. SIMON, *Organizations*. New York: John Wiley, 1958. Copyright © 1958 by John Wiley & Sons. Reprinted by permission.

MARQUIS, DONALD G., "Ways of Organizing Projects," *Innovation,* Project Management Publication, 3, American Institute of Industrial Engineers, 1969.

MEREDITH, JACK R., and SAMUEL J. MANTEL, JR. *Project Management: A Managerial Approach.* New York: John Wiley, 1985.

VAZSONYI, ANDREW, "L'Histoire de Grandeur et la Decadence de la Methode PERT," *Management Science*, 16, no. 8 (April 1970), pp. B-449–B-455. Copyright 1970, The Institute of Management Sciences.

OTHER READINGS

BRUNER, JEROME S., "The Conditions of Creativity," in *On Knowing: Essays for the Left Hand.* Cambridge, Mass.: Belknap Press of Harvard University, 1962.

CLELAND, DAVID I., and DUNDAR F. KOCAOGLU, *Engineering Management.* New York: McGraw-Hill, 1981.

GEORGOFF, DAVID M., and ROBERT G. MURDICK, "Manager's Guide to Forecasting," *Harvard Business Review*, January-February 1986, pp. 110-20.

HAGAFORS, ROGER, and BERNDT BREHMER, "Does Having to Justify One's Judgements Change the Nature of the Judgement Process?" *Organizational Behavior and Human Performance*, vol. 31 (1983), pp. 223-32.

HAJEK, VICTOR G., *Management of Engineering Projects*. New York: McGraw-Hill, 1977.

HARRISON, F.L., *Advanced Project Management*. New York: John Wiley, 1981.

HUNT, PEARSON, "Fallacy of the One Big Brain," in *Harvard Business Review: On Human Relations*, ed. New York: Harper & Row, Pub., 1979.

LAWRENCE, PAUL R., HARVEY F. KOLODNY, and STANLEY M. DAVIS, "The Human Side of the Matrix," in *Matrix Organization and Project Management* (Michigan Business Papers, 64), eds. Raymond E. Hill, and Bernard J. White. Ann Arbor, Mich.: University of Michigan Press, Division of Research, Graduate School of Business Administration, 1979.

LAWRENCE, PAUL R., and JAY W. LORSCH, *Organization and Environment: Managing Differentiation And Integration*. Division of Research, Graduate School of Business Administration. Cambridge, Mass.: Harvard University Press, 1967.

LOYE, DAVID, "The Forecasting Mind," *The Futurist*, June 1979, pp. 173-77.

MAKRIDAKIS, SPYROS, and STEVEN C. WHEELWRIGHT, *Forecasting: Methods and Applications*. New York: John Wiley, 1978.

MARTIN, CHARLES C., *Project Management: How to Make It Work*. New York: AMACOM, 1976.

NADLER, DAVID A., and EDWARD E. LAWLER III, "Motivation: A Diagnostic Approach," in *Perspectives on Behavior in Organizations* (2nd ed.), eds. J. Richard Hackman, Edward Lawler III, and Lyman W. Porter. New York: McGraw-Hill, 1983.

ROBBINS, STEPHEN B., "Reconciling Management Theory with Management Practice," *Business Horizons*, Fall 1977, pp. 38-47.

RUSKIN, ARNOLD M., and W. EUGENE ESTES, *What Every Engineer Should Know About Project Management*. New York: Marcel Dekker, 1982.

SILVERMAN, MELVIN, *Project Management: A Guide for the Professional*. Cliffside Park, N.J.: Atrium Assoc., Inc., 1984.

SILVERMAN, MELVIN, *The Technical Manager's Survival Book*. New York: McGraw-Hill, 1984.

Chapter 4   Forecasting: Tieing It All Together

PERSONAL
MANAGEMENT
STYLE:
THE INDIVIDUAL
SYSTEM

PEOPLE MANAGEMENT:
THE PROJECT SYSTEM

PROJECT OPERATIONS:
THE PROJECT SYSTEM

MEASURING:
THE ORGANIZATIONAL SYSTEM

FORECASTING:
THE ORGANIZATIONAL SYSTEM

FOUNDATION: PROJECT MANAGER'S AUTHORITY

THE CASE OF THE "EASY" PROJECT

The Cast

Mildred Monsey, Project Manager, Widget Project
George Peabody, Chief Engineer
Mary Wilson, Purchasing Manager
Will Opson, Manufacturing Manager
Harry Sloan, Manager of Project Management

Late one Friday afternoon, Mildred carefully shut the door to her office (indicating to any casual visitor that she wanted to be left alone), sat down at her desk, and began to review the events of the past week. Jim Mong, the previous project manager on the Widget Project, had been promoted into a line management job at one of the company's divisions several hundred miles away. She had been given the Widget Project to finish up by her boss, Harry Sloan. He had called her in several weeks ago, told her that because of the great job she had done on the last two projects that she had managed, he was giving her Jim's Widget Project. He said that this one would be easy since most of it was done and all she had to do was "keep it on the tracks."

As a first step, she requested all the previous documentation. It was almost impossible to make any sense out of what she received. There was a minimum overall plan and a lot of internal memos, but there wasn't any central file that she could analyze. After spending several days trying to read it, she called a meeting with the team managers, George Peabody, Mary Wilson, and Will Opson. It was a very interesting review.

The meeting started about half an hour late. Mildred wondered if that was an indication of her team managers' personal commitment. The first person to speak was George.

George: Well, the best that I can report is that the tooling designs were several weeks behind schedule. There were some problems with the way the clamping devices had to fit on some of the newer components. I had told Jim Mong, the previous project manager, about it and requested some contingency funds in order to hire some outside temporary engineering designers. As I recall, Jim wasn't too concerned then because it was only the beginning of the project and he said that the project had plenty of funds. Jim said that if things got a bit tight, perhaps some overtime would fix it. Nothing was done about it then, so we'll need some additional money and time now.

Mary: It's been very rough providing purchasing support for this project. The design engineers had gotten some telephone quotations on some special motors for the Widgets several months ago without going through Purchasing. You know that they're not supposed to do that. If they asked us to do that, we'd have a record of what happened in order to hold the vendor to any promises made. Anyhow later, when the engineering requisitions arrived in Purchasing, my people found that they couldn't get either the price or the delivery that those engineers said were quoted to them.

There's another problem too. The vendors listed on the engineering purchase requisitions have never been evaluated by Quality Assurance and were, therefore, not yet qualified to supply the Widget Project. When my people reported that to the engineers, they said that the vendor was very well known and that we "shouldn't worry about it!" I had mentioned this to Jim, but Jim told me to place the orders anyhow, since "all vendors are a little difficult to deal with. And you know how picky Quality is anyhow. They'll go along. It will all work out."

Will: I'm not even sure when the manufacturing drawings will arrive since George said Engineering was delayed. And my Production Control manager just told me that he had gotten a new bill of materials showing some special motors.. There were no provisions made for position stands to hold those special motors during assembly, and they each weigh about 60 pounds. And finally the Manufacturing division was facing a union negotiation next month. All of this means that I can't give you any idea about when I can start making widget parts or assembling them.

By the way, my schedules have been changed by all the other project managers so many times that I've decided not to do anything for any project until everything for that project was actually on hand. That means I can't estimate delivery schedules until the widget prints are on hand and all the parts are in inventory.

Questions

1. What should Mildred do now? What is the sequence in which it should be done? Who should be involved?
2. What kinds of tools would you use if you were Mildred?
3. How is this different from starting a new project?
4. Are there any general rules that cover the timing for this kind of management activity?
5. Has this ever happened to you? What did you do?

1.0 ASSEMBLING FORECASTS

The bar (or Gantt) charts or PERT networks are among the more popular techniques used in forecasting technical projects. Bar charts are primarily concerned with elapsed time and PERT with time and achievement. (See Marciariello, 1978, for more details about this.) The relationships among various tasks and the "how to do it" which each task should cover are usually described by a *work breakdown schedule* (WBS). In Chapter 3, we briefly touched on the work breakdown schedule (WBS) and described it as a clearly documented set of interrelationships among-project tasks. We will continue to explore the WBS here, since the process of developing it is often very important to the end result. That development process supports communication and integration. It requires personal commitment from the various team managers. One method that provides these benefits and provides the WBS as an end result is the *coordinated forecasting process* (CFP). It ties together the many forecasting techniques such as bar charts and PERT networks.

To review briefly the WBS summarizes the outputs of the task hierarchies (bar or Gantt charts), network-time relationships (PERT), and costs (including knowns and unknowns). It shows technical achievement, time, and cost. In some cases, the work breakdown schedule (WBS) is somewhat like a bill of materials used in ordinary production planning. A bill of materials has a sequence that follows how the product is to be assembled. For example, start with raw metal, then we machine it, then build subassemblies, and finally build the total product. A WBS can define the project in a similar hierarchal chart form showing initial design, contracting, manufacturing, and shipment. These are typical elements in many projects. It outlines *how* the project is to be completed. It also includes a task hierarchy. However, it goes beyond the bill of materials analogy since it also has time and cost considerations. Each work package in the WBS has a minibudget that includes a complete description of the inputs, the expected time and cost, the interim checkpoints, and the output for that package. Since the tasks include "milestones," they become very useful indeed as a measuring and control tool.

There are several typical kinds of work breakdown schedules that can be used, depending on the situation. There is a *task-related WBS*, which I've used when the

project has a high proportion of unknowns; a *skill-related WBS*, which is less costly than a task-related WBS; and a *staff-related WBS*, which is used when the project is very similar to others you have managed or when the project team is very experienced and committed. If these types of work breakdown schedules don't exactly fit your situation, there should be sufficient information here for you to build your own modification(s).

1.1 TASK-RELATED WORK BREAKDOWN SCHEDULES

A *task-related WBS* most closely resembles a bill of materials. All of the subtasks must be completed before the next higher task can begin. In effect, this is oriented to a series of deliverables that controls each step of the project. It's probably the safest WBS because of the step-by-step approach. The project manager will undoubtedly have the most control using this WBS, but like most things in life, this approach has a disadvantage. It's very costly because it doesn't use people optimally. To illustrate, let's assume that the project is building the company's first space ship. We want to be sure that each work package has been completed and accepted before the next package is worked on. Thus, we choose a WBS that is task-related (see Fig. 4-1).

For example, the work to be done in the life support work package (task 1.1.1) must be completed before construction on the nose assembly (task 1.1) can begin. This kind of WBS is well suited for projects that have a large number of unknowns or that are very novel. Now you can see that if the most expensive element in your project is the human technical input, as so often occurs on novel or unknown projects dealing with many unknowns, this WBS design would be very costly because these inputs are not used optimally. For example, assume that we're using a task-related WBS and a particular engineer is needed for only a week on a particular task. When the engineer's portion of the work is done, he or she would no longer be able to charge the work package and would either be reassigned to another concurrent task, another project, or else charge overhead until his or her services are needed again. If there are no other tasks immediately available, overhead costs in-

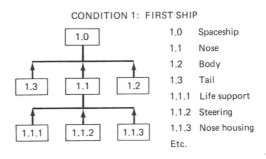

CONDITION 1: FIRST SHIP

1.0	Spaceship
1.1	Nose
1.2	Body
1.3	Tail
1.1.1	Life support
1.1.2	Steering
1.1.3	Nose housing
Etc.	

Figure 4-1. Condition 1: First ship

crease. Conversely, if there is another project available and the engineer is assigned to it, he or she might be unavailable if quickly needed for some unforeseen correct-ive actions. This sometimes happens. Conversely, if the particular engineer charges overhead, he or she would be available for emergencies, but the overhead charges would mount up whenever an emergency doesn't exist.

To briefly review the disadvantages of the task-related WBS: Various workers might not be available when needed if there are no other available work packages *at that time* to which to assign them and if they must be available for emergencies. This results in a less than optimal use of people for the sake of optimal task comple-tion.

This WBS does have advantages, however. It is probably the most reliable, since it requires task completion. It uses milestone "controls" (see Chapter 3 for a description of milestones) at the beginning and end of each work package, assuring that the work is actually completed satisfactorily before the next task is opened for charges.

1.2 SKILL-RELATED WORK BREAKDOWN SCHEDULES

If the project has a moderate set of unknowns and if the project manager is familiar with similar kinds of projects that have already been completed, a less ex-pensive WBS that increases the use of human skills can be used (see Fig. 4-2). In the *skill-related WBS,* as an example, all of the electrical engineering, the mechani-cal engineering, or other specialized skills would be grouped under the various team managers. The detailed WBS under these major tasks would then be skill-related. The WBS is then very similar to that of an equivalent functional organization. (Again, this assumes that one of the most expensive costs is human services.) The disadvantage is that the milestones are much more difficult to define and therefore subject to greater personal interpretation. The deliverable milestone can't be used to open and close work packages because those work packages are skill-oriented, not task-oriented. In this kind of WBS, there will probably have to be more design re-

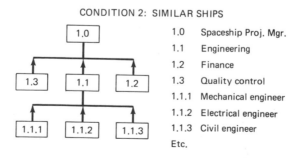

Figure 4-2. Condition 2: Similar ships

views (see Appendix 3) or project status review meetings. With this WBS, team managers are responsible for controlling and reporting progress on their own specialized skill-related work packages. This means more coordinating with other team managers and more reporting on progress to the project manager. It places a greater load of responsibility on each person.

Using a skill-related WBS is similar to establishing another *temporary* functional organization. This can lead to increased commitment, since all the team managers would probably be involved throughout the project life cycle. Most conflicts among team managers should be resolved during the initial forecasting meetings and charter revisions. Therefore, this WBS can promote a closer commitment. The management load on the project manager increases, however, because there are often many more interfaces at one time.

A skill-related WBS is used for projects that are more completely defined. This definition could include not only the new products or services to be developed but also the relationships among the team managers.

1.3 STAFF-RELATED WORK BREAKDOWN SCHEDULES

There is a WBS design that provides the advantages of detailed control over work packages such as achieved with a task-related WBS and the decreased costs of a skill-related WBS (see Fig. 4-3). This design is called the *staff-related WBS* and has three major differences from the design of the other two work breakdown schedules:

1. Staff size: It assumes that the project is large enough to have a semipermanent project management staff of several team managers.
2. Materials: It provides a centralized control over materials.
3. Contingency funds: It segregates all contingency funds into one work package.

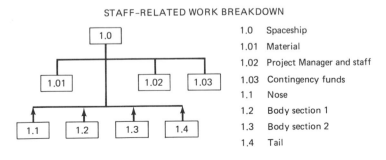

STAFF-RELATED WORK BREAKDOWN

1.0	Spaceship
1.01	Material
1.02	Project Manager and staff
1.03	Contingency funds
1.1	Nose
1.2	Body section 1
1.3	Body section 2
1.4	Tail

Figure 4-3.   Staff-related work breakdown

If the project is large enough to have a supporting project management staff, this staff works throughout the project life. There are no milestones to be delivered by this staff. They provide a management service. Since these team members report directly to the project manager, the staff is just like a minifunction with the usual budgetary and review controls that any organizational function has. It operates as a "permanent" structure, even though it's really a "temporary" structure that is to be dissolved when the project ends.

Let's assume that the project has a material content that is more known than the labor content. A single work package could be set up just for materials. When the project is costed from the bottom up, one of the tasks is to build an imaginary bill of materials for the new product and attach estimated materials costs to each product component. Then the costs are estimated for subassemblies and finally for the total assembly. This is a costed (estimated) bill of materials. Placing the costed bill of materials for the product in one work package eliminates having any material costs out in the rest of the WBS. Only labor costs should appear as charges to the project in the other work packages.

When the various work package managers need material purchased outside the company or from the company's inventory, they issue requisitions against the central material work package. The costs, after being matched against the costed bill of materials and approved by the project manager or his or her designee, are charged against the project immediately as committed costs. The materials would then be obtained as directed from organizational departments such as Purchasing or Inventory. If they are purchased from vendors, Purchasing is responsible for the efficient buying of materials against the costs shown on the requisition. Whether the materials are specifically purchased or taken from materials inventory, the various work package managers are primarily responsible for the efficient use of technical labor. If scrap or rework occurs, those costs are easily visible because additional procurements must be justified through additional material requisitions. Since all the materials are purchased under one work package number, that number has the estimated materials as a standard. Any deviations in quantity, usage, or cost easily becomes noticeable. In any event, the accountability for materials becomes very straightforward. (More details concerning this are discussed in the next chapter.)

This WBS also can provide for establishing one work package or task for all the contingency funds in the project. In some ways, contingency funds are a function of potential risk. Therefore, they're somewhat similar to profit. When the unknowns are high, contingencies should be equally high. When the project is similar to others and the problems to be solved are fairly well known, contingency funds should be lower. But contingencies are not linear; they vary depending upon the project life cycle.

1.4 HANDLING CONTINGENCIES

Contingencies are intended to cover the things that we don't know. In some cases, they might be called a *safety factor*. For example, it is the rare designer who will use the basic stress loads in the textbook to design a building. There is always a safety factor that is chosen subjectively and built in to reduce the actual stresses below those that the book says the particular materials should be able to withstand. Contingencies are similar to this. They obviously decrease as we move from the greater unknowns to the lesser ones. For example, contingencies *decrease* as the project

A. Moves from feasibility to design and then to manufacturing or from estimating labor to estimating materials.

B. Moves from the short term to the long term, usually. There is an apparent conflict here, of course, since most people would reverse this, but uncertainty is higher in the short term than in the long term for projects. Therefore, contingencies should theoretically be higher in the project's short term. How can this be minimized? One way is to use feasibility studies or several design reviews in the beginning stages of the project. Then, if uncertainty remains high, use contingency funds, restructure or recost the project, or, as an extreme measure, stop the project.

In other words, most project managers would usually tend to increase contingencies as they forecast further ahead for labor costs ("What will happen at the next union negotiation?") and raw materials costs ("Will tungsten for the widget go up next year?"), since many people tend to increase their estimates when they're not sure about something. But logically they should really *decrease* contingencies for the work packages that occur later in the project life cycle simply because most of the technical unknowns occur in the project's beginnings. And technical forecasts come first. It depends on the decision maker's subjective balance of emotions and logic.

There is almost a constant balancing act among the contingencies associated with technical, time, and cost objectives. Yet there is another way to resolve contradictions in setting the amount of contingencies: Decrease the size of the work packages in the beginning of the project. There is less room for contingency *usage* before the beginning work packages end. Since uncertainty is high in the initial stages of a project, decreasing the size of the initial work packages and putting more measurable milestones in them decreases contingencies.

As noted before, if the first few work packages don't do well, it might be necessary to reforecast or reiterate the whole project plan. When this happens, it's because additional information has been acquired. This means less uncertainty. You may find that the contingencies you forecasted for increased labor due to that upcoming union meeting or the increased cost in tungsten due to an economic problem in the supplying countries will be omitted by the reiterated forecast for the revised project.

It's often a reasonable step to segregate all the contingencies into one work package, which, of course, is appropriately labeled. When a particular team manager then requests additional funds because of a possible overexpenditure in a particular work package, the request is similar to internal impact statement issued to the project manager. The project manager gets the team managers together to review it and to determine the effect on the project. If everyone agrees, the funds are transferred in accordance with the internal impact statement. If they don't agree, the project manager will perhaps have to develop an impact statement for top management's approval, because the team members are explicitly saying that they don't believe the existing contingencies are sufficient to cover the rest of the project if this internal impact statement is funded. This could happen when the project has enough funds or time at one point in the project, but doesn't expect that there will be enough time later. Top management should be notified as soon as this is discovered, because the project expects to exceed one of the three "golden limits."

This process helps to bring problems up as soon as they occur. The discussions can be interesting, since everyone has a stake in minimizing the use of the contingency funds ("If we give them out now, will there be any left for me later on?") and in expending them ("If we don't allow some funds to be spent now, how will that work package and the project ever be completed?").

2.0 GETTING STARTED

There are two major assumptions in using all of these forecasting techniques. These are:

1. The forecaster of the WBS, the PERT network, or whatever process used has some knowledge of the available project parts, how they should generally fit together, and whether those parts are acceptable.
2. The forecaster is able to obtain agreement among the project team members on a specific plan.

These could also be self-fulfilling prophecies, since there are few projects that deal with completely unknown ideas and products, and project teams eventually *do* get things accomplished. But let's look at these two assumptions a bit closer.

In most cases, the project manager and his or her team probably know about plans and systems that worked in the past and have some kind of a beginning point for the project. The first assumption of knowledge about similar projects is usually justified, since the team managers concerned will always have an adequate technical background. If they need more detailed technical inputs, they can get them from their people. Team managers need an overall familiarity with the technical considerations in order to evaluate the potential solutions provided by the team members during those first crucial months of the project. Therefore, the first assumption that few projects deal with totally unknown ideas and products is usually a safe one.

However, the second assumption rests on less secure foundations. Assuming that no laws of nature are being violated and the project has adequate time and funds, getting things accomplished as well as is possible depends on the personal commitment of the team manager and team members. Otherwise project forecasts can too easily be modified or abandoned at the first sign of problems or internal disagreement. For example,

One engineer manager described the importance of this commitment in the following manner:

A work package is a commitment made and agreed to by two individuals "eyeball to eyeball," so to speak. It is a personal thing and has nothing to do with charts, graphs, or other inanimate pieces of paper. You don't have to make any commitment you don't agree to—normally. In very unusual circumstances your boss, and only your boss, will make a commitment for you, but this is the exception, not the rule. Only the parties to the commitment may agree to a change or cancellation of the commitment. . . .

Of course, this commitment presupposes that the necessary planning has been done, i.e., the tasks have been adequately defined so that the personal commitment can be made. . . .

If you can't commit to what the other person is asking you for, then the two of you had better redefine the task in such a manner that a commitment can be made. . . . (Cleland & Kocaoglu, 1981, p. 44)

Obtaining the necessary degree of personal commitment is difficult, but it can be obtained if the forecasting processes include the people responsible for achieving project goals and if the inputs received tie the forecast together. This can be done using a process called the *coordinated forecasting process* (CFP).

3.0 COORDINATED FORECASTING PROCESS

The CFP is intended to get answers from various project team managers to the typical project questions: What kinds of tasks have to be done to accomplish interim goals that will eventually result in the overall project goals? How will those tasks be measured? Who will do the work? Who will do the measuring? When project team managers participate in developing the forecasting process and the result obviously reflects that participation, the forecast more easily becomes a personal commitment. This also helps to resolve the next round of questions, such as, When will it happen? How will we know it when it does happen?

CFP works by defining which tasks and responsibilities are accepted by which project team manager (and even team participant, if necessary), by defining what the milestones or measurements will be, and by coordinating these tasks into a combined team effort. The process usually takes a minimum of three sequential meeting, as follows (see Fig. 4-4):

Meeting 1: Organization—Defining tasks, organizing the project team, and developing preliminary responsibilities.

Meeting 2: Forecast Development and Coordination—Defining interfaces between each manager's work packages, summarizing all forecasts into a project forecast, developing contingencies and project operating procedures, and designing an appropriate work breakdown schedule.

COORDINATED FORECASTING PROCESS

- Meeting 1: Organization
  Defining tasks
  Organizing the project team
  Assigning preliminary responsibilities
- Meeting 2: Forecast Development and Coordination
  Analyzing forecasts
  Defining interfaces
  Milestones
  Summarizing all forecasts
  Developing contingencies
  Overall W.B.S.
- Meeting 3: Project Presentation
  Accepting criticisms
  Modifications
  Changes as necessary

Figure 4-4. Coordinated forecasting process

Meeting 3: Project Presentation—Presenting the forecast for approval and accepting criticisms, modifications, and changes as required.

CFP is an integrating forecasting process that varies with the amount of detail. It can be adjusted according to the situational complexity or the organizational size. It simplifies the process of describing the overall project interactions or networks, and it shows how the project is to be managed. It even partially documents how managers absorb uncertainty when making forecasts. It supports interactive constructive criticisms, modifications, and improvements suggested by other team managers. When the organization's top management has critiqued the forecast, it usually eliminates after-the-fact criticism, such as, "Why didn't you think of—?"

CFP defines work activities as well as how those work activities are related to one another. It includes conventional forecasting techniques such as Gantt or bar charts and PERT networks, and their many modifications. It's strength lies in the requirement for each project manager (and contributing team managers or project participants) to define and be committed to achieving his or her documented project goals.

The process starts with the project charter and involves the project manager and the first line of team managers assigned to the project. As already discussed, the charter includes the project description and the project manager's suggestions for various work packages. The manager's suggestions are noted in two documents: (1) the responsibility diagram and (2) the project flow diagram.

The *responsibility diagram* is intended to define the team managers who are to participate in the forecasting process (see Fig. 4-5). It is completed during the first meeting. It shows who "owns" a particular work assignment and which of the other team managers is to assist in the forecasting and completion of that package.

The *project flow diagram* determines what the work packages are and how they are to be tied together (see Fig. 4-6). It also defines the milestone input or output, documents the particular manager's confidence or uncertainty in completing

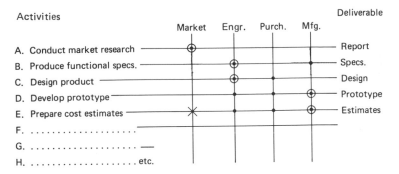

Figure 4-5.   Responsibility diagram

PROJECT FLOW DIAGRAM

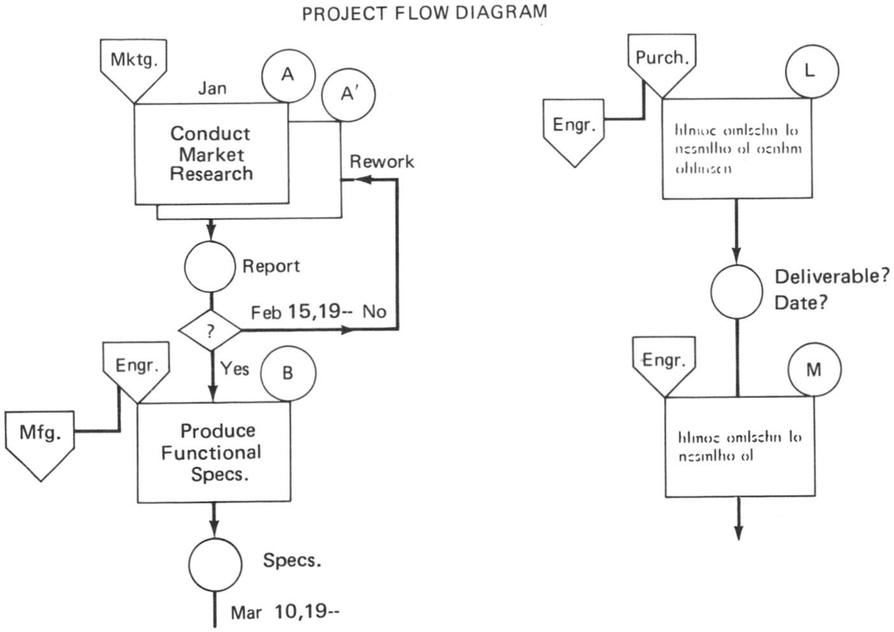

Figure 4-6.   Project flow diagram

a work package, and shows the scheduling of design reviews and important project accomplishments.

Three meetings are needed to complete this process. The first two meetings involve only the project manager and the first-line management team. The third meeting involves management and anyone else interested in reviewing the project in addition to the team. Meeting 1 defines individual project responsibilities and outlines the general flow diagrams for the initial project plan. This is similar to a "kickoff" meeting, during which managers have a chance to critique the charter and provide their inputs. Later during meeting 2, which occurs several days or weeks later (depending upon the urgency of the project), the project management team reviews all the detail work packages and the total forecast and matches their forecast against overall project requirements of technical achievement, time, and costs.

3.1 MEETING 1

At meeting 1, the project manager distributes the project charter, any specifications, the preliminary work package description that was developed as part of the initial project "envelope," and the project definition and authorization (see Fig. 4-7). The group's first item of business is to review this information and modify it as required. If there are to be changes in the work package descriptions, sizes, or relationships, this is the time to suggest them. At this point, the group is acting as a total committee. In order to preclude difficulties in the definition of responsibilities among team members, the project manager also prepares and distributes the task assignments that describe each team manager's responsibilities. These task assignments, which have already been described in Chapter 3, should be reviewed and modified as required in one of the first committee acts.

3.1.1 Responsibility Assignment

At meeting 1, the project manager also outlines a generalized series of tasks that he or she thinks should be completed during the project (see Fig. 4-5). These can be a first design of a WBS or even just a listing of tasks, since this series is the starting point for the forecasting process. This initial list is revised and accepted by the team managers acting as a committee. By the way, this first meeting of a CFP can also be part of the bottom-up meeting that is intended to organize the project team. The project manager then asks which of the managers at the meeting is to be primarily responsible for task A (see Fig. 4-5). Assuming that task A—conducting market research—is accepted by the project marketing manager, a circled dot is marked on the responsibility diagram by the project manager. This means that the project marketing manager is a member of the team working on that task, indicated by the dot, and that she is in charge of it, indicated by the circle around the dot. The marketing manager then determines which of the other managers present are needed to assist in the planning of the task. Of course, if no one accepts primary responsibility for a particular task or if the project marketing manager (as in this example) requests the help of another team manager and it is refused, there should be an immediate discussion and clarification of the various responsibilities of the dispute. It is often possible to solve potential future disagreements in this preliminary meeting. Figure 4-5 indicates that the marketing manager has said that she can handle task A alone, so there is no problem.

PROJECT DEFINITION AND AUTHORIZATION

Date: 3 March 19xx

Project Title:	M-15 Diagnostic Chair
Description:	Provide measurements of patient's vital signs when patient is seated in chair.
Objective:	Measure blood pressure, temperature, respiration, weight, and electrocardiogram in accordance with governmental regulations # xxx.
Results:	Provide 6 test chairs that are fully operational within 12 months. Interim test requirements according to company standard # xxx. Chair response time should be no greater than 1 minute.
Project Budget Estimate:	Direct costs = $750,000 Project similar to last year's M-14 chairs

Figure 4-7. Project definition and authorization

Let's assume that the project engineering manager has accepted the responsibility for task B and requests assistance from the manufacturing team to accomplish task B—produce functional specs. The manufacturing manager is asked to assist in preparing the budgets for that work package, under the direction of the engineering manager. If the manufacturing manager agrees, the project manager notes the primary responsibility as belonging to the engineering manager. This is shown on the responsibility diagram as a dot with a circle around it for the engineering manager and a dot for the manufacturing manager (see Fig. 4- 5). If the task requires no other managers, the assignment is clear. Again, if a request for assistance is not accepted, the project manager must immediately determine the reasons for the rejection by either referring to the form of the agreed-upon task assignment or else through a redefinition of the particular task. This should resolve any problems without delay. In the event that the dispute cannot be immediately resolved, the project manager puts an X on the diagram in order not to hold up the meeting further. For example, task E has been accepted by the project manufacturing manager. He has requested assistance from the project marketing manager, who has stated that she felt her help wasn't really necessary. If this dispute cannot be immediately resolved, there can be another meeting with the project manager and the two managers concerned to resolve the problem. It's not necessary to take up the time of all the meeting participants to resolve a problem between only two of them. For every task, the responsible manager (say, the marketing manager for task A) must determine what the particular deliverable milestone is to be. In this case, it's a report (see Fig. 4-5). (Remember, though, the engineering manager, the manager of the next work package—produce functional specifications—must define the acceptance criteria of the deliverable milestone for the marketing manager. This process always demands cooperation.)

As another example, assume that task D—develop prototype—depending upon the particular organizational design, can either be handled by the engineering manager or the manufacturing manager. In the example shown in Fig. 4-5, the manufacturing manager has accepted it, with support from purchasing and engineering.

Meeting 1 is thus supposed to redefine initial tasks, to assign them to team managers, and generally to close major gaps in the provisional plan (or envelope) outlined by the project manager. It also builds personal commitment by the various meeting attendees. The responsibility diagram also indicates what the expected deliverable or output of a work package will be, according to the responsible team manager. As noted before, in our previous example of the marketing manager accepting task A, the deliverable is a marketing report. The deliverable for task B is specifications, and so forth.

This initial kickoff or bottom-up summary project plan (developed by the project team) holds specific team managers responsible and identifies the jobs that have to be done and the deliverables that will be forthcoming at the end of each task. It also coordinates the interests of all the managers participating in a specific task toward an agreed-upon goal.

This sequential series of meetings should coordinate the effort and thinking of a project group very quickly and effectively. It closes many of the initial project gaps by clearly stating the objectives, develops and aligns the interests of various managers within the project, identifies potential problems and rewards, provides an environment of cooperation, and supports creative decision making since each participant is a valued contributor.

3.4 REVISIONS AND CHANGES

CFP can be used at any time, not only during the initial phases of a project. If the project manager is changed during a long life project, it is wise for a new project manager to do a full analysis and another CFP when taking over. This is not only to become familiarized with the project but "also to clearly establish his performance baseline so that he cannot be held responsible for delays and over-expenditures of the previous management" (Harrison, 1981, p. 185). CFP ensures a smooth transition.

Technical, time, and cost changes to forecasts should always be documented and controlled. Therefore, if a revised CFP is to be completed for these kinds of changes, there should be a change control implemented similar to those used on drawings. For example, the third revision should have a cover document explaining the changes from the previous revision. Of course, there should be a similar process followed internally if only a particular task within the WBS is changed. Documentation of changes should do the following:

1. Identify the change and how it varies from the last scope.
2. Forecast technical, cost, and time changes.
3. Identify interactions with other work packages.
4. Ensure that all other managers are informed.
5. Provide for implementation.

At the conclusion of the project, it would be instructive to review all the changes in order to improve the forecasting process the next time. Those changes that affect the overall technical, time, or cost objectives should be classified as major changes. Those that are within specific work packages and do not affect overall project goals are classified as minor changes.

3.4.1 Major Changes

Major changes affect one or more of the three important criteria of any project appreciably—the contracted technical achievement, the total time, or the total costs. A change like this cannot be handled within the project; it requires the approval of management. It also requires an impact statement (see Chapter 2 for a description of the project charter and the impact statement). The documentation follows the regular change control format noted above with this major difference: It has an "automatic approval and funding unless stopped" clause, or some other "automatic release" clause. We have discussed this before, in Chapter 2, Section 4.4.1.

Some organizations establish a Change Control Board to approve any changes after the first formal design review. Usually a weekly short stand-up meeting (i.e., no chairs means short meetings) at some fixed time is chosen for the Board to convene and act.

3.4.2 Minor Changes

Minor changes require the coordination of the project manager and the particular work package manager. An impact statement is written by the work package manager if the technical, time, or cost goals of *that* work package are to be exceeded. The project manager must deal with these changes, since they're within the total project perimeter. The amount of detail required usually varies with the life of the project. During the initial phases, when not much has been done and the project is almost in a definition stage, the documentation can be minimal. However, as the project progresses, even these minor changes require a more detailed and formalized approach because uncertainty should have dropped.

4.0 REVIEWING IDEAS ON FORECASTING

Forecasting is providing present solutions to future problems. The forecasting techniques covered here are not infallible. The future, obviously, is not completely predictable. These techniques help in selecting the better current alternative solutions. But there are always unknowns, and these are the main concerns of project managers. During early training in school and college, there is an emphasis on logic and consistency, because most predictions deal with known laws and relationships. Management in general and projects in particular are only partially concerned with this logic and consistency since they are, by definition, involved with less predictable elements of human emotion and creativity. It's difficult and much of our early training may no longer seem applicable, but there are some general rules that can help in managing a changing situation:

1. When uncertainty increases, produce more detailed breakdowns of goals ortargets, specifying outputs or milestones.
2. Be sure that forecasts, work packages, milestones, and so forth developed by each manager are presented to the project team for review. This leads to higher accuracy in forecasting.

   The results showed that having to justify one's judgments will lead to higher consistency in those judgments when task predictability is low. . . . The results are interpreted as indicating that justification may lead to an analytical mode of functioning in judgment behavior. (Hagafors and Brehmer, 1983, p. 229)

3. When there are similar projects, prepare forecasts that duplicate as much of the past project's success as possible.

4. Have people who are not closely allied with the project review it quickly and estimate its length. This could be one task for meeting 3 of the coordinated forecasting process.

5. Provide a clear concept of the three project goals (technical achievement, time, and cost), the project charter, and specific managers' responsibilities. Then allow the project team to question them, thereby supporting personal commitment.

6. Don't use forecasting techniques that are more complex than you need. If you don't need a computer to control PERT because you have less than 100 events to control in your network, you may not need to use PERT at all for control purposes. All you probably need is the networking part because it requires careful analysis of relationships. The critical path analysis and reanalysis as the project goes along may be superfluous. The development of the network alone will be sufficient.

7. Try to place the same management emphasis on the costs of underestimating as there are on overestimating, since forecasters (as do most people) tend to respond to the direction that will result in greater personal gain.

8. The most difficult parts of all projects are those involved with an adequate definition of the project scope. If there is any confusion on this point, use a feasibility study to clear it up. There has been more money made and lost during initial design than during any other project phase. If you don't have enough time for a feasibility study, try multiple design approaches, and review them all during a preliminary design review.

9. Don't treat all parts of the project equally. Generally, the first or design parts are more important, but also pay attention to high-cost items. Usually 20 percent of the project design accounts for 80 percent of the cost.

There are always organizational and personal uncertainties. The types of uncertainty requires different kinds of thinking from the kind of thinking usually used in purely technical operations. Try to learn the different modes of thinking. Rational, logical thinking is typical of the schooling of many technically trained managers, but there are other more intuitive, emotional thinking patterns that may have to be learned. For example, Bruner suggests that

> A willingness to divorce one...elf from the obvious is surely a prerequisite for the fresh combinatorial act that produces effective surprises. . . . One final point about the combinatorial acts that produce effective surprise: they almost always succeed through the exercise of technique. (Bruner, 1962, p. 353)

Creativity might possibly be learned, and managers who can use this intuitive thinking pattern can often improve their forecasting abilities. There is a well-known incubation process that states,

1. Work as hard on the problem as you can until you cannot come up with any more answers that will satisfy you.
2. Push the problem completely aside and work on something else.
3. Keep a notebook handy in case something pops up. (That's a good idea in any event. A note pad next to the bed can be used to jot down those brilliant ideas that would otherwise fade with the morning mist.)
4. If the solution that you like doesn't appear within a day or so, repeat the process. It will eventually come to mind.

5.0 SUMMARY

We have developed some techniques for tieing the forecasting techniques into a system. The coordinated forecasting process is an important one that can coordinate the team manager's efforts, demonstrate authority and responsibility patterns, and support personal commitment. The resulting work breakdown schedule, which can be described as an upside-down project flow diagram, is the summation of all forecasts.

In the next chapter, we develop the second major organizational system, measuring. It includes the measuring systems and devices that provide feedback for the forecast.

My Suggested Answers to the Case Study

1. As far as Mildred is concerned, this is a "new" project. She should complete a top-down charter for the *rest of the project* as if that were the total project. Then she can call a bottom-up meeting with her team leaders to develop a new project forecast and present the new project forecast to Harry. If the Widget Project is a major one, a design review meeting might even be appropriate. This is not a project at present. It's an uncontrolled and almost unrelated series of tasks.
2. The usual ones, such as
    a. Task assignments—carefully outline each team manager's job and responsibilities.
    b. Coordinated forecasting process—meeting to forecast the rest of the project.
    c. PERT network—to be sure that everyone understands the relationship among project tasks.
    d. WBS—carefully defining milestones.
3. It is no different from starting a new project.
4. Yes, this should be done whenever there is a requirement for an impact statement or if there is a change in any of the team managers on the project manager's staff.

BIBLIOGRAPHY

BRUNER, JEROME S., "The Conditions of Creativity," in *On Knowing: Essays for the Left Hand*, ed. Cambridge, Mass.: Belknap Press of Harvard University, 1962.

CLELAND, DAVID I., and DUNDAR F. KOCAOGLU, *Engineering Management*. New York: McGraw-Hill, 1981.

HAGAFORS, ROGER, and BERNDT BREHMER, "Does Having to Justify One's Judgments Change the Nature of the Judgment Process?" *Organizational Behavior and Human Performance*, vol. 31 (1983), pp. 223–32.

OTHER READINGS

GALBRAITH, JAY R., "Organization Design: An Information Processing View," in *Organization Planning: Cases and Concepts*, eds. Jay Lorsch and Paul Lawrence. Homewood, Ill.: Richard D. Irwin, 1972.

HAJEK, VICTOR G., *Management of Engineering Projects*. New York: McGraw-Hill, 1977.

HARRISON, F. L., *Advanced Project Management*. New York: John Wiley, 1981.

LAWRENCE, PAUL R., and JAY W. LORSCH, *Organization and Environment: Managing Differentiation and Integration*. Cambridge, Mass.: Harvard University Press, Division of Research, Graduate School of Business Administration, 1967.

MACIARIELLO, JOSEPH A., *Program Management Control Systems*. New York: John Wiley, 1978.

NADLER, DAVID A., and EDWARD E. LAWLER III, "Motivation: A Diagnostic Approach," in *Perspectives on Behavior in Organizations* (2nd ed.), eds. J. Richard Hackman, Edward E. Lawler III, and Lyman W. Porter. New York: McGraw-Hill, 1983.

SILVERMAN, MELVIN, *Project Management: A Guide for Professionals*. Cliffside Park, N.J.: Atrium Assoc., Inc., 1984.

Chapter 5   Measuring: The Second System

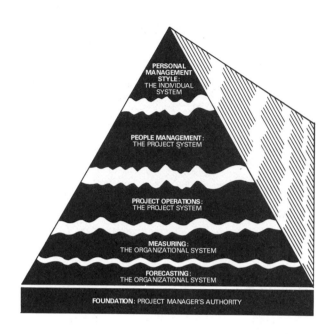

PERSONAL MANAGEMENT STYLE: THE INDIVIDUAL SYSTEM

PEOPLE MANAGEMENT: THE PROJECT SYSTEM

PROJECT OPERATIONS: THE PROJECT SYSTEM

MEASURING: THE ORGANIZATIONAL SYSTEM

FORECASTING: THE ORGANIZATIONAL SYSTEM

FOUNDATION: PROJECT MANAGER'S AUTHORITY

WILLIAMS AND COMPANY

Cast

Roger Gunther, Project Manager, Turtle Project
Alex Bedder, Vice President, Engineering
Joyce Fedders, Chief, Quality Assurance
Angela Mission, Chief, Purchasing

The Turtle was an armored heavy-duty bulldozer that was intended to re-trieve disabled army tanks and bring them back to repair depots. It was an adapta-tion of the company's very successful commercial model of bulldozer. One of the major modifications was the winching mechanism that was to be operated by only one person from the cab of the Turtle. That winching mechanism was the result of a revolutionary engineering design, which Williams and Company had used to convince the Army to award it the Turtle Contract.

Roger Gunther, the manager of the Turtle Project, was on his way to the management review meeting. This was a meeting held on each major project on a regular quarterly basis. Dr. Alex Bedder, the Vice President for Engineering, was the chairman. The project manager and all the immediate supporting managers were to attend and report on progress, problems, opportunities, changes, and anything else that would affect the project. Roger was very confident as he approached the meeting room. He had carefully gone over the progress of his project with his staff and everything seemed to be under control. Expenditures were about 10 percent under budget, all of the milestones were being met, and the critical path on the PERT chart was still achievable.

Roger made his presentation without any interruptions from anyone at the meeting. He had prepared charts showing expenditures versus budget and achievements versus forecast. He stated that this project was coming along well because of the excellent cooperation that everyone on the project had shown, and he predicted that the Turtle would probably be completed within the next two years as scheduled and might even come in about two months before that because it was going so well.

The next person to speak was Angela Mission, chief of the purchasing team for the Turtle. She said that she was a bit surprised to see that Roger showed the project on schedule. She had expected to receive a vendor approval report from the quality assurance team (headed by Joyce Fedders) four weeks ago, and since it had not arrived yet, she had been unable to place purchase orders for the winch main castings with the vendor. This would cause a delay in receiving the castings. That delay might be slightly alleviated if Production worked overtime on them when they came in. This assumed that the castings would be perfect upon arrival from the vendor. She said that if Roger wanted to be sure of that, he might want to fund an additional purchasing coordinator to stay at the casting vendor's plant to make sure the castings were tested and shipped on time. She estimated that the cost for the coordinator and the overtime in production would be about $15,000.

Roger looked at Joyce. She seemed to be intently looking at a report that she had spread out in front of her on the desk. Alex asked Angela if the winch castings were on the critical path. She said that they were. He then asked Joyce if she had any additional information to offer at this time. Joyce said that she would discuss that right now, but Angela interrupted with another comment.

Angela: I'd like everyone to know that the time expended by my purchasing team in waiting for the vendor approval report for Roger's project was charged to the Turtle Project. I now need some additional funds just to buy the castings even if the approval is received today. Those costs are, of course, in addition to the $15,000 I mentioned a moment ago.

Roger: Why so?

Angela: My people have schedules to meet. They charge their time to the projects that they are working on so that we can be sure that they respond to the various needs of the different project managers. We have a procedure in purchasing, and they are supposed to follow those procedures. If you have allocated the time for my people to work on your job, they'll do it. What am I supposed to do with them if you're not ready on time? Fire and rehire them when you are ready?

Alex: Well, Roger, it looks like your project is beginning to show some problems. Joyce, perhaps you can shed some light on this situation.

Joyce: Alex, I told Roger when we started this Turtle Project that there wasn't enough time allowed for my people to visit the vendor, supervise a complete vendor evaluation, and report back with any corrective action. As it turned out, I was right.

Roger: But you reported last week on your PERT input that you were on time and you had delivered the report to purchasing.

Joyce: Yes, I did. Even though we didn't have enough time budgeted to complete the vendor evaluation report, we managed to get most of the work done. There were a few small corrective actions that the vendor still has to complete before we give him a final approval. We've been working very closely with him, and they should be finished shortly. I've told this to Angela but she refused to issue the purchase orders until the report was delivered to her, and we just haven't been able to get every little detail finished yet.

Alex: Joyce, we seem to have several problems here. Perhaps, to start with, why would you accept a budget that you felt was inadequate?

Joyce: Alex, I don't seem to have a choice. According to Roger, it's either "This is your budget and no arguments" or else!

Roger: Everyone always adds a "safety" factor to their budgets. If I let everybody get away with it, our costs would be astronomical, and we never would have gotten the Turtle contract.

Joyce: There's another point that is related to these winch castings. Even if we do spend the extra funds on the purchasing coordinator and work overtime in production to machine them, the pressure test stand won't be available for six months, since our other projects are using them. I was talking to the supervisor of the Test Stand department, and he said that he had discussed this with you, Roger, two months ago. Won't this affect our delivery schedule?

Roger: Well, it might, but that problem won't be facing us for another several months. We can work on it when it gets here. Maybe we can reschedule the test stands then.

Alex: Roger, it appears that your project has a few unsolved problems that haven't been fully presented here. I want a complete recovery plan on my desk in one week. This meeting is over.

QUESTIONS

1. What can be done about Joyce's inaccurate reporting? Is Roger doing the same thing? Why?
2. Do you believe Roger's statement about everyone inflating their estimates? What can a manager do about that?
3. Should time be charged to a project if there is no work being done on it? Why? How can that be controlled?
4. What changes should Roger make? What changes should Alex make?
5. Have you ever been confronted with a similar situation? How did you handle it?

1.0 INTRODUCTION

The usual backbone of most corporate measurements is an array of various financial reports produced by the Accounting department. In many countries, the publication of financial data of differing complexity has become a legal requirement. For example, every publicly owned company in the United States is required by law to provide certified financial reports to its stockholders at least on an annual basis. Those measurement reports are in fairly standardized formats. They use double-entry bookkeeping techniques to produce balance sheets and profit-loss statements. The balance sheet shows the company's financial status at a point in time, usually at year end. The profit-loss statements show the differences between the revenues the company receives and the expenses it incurs over time, which is also usually a year. These reports are almost useless for operating management purposes, since they indicate the financial situation of the past. Management wants to know how well they are doing right now. It's not the same thing.

Unfortunately, there are not enough kinds of accounting measurements that are useful to help management decision making that can affect the future. The few that are, report progress of more limited objectives such as technical achievements, time used, and current cost analysis. But even these few reports are more useful to the functional groups with their ongoing organization structures than they are for projects. Functions operate on a more or less consistent pattern of uncertainty over time, and the measuring systems consistently feed data back on a similar pattern. For example, monthly or weekly reports reflect a linear world that is not applicable to most projects. Projects have a pattern of nonlinear uncertainty absorption. If measuring systems are to be helpful to projects, they have to produce reports with an initially small time base that gradually lengthens. Nonlinear reporting is not the norm, and this can be a major consideration in designing project measurement systems. A quick glance back at Fig. 3-5 illustrates this requirement for nonlinear flow.

1.1 NONLINEAR TIME REPORTING: UNCERTAINTY AND REPORT FREQUENCY

Therefore, a linear reporting system that provides a periodic report on the first of every month may be useful for functions, because the uncertainty levels to be absorbed by the various managers are about the same, month after month. (Of course, this assumes a cyclic kind of function with no major changes.) This is not typical for most projects. Projects require fast reporting in the project beginnings when uncertainty is fairly high and is being absorbed by the various project managers as they make their nonrepetitive decisions. The problem gets even more complex when you consider that there are relatively small expenditures at the beginning of a project, which rarely appear to be significant in most financial reports. However, the effect on the eventual success of the project of these initially small expenditures can be high.

For example, a feasibility study at the beginning may be relatively inexpensive yet produce great results. If we assume there are limits to how much uncertainty can be handled by a person during a given time period and the level of uncertainty being faced by team and project managers during the beginning stages of a project is high, the time period for reporting has to be shorter. This assumes that when uncertainty is high, reports must be made more frequently, and that means shorter time periods. When uncertainty drops, reports can be made less frequently and can even cover larger time periods. Since reports cost money, it's always desirable to minimize them as uncertainty drops since their usefulness declines. In other words, as uncertainty drops with the project age, the time period for reporting can be increased. The description is that of a geometrically increasing time period between measurement reports while uncertainty drops. However, the uncertainty curve over time is just the opposite of that of cumulative costs. Cumulative costs start low and go up geometrically. That's very different from equivalent functional organizations, in which uncertainty is relatively steady over a given time period and cumulative costs usually increase linearly.

Therefore, measurement techniques for projects are different. They should consider the differing levels of uncertainty to be handled, the amount of cumulative costs to be incurred, and the consequent differing frequency of reporting.

1.2 THE IMPORTANCE OF REPORTS

To review a bit, it's obvious that if my model of uncertainty decreasing with time is valid, the early decisions in most projects have more effect on project success than those made later on. More money has been made or lost in the research or Engineering department than in consequent activities of the manufacturing and service shops. When the research or the engineering design is completed optimally, the potentials for major losses later in the project are decreased tremendously if our decreasing uncertainty curve is reasonable. Achieving the optimal design effort may usually not show an appreciable expenditure of time or funds. If the project's uncertainty curve is as suggested, costs do not become *cumulatively* large (generally) until the designs are completed and major purchase orders are placed. Thus, the importance of time on a project is also nonlinear. The time initially spent in engineering and research is more important than that spent later on. Of course, if the design effort is a major part of the total project costs, these statements may not entirely apply. However, it's the relationship of uncertainty to cost that is the major concern. This applies to all projects.

It's an interesting circumstance that if the time and resources are spent very well, no one may even notice how well it was done. If problems are initially eliminated through effective design and management, the project will probably never suffer with them later on. One doesn't see problems that are eliminated. Therefore, while the cumulative time and costs being spent at the project beginning are relatively minor, the project manager should be very concerned with the effects of these smaller expenditures in the beginning. The measurement system has to reflect this concern.

If this uncertainty curve is typical, it can readily be seen why regularly delivered financial reports are almost useless to the project manager. Regular reports delivered in the project beginnings will probably show small expenditures of time and cost. Subsequent reports may show much greater spending, and while that might be of great interest to a functional manager, it's of little interest to the project manager, since it merely reflects the vital technical, upfront decisions that were made long before. As noted before, the cumulative cost curve for projects is almost the opposite of the uncertainty curve. This leads to a high frequency of measurement and reporting during the initial stages of a project with not much cost incurred. It also means a decreasing reporting frequency with project life when most of the funds are actually spent. It's practically a violation of regular financial reporting and reporting this way can be expensive.

Every measuring system or reporting technique costs something. And no matter how important these data may be, there's always a balance between a manager needing to "know everything" and "paying for it." Regular financial reports are inadequate, so, to minimize formal reporting costs, project managers have many project meetings during the initial high-uncertainty phases of the project. Meetings provide fast feedback or measurement as problems are defined and solutions are noted or corrective action started early. But there is a disadvantage too; meetings may cost more than reports do. The time people spend attending them is more expensive than written reports are because everyone may be at meetings but only one person writes a report. But it does gain prompt information, which is usually more valuable than the meeting cost. This doesn't apply later in the project. Then, reports are used because of lowered uncertainty. With lowered uncertainty, meetings become relatively expensive and their frequency should decrease dramatically.

Since each project undoubtedly has different absolute levels of uncertainty (with the level of uncertainty being partially defined subjectively by the project and team managers), reporting frequency is, therefore, dependent upon the needs of each of those managers. Responding to this need for different reporting frequency is almost an impossible task for most Accounting or Data Processing departments. Consider this: "Charley, I'd like your data processing people to provide progress reports to me about my project twice a week for the first eight weeks, then once a week for the next eight weeks, then every two weeks, etc. and so forth. By the way, I'll let you know if that frequency of reporting is to be increased when we run into trouble! And oh yes, the other team managers may want a different report frequency." Can you imagine Charley's response?

Reporting techniques don't stand alone. They must be closely linked to the forecasting techniques and the management strategy. As you recall, "strategy" was defined as the predetermined answers to expected potential variances or differences between forecasts and measurements. Project management strategy is based in the next two project-related systems: Project Operations and People Management. The last system, Personal Management Style, deals with the behavior of the project manager. In this book, we'll deal with measurements and variances from forecasts that occur during those measurements as if they stood alone. That's for the sake of clarity. In reality, they are interactive.

2.0 THE FIRST MEASUREMENT: TECHNICAL ACHIEVEMENT

The first and most important measurement is technical achievement, and this is measured primarily by the delivery of an appropriate "milestone." For example, let's assume that the *end result* of the technical achievements of a project have been fairly well defined during the initial stages of the project. A final acceptance test has been defined by the customer or client and developed during the design of the charter. This test measures the project's technical achievement. But the project manager is not prepared to start the project and "hope" that the technical achievement will be met. Progress must be monitored and reports received as the project moves along. The milestone is one of the more important monitoring devices.

As previously defined, the adequacy or the acceptance of a milestone is determined by the receiver of that milestone. For example, let us assume that the final test data are a milestone. The project is then accepted only when the test results are approved by the customer. During the project life cycle, milestones are also used to begin a work package. Each "deliverable" is from a team manager who is responsible for the previous work package. It can also be an interim "deliverable" to Quality Assurance, the project manager, or anyone else as a measure of progress. And finally, it can be an ending "deliverable" to show that the work package is completed. When the project is begun, the first milestone may be a work release form approved by the project manager. A work release form is delivered to the managers of the first work packages. After they approve the work release forms as adequate, the work starts.

Milestone deliveries ensure cooperation between neighboring work package managers. The manager of the preceding work package, that is, the one delivering the milestone, must satisfy the needs of the next work manager package, otherwise the preceding work package is incomplete. Therefore, it is in every team manager's best interests to deliver and/or receive an acceptable milestone. If the tangible deliverable, the technical milestone, is defined by documented test data, a technical achievement has occurred when the documentation is accepted, not when the test is finished. When these kinds of milestones have been defined well, there is almost a digital response, since the data are either in an acceptable format or not. "Has the test been completed? Good, did the product meet the specifications or did it fail? Where is the milestone, which are the acceptable test results?" The use of milestones as deliverables makes technical measurements the most straightforward kinds of measurements.

3.0 THE SECOND MEASUREMENT: TIME

Time is not really a linear measurement for projects. As said before, the time expended at the beginning of a project is usually much more valuable than that spent at the end. This valuable time is very carefully measured in small doses at the beginning. There could be many meetings for this purpose, possibly on a daily basis. The technical solutions developed then affect the whole project. However, there could be exceptions to this general rule. If the project is an all-or-nothing-at-all project (like a new parachute), this concept of time being valued differently might not apply. The uncertainty level may not drop with time if you are designing a new parachute. It will drop precipitously when the parachute is tested, at the end of the project. It's like a total yes or no answer in that case. But valuable beginning time and less valuable ending time are typical because there are usually interim technical achievements made as most projects go along. For example, if the project is designing and building a new diesel engine, uncertainty will drop a little after the first test of an injector nozzle, and so on.

When important technical goals are achieved early in the project, the project will have a high success potential. If not, the project should be stopped or else it will probably fail. And, until there is an indication of potential technical success, cumulative spending should be low.

3.1 PROJECTS AND FUNCTIONS: COMPARING REPORTING

Since functions, like the Engineering or the Accounting departments, are a continuing organization effort (as noted before, almost "eternal"), their reporting techniques and measurement systems are usually based on some extended time-based level of effort rather than large and very specific deliverable milestones. Their milestones are not delivered irregularly but rather continually through the entire year. Regular periodic reporting of technical progress and expenditures versus forecasts are their reasonable management measuring techniques. A management review is a regularly scheduled affair.

Functions can also rely upon long-term and relatively predictable human interactions. Therefore, functions develop a culture and a definition of an acceptable work pace that has a longer range outlook. They also handle a middle range of problems, minimizing the ups and downs of the work load with as little disruption as possible.

It's not so for projects. The project culture has to be reinvented to a great degree for each project. Problems vary greatly in magnitude, occasionally requiring extremely heavy work loads. Team participants no longer contributing to the project's goals are reassigned back to their functions. The specifications are not at all the same. There is less time for development of a project culture and getting team participants to cooperate with it.

The slower acculturation of a project team is partially minimized since projects usually deal with relatively small groups. This can speed things up. For example, if the *direct staff* (i.e., those team managers reporting directly to the project manager) on most projects exceeds 10 people, it's a very large project. With a span of control of 10, four organizational levels can have 10,000 people. Few projects have that many levels or are that large today. Fewer people in a group can tend to increase communications and support cooperation. A smaller staff allows the project manager to have each team manager supervise and feed back measurements against the forecast for the area under his or her control. This is not as troublesome as it might appear, since there are only a few work packages that should be open for charges at any one time, and this simplifies and limits reporting. These limited measurements are then summarized as necessary by the project manager (whose job, anyway, is primarily managing the interfaces among team members) and a timely total project report can result. This might be the basis for corrective action, if required. Small teams make meetings easier. The design and development processes in the bottom-up meetings, the coordinated forecasting process, and the close working arrangements possible in a small group also help in developing a supportive culture.

4.0 THE THIRD MEASUREMENT: COST

As noted before, project cost measurements need a different feedback schedule report content than function cost measurements do. In my opinion, most standard organizational cost systems are not effectively applicable for projects. Cost systems that are based on historical financial data help very little in forecasting the future unless one assumes that the past will be repeated. Of course, past experience is a valuable guide but using this kind of data is almost like trying to steer a boat by watching the wake in the water. Historical data may be necessary for the Accounting department's profit and loss statements, but it is less valuable in controlling project operations. History doesn't provide what is really needed: assistance today in projecting the unique tasks to be completed *in the future*. The most effective project managers are almost always concerned with answering the question, ''Based upon what we know *now*, what do we do next?'' rather than, ''Did we do all right in the past?'' It's an existential or present-time attitude concerned with trying to solve the problems of today to minimize those of the future. One of the more important (and possibly destructive) cost concepts that functions use in the budgeting process is the absorption costing technique, which involves the allocation of overheads or other corporate fixed costs, downward to the project.

4.1 ABSORPTION VERSUS DIRECT COST
SYSTEMS

All cost measurement systems are not equal. We briefly touched on the subject of responsibility accounting before. That's defined as assigning responsibility for financial results only to people who are responsible for the expenditure of the costs concerned. Let's assume the original forecast of technical, time, and costs flow as shown in Fig. 5-1, from the functional support groups to the various project

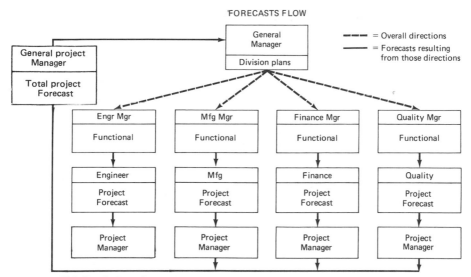

This flow is shown in one direction only for descriptive purposes. There are usually
many revisions or back-and-forth adjustments before submission to top management.

Figure 5-1.   Forecasts flow

team managers, then to the project manager for consolidation and up to the general
manager. Under responsibility accounting concepts, the project people are charged
only for those variable costs that are in their forecasts.

There are some organizations that operate under an *absorption costing system,*
in which a charge is made for *overhead* or *burden* as a function of some variable
cost that is incurred directly. For example, with an absorption cost system, there
might be a burden charge assigned to some variable such as labor, materials, or
units produced. The overhead usually consists of the time-based costs associated
with operating some manufacturing facility. This overhead is then distributed according to some direct variable such as direct labor or the sum of labor plus material
called *prime costs.* Often there is even an allocation or down-loading of central corporate costs, called *general and administrative costs,* as a function of labor, materials, or some combination of the two.

The theory is that these costs should be absorbed by the users in accordance
with how much they use the services that these costs represent. On the surface, it's a
reasonable idea. One is charged according to the benefits received from some central corporate source (''Well, you people need Central Purchasing, don't you?''
''Somebody has to pay for all of these corporate services, don't they?'') But it fails
because those who pay don't control, somewhat like taxation without representation. These overhead services and corporate charges are under the control of some
other manager, not the project managers who have to absorb them into their
operating budgets. When this happens, there's no freedom of action for the project
managers, and consequently no authority over them. The theory is sound for setting

prices but not for costs. A customer, as the final absorber of "uncertainty," has the choice of paying your price or going to someone else. In effect, in a free market, the customer accepts the responsibility by buying your products or else buying someone else's. The manager in an organization where overhead and general and administrative costs are assigned downward usually doesn't have that choice. The total amount of expenditures for overheads and corporate expenses is not controlled by the "absorbers" with an absorption cost system. When those expenses go up, the amount charged (per unit of variable costs) goes up automatically. That is not constructive, to say the least. Therefore, we'll turn to accounting concepts that are more applicable, such as responsibility accounting.

Responsibility accounting is a mainstay for *direct costing* systems. *Direct costs* are the variable production costs of the product. If they are dependent upon each product, it's a safe bet that the costs are direct. Fixed or time period–based manufacturing costs, such as overhead, and those due to corporate activities not directly associated with the product or service being produced, such as general and administrative costs, are charged against revenue immediately rather than as a product cost. They are not assigned to the units produced but the total revenue received during the time period that they are incurred. Except for the variable overheads associated with each employee such as insurance, and so forth, there are no fixed overheads or burdens or general and administrative costs assigned downward. Manufacturing overheads and budgets that are associated with a time period of some kind are the consequent responsibility of the managers who run the factory or provide the manufacturing capability. In other words, general overhead is the responsibility of the plant manager who is in charge of the physical facilities. Similarly, corporate general and administrative costs become the responsibility of the corporate officers who control them. For example, the company president can affect real estate or state income taxes by moving the operation to another location where those taxes are lower. (See Horngren- 1972 for details.) Therefore, those taxes should be in the company president's budget.

The general rule should be, "If you can't change it, it doesn't belong to you!" In these systems, instead of loading costs downward, "contributions" (which are the difference between the revenue that the project brings in and the direct costs that it incurs) can be sent upward. If those "contributions" are sufficient to pay for the overhead and the general and administrative costs, that would be fine. If there is anything left over after those costs are paid, it's profit before taxes. If the "contributions" upward are insufficient to pay for overhead, there are two sequential strategies to follow: reduce overhead to match the contributions received or try to raise contributions by cutting costs and raising sales. The same alternatives apply to general and administrative corporate expenses.

With direct costing, total revenue is first distributed for direct costs incurred, such as labor, variable overhead, materials, and so forth. Then the difference between revenue and direct cost is used for the total period costs such as overheads, corporate expenses, and profit. When cost measurement systems do not provide this clear measurement of the flow of revenue and costs, the measurements should not be attributed to the managers who provided the original forecasts. If that happens, the variances confuse rather than assist effective management corrective actions. Responsibility accounting and direct costing provide clearer measurements upon which to base management actions.

This applies even when projects are funded internally and therefore provide no contribution. With no revenue, there can be no contribution. In this example, there should be no question that projects should be measured on a stand-alone, direct cost basis, since there will be no contributions to pay for overheads, administration, and so forth. Therefore internal projects are measured by a discounted cash flow technique that compares the values of some expected future increased stream of revenue against the project's present or direct costs. But either in internally or externally funded projects, the responsibility for fixed overheads and other corporate costs cannot be assigned to the project manager. If clarity of measurements and responsibility is desired with a resulting optimum project management operation, direct costs should be used.

Direct cost systems measure and report only those costs that are directly applicable (or variable) to a particular project. Since the forecasting and expenditure of funds *within* a work package are the responsibility of the team leader who is managing it, responsibility accounting then is automatically a part of internal project cost reporting.

Overheads, corporate cost "downloads," or other nondirect costs that *cannot* be changed by the project manager that are allocated to a project, as in absorption systems, are almost invariably a detriment for the project manager. When motivation in technical people is primarily dependent upon their expectations for the personal achievement of future rewards, this would definitely be a demotivator. The demotivating influence applies also to functional managers but not in the same degree. They don't have the project manager's disadvantage of only having one time through to complete the project. With enough time and repetitive reviews of periodic budgets, functional managers often find that it is possible to adjust their next year's forecasts upward, which, of course, then allows for absorption of the possible increases in these kinds of nondirect costs. A direct cost system supports improved motivation in managing projects, because there is a clearer connection among the forecast, the measurement, and a consequent management strategy to handle variances between the two.

To briefly review, using direct costing techniques means that projects are responsible only for costs that they can control. When projects provide products or services that are sold to outside firms, those projects either have to generate sufficient differences between revenue received and direct costs to pay for the corporate costs, or else uncontrolled (overhead and general and administrative) costs must be reduced. Reducing overhead, general, and administrative costs is not the job of the project manager. It's the job of the functional managers who are responsible for the costs. Of course, when there are many projects and the individual project contributions are totaled into one overall contribution, there may then be sufficient funds for this purpose. The manufacturing management responsible for overheads and corporate management responsible for those general and administrative costs have to live within the total contribution generated.

For example, if the project's direct costs start to exceed the total forecasts for the costs and variable overhead that is directly associated with those direct costs, such as insurance, social security, and so forth, the responsibility is clearly the project manager's. However, when the total contribution is calculated by deducting all the project's direct costs from all project revenues, and there may be insufficient contributions to cover manufacturing overheads and corporate or administrative costs, it is not a direct responsibility of the project manager if direct project costs are within forecasts. It is usually the responsibility of others if they negotiated the selling price of the project.

> Absorption costing is much more widely used than direct costing, although the growing use of the contribution approach in performance measurement and cost analysis has led to increased use of direct costing. . . .
>
> Direct costing has been a controversial subject among accountants—not so much because there is disagreement about the need for delineating between variable- and fixed-cost behavior patterns for management planning and control, but because there is a question about its theoretical propriety for *external* reporting. (Horngren, 1972, p.311)

External reporting is *not* a major concern of project managers, and in many cases, not that of most middle level functional managers either. However, it is a major concern of the organization's top management because the external financial reports are one measurement of the success of the total organization. But it's a fairly straightforward accounting calculation that can convert the data from a direct cost system to an absorption system. This can be done quite easily in the Accounting department without inconveniencing (or confusing) operating managers.

Direct cost systems have been around for many years and are quite familiar to most accounting groups. Those groups may have a reluctance to move from absorption costing for *financial reporting* reasons, but occasionally I think that there may be other reasons for the lack of widespread application of direct cost systems. While the basic premises and advantages of direct costing are clear and easily understood, there is often a top-down functional management reluctance to implement this financial system. Direct costing clarifies the qualitative differences in the decision-making processes of various managers. Therefore, the system might be interpreted as a technique that helps to differentiate those managers who learn and apply from those who don't.

Although the use of direct costing is often a goal stated by many top level managers, those top level managers in larger organizations often reached their positions not by taking risks but by avoiding risks. They succeeded through not making mistakes and surviving, by achieving seniority. It is a rare top manager who allows major changes in forecasts, because now "we know more." In contrast, project and team managers should always be willing to change as new data are presented. Project and team managers are almost always willing to iterate in order to modify past forecasts. In some functional or conventional organizations, these iterations might be interpreted as "mistakes." They're not, of course. They're just corrections as unforeseen problems are being resolved. Therefore, clarity in accounting measurements is useful, depending on who (the level of management) or what (functions versus projects) is being measured. Returning briefly to the Accounting fraternity, we even find that they might be at fault, according to these authors:

> Whether managers are familiar with accounting systems or not, they can find it useful to reflect on the methods and assumptions used in the preparation of accounting data. Among the more crucial are the following:
>
> 1. Accountants live in a linear world. With few exceptions, cost and revenue data are assumed to vary linearly with associated changes in inputs and outputs.
> 2. The accounting system often provides cost/revenue information that is derived from standard cost analyses and equally standardized assumptions regarding revenues. These standards may or may not be accurate representations of the cost/revenue structure of the physical system they purport to represent. . . .
> 3. Remember that assignment of overhead cost is always arbitrary. The accounting system is the richest source of information in the organization, and it should be used—but with great care and understanding." (Meredith & Mantel, 1985, p.54)

4.2 OTHER MEASUREMENTS: COMMITTED VERSUS ACTUALS

In the previous chapter, the staff-related work breakdown schedules separated the forecasts of materials from those of labor by establishing a work package for materials only. The format for that work package followed a projected bill of materials. This can be invaluable when trying to implement fast, responsibility accounting reports applied to expenditures for materials. The usual process is to write a requisition, then have Purchasing issue a purchase order or else have Inventory Control draw it from stock, receive the material, then wait for the invoice before charging the project. This results in measurement delays that could cover extended time periods.

However, when work packages for materials are separated from those of labor, measurement becomes easier. Any team manager can then start a material requisition if the materials are to be used in a work package that he or she is managing. The material requisition is matched against the progress in the particular work package where the material is needed and the bill of materials forecast in the material work package. If there is a match, the requisition is approved by the project manager, and that requisition cost is immediately charged against the project. If there is not a match, there is an immediate investigation into the variance. If the variance is justified, because of a change in design made after the work package was estimated and/or because of an increase in a vendor's price, contingency funds may have to be added to the materials work package at this point if future cost reductions cannot be forecasted. The total requisition costs are *committed costs* rather than *actual costs*, but they are just as effective for control as subsequent actual costs. The project manager could use those committed costs as a measurement against forecasts as well as any later actuals.

The materials requisition is then sent over to Purchasing, where a firm cost is negotiated. It can also go to Inventory Control if the material is unallocated and in stock. If there is a variance between the buying cost and the requisition cost, the variance is charged to Purchasing. Usually, when there is a positive variance—the buying cost is less than the requisition cost—Purchasing gets the credit in its variance account. When there is a negative variance, Purchasing is usually reluctant to issue a purchase order. This decreases the positive amount in the variance accounts and Purchasing, therefore, has a tendency to return to the project and attempt to investigate the difference. As far as the project manager is concerned, this behavior is just what is wanted. When this happens, it is possible to investigate the potential variance, possibly start corrective action in engineering or wherever the requisition came from, and thereby change the requisition to minimize the negative variance. An alternative is to increase the requisition funds using additional contingency funds. The important idea is that this variance is promptly investigated. The variance accounts are analyzed monthly to determine how well forecasts are being met. These kinds of accounts definitely support the idea of responsibility accounting.

4.3 SUMMARIZING THESE IDEAS ON ACCOUNTING

The relatively short life cycle, the usually decreasing uncertainty curve, and the increasing curve of cumulative expenditures of costs of projects require many measurement techniques that are different from those of functions. These techniques result in systems that more clearly fix responsibility and provide control over costs to those managers who can change them. Project budgets are sometimes considered to be a duplication of the functional organization's budgets. In some cases, it's almost as if everything were forecasted twice with the project budgets just arranged differently, in parallel but not exactly the same. This different, yet parallel, arrangement provides project managers with very relevant measurements, such as those where work packages are independent of one another. It's almost a necessity because

> The parallel accounting systems provide independent accounting controls that are consistent with the characteristics of the work in each type of unit and that recognize the partial autonomy of each organizational subunit. (Lawrence, Kolodny & Davis 1977-P.46)

Going a step further, not only is the total project budgeting system a partial duplication of the functional organization's budgeting system, but the budgets for the work packages witihin the project also duplicate the total project budget. There are budgets within budgets or, in other words, forecasts within forecasts. The total project may have three separate, related forecasts: those of direct labor, support services, and material (whether purchased or taken from inventory). The work packages for labor will be started and completed through the delivery of some tangible milestone. The work packages for support or staff services and materials are ongoing and will not be "completed" because there is no deliverable. Therefore, these support services have to be controlled by the project manager acting temporarily as a functional manager. Control exerted over materials through a requisition charge and a committed cost provides fast, effective feedback.

The beginning of projects is usually typified by smaller work packages because of the higher level of uncertainty. In other words, these packages have smaller total costs budgeted. They are reviewed often. Also, the deliverable "milestones" under the control of the work package managers are probably more important at that time in the project cycle. However, the design of the reporting techniques and measurement system is more than a conceptual exercise. It requires implementation. Sometimes this implementation is accomplished by the same project team that will run the project. This team not only develops the project, but it also helps to develop the systems that will be used to manage it (remember the bottom-up meeting?) and even to implement those systems if necessary.

5.0 IMPLEMENTING: DEVELOPING INFORMAL PROJECT REPORTING SYSTEMS

Fast information is always better than slow information. Reports concerned with the beginning phases of a project are concerned with higher uncertainty levels than those later on. Therefore, the design criteria for the timing of project reports become obvious: The faster they are delivered after any time period, the more useful they are. In management, fast reports are often dependent upon "educated guesses" about the unknowns that are not available at the time the report is created. In accounting, an educated guess about unknowns is called an *accrual*. Learning how to become an "educated guesser" or a better than average estimator of accruals is fairly easy. If the data always come from the same sources, repetitive educated guessing (or accruals) eventually provides quite amazing accuracy.

The "guesser" doesn't act on the first few reports that are produced but compares the fast report with the *actuals* that are produced weeks later from the regular sources of data, such as data processing or accounting. Most errors or differences are repetitive, and when these are recognized and compensated for, the guesser will be able to produce astonishingly accurate reports. In other words, the first few fast reports may not be very accurate at all when they are compared with the later, more accurate actual reports. But after comparing these first fast reports with the more accurate actual reports, repetitive errors can be accounted for and factored into future fast reports. Thus the accuracy of the fast reports is increased.

"Fast" reporting is easily implemented, since all it requires is some repetitive note taking or educated guessing about what's happening. When there is a choice between speed and accuracy in reporting, go for speed every time; but go for speed in a disciplined way. Learn where the repetitive problems are and compensate for them. In other words, don't make any major decisions until there has been a chance to debug the first few reports and find the errors in the fast reports. Set up reporting periods at minimal time intervals and then have them condensed as the project progresses. A reporting period of one week may be satisfactory during the inception of the project when uncertainty is high, subsequent weekly reports just pile up later in the project when the design is completed and it is in manufacturing. Then the reporting frequency can be a month or even longer.

6.0 GETTING IT ACCURATE: THE ESTIMATE AT COMPLETION

The *estimate at completion* (EAC) is an important measurement tool that is useful to track all aspects of a project: technical achievement, time, and cost (see Fig. 5-2). Most reports are concerned with the differences due to comparisons of actual measurement against the forecast, that is, the *variance*. This is also valuable information, but the real concern of any project is with the future, *not* altogether with what is happening now or with what happened recently, but what will we end up with, what will it look like when it's delivered, and when will it be done? When the team manager responsible for a given work package or task is required to provide an estimate to complete (ETC) at the same time that the actuals are reported, it is possible to easily monitor the actual progress toward a goal which is the *estimate at completion* (EAC). Being concerned with the estimate at completion, the manager *receiving* the report just adds the ETC to the actual cost, (EAC = ETC + actuals) to determine the predicted end point at that time.

When the ETC is being forecasted, the particular work package manager is reforecasting the remaining portion of work that has to be done in that particular

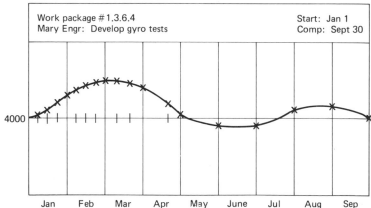

PLOT OF ESTIMATE-AT-
COMPLETION OVER TIME

Figure 5-2. Plot of estimate at completion over time

work package. This is not the same as requiring a report of the work completed or expended versus the plan. When reports include the estimate of the percent completed rather than a report of the actual work to be accomplished, misleading conclusions are often drawn. For example, let's assume that two work packages are reported to be 75 percent completed. This information would not be very helpful, since package A has only 50 engineering hours in it and will be finished in two days, whereas package B has 5,000 hours in it and will take the rest of the year. It is a repetitive requirement for each responsible task leader or team manager to reestimate the amount of work, costs, and so forth that is required to complete the tasks or to deliver the appropriate milestone. Percentages don't apply to projects.

The estimate to complete is another way of updating the forecast. It requires revising and re-estimating future tasks in every reporting period. When reporting of the estimate to complete is repeatedly done in an open forum such as in a project meeting, people improve in the accuracy of their estimates, according to research (See Hagafors & Brehmer, 1983). They become better forecasters (and, of course, better managers) since they have to justify their cognitive processes to their peers and thereby learn what parts of their forecasts are sensitive to changes and which are not. Changes in the forecasted EAC over time modify present actions or existing forecasts that influence actions taken in the future. And this can be done *now*. The ETC is produced on a nonlinear time basis, that is, smaller time increments in the beginning and longer at the end.

There is an additional advantage gained in using the ETC. When an ETC concerning the same work package shows a widely varying EAC from report to report, it indicates that many unforeseen things are happening or that the work package manager doesn't understand something. A sharp rise in the EAC indicates problems and can justify writing an impact statement if the new EAC exceeds project time or funds. Conversely, if it drops precipitously, there must have been a breakthrough of some sort. It's a very useful measurement tool.

The ETC and EAC are even appropriate when a materials work package is used. In this case, it is acceptable to use an arithmetic difference between the original total work package cost (the original EAC) and the amount of costs that have been already committed. This figure is really the total forecasted material in the work package less the requisitions issued. Of course, even this might change if the bill of materials on the project was modified or contingency funds were added to the work package.

Figure 5-2 demonstrates the varying shape of the estimate at completion (EAC) over the life cycle of a work package. The original EAC for the work package 1.3.6.4—Develop Gyro Test—was 4,000. During February, Mary reported an ETC that resulted in a higher EAC than the original 4,000. That ETC was added to the actual to get the EAC. She seemed to be in trouble. The trouble was solved, and during May and June, she seemed to be coming in below the 4,000. During August a few minor problems surfaced, but she finally ended on target. This reporting technique helps the project manager watch Mary's progress as she solves problems and makes decisions.

7.0 OPENING AND CLOSING WORK PACKAGES

As a quick review, work packages are specific bundles of achievements, amounts of time, and groups of costs that accompany specific parts or predicted accomplishments of the project. Work packages in a project are usually related to one another in some sort of hierarchal fashion and are intended to control various parts of the project. Conversely, when the *entire* project is opened for charges and the historical cost systems show expenditures against plan, it provides no useful management information for the manager. It's merely a comparison of expenditures versus forecast. There is no way to determine when and if a particular work package is being worked on and if incorrect charges have been incurred. Historical reports show only the forecast and total expenditures against it.

Establishing opening and closing work packages *within* the project breaks down the work into manageable tasks. It fits the concepts of responsibility accounting, limits potential overruns only to those work packages open for charges, and helps to point out internal inconsistencies in scheduling within the project itself.

The project cost system must support opening work packages when they should be worked on and closing them when they are completed. It must also provide that a closed work package will not accept any more charges. If charges are attempted against a closed work package, they should be rejected and will, then, quite properly become a recharge to and a consequent concern of the functional manager in whose department they originated. This eliminates a lot of search-and-destroy missions for bad project charges for the project manager. The milestone provides a control to support the correct use of this open/close ability.

We have defined *milestone* as a tangible deliverable. It is *not* some significant event that is intended to occur sometime within the project life cycle unless someone has actually defined how it will be objectively measured. Defining milestones as deliverables decreases some of the problems with measurement of achievement in historical reporting systems, since at least two people on the project—the milestone deliverer and the receiver—know if a delivery has been made and if it has been made to schedule. It is in the best interest of both the deliverer and the receiver if a milestone is mutually defined during the project forecasting phases, by establishing what that definition will be and how they both will know it has been delivered. This tends to decrease conflict when work packages cross functional boundaries. Milestone receivers want to receive them as soon as possible because then their work packages are open for charges. In effect, they begin to informally monitor the progress of the milestone deliverer. Milestone deliverers cannot hold onto a milestone too long because their work packages will be overexpended. An acceptable milestone, then, is approved by both parties before the next work package is opened. Since the accepting of an opened work package implies the ability to work, few milestone receivers will approve opening their work package unless there are adequate inputs. If the input was inadequate and they said that it was adequate, it means that they will have to perform extra work within their own work package to make the accepted milestone useful, since the prior work package was closed. Few receivers want to do that. In summary, the project work packages should contain only direct labor costs that produce some tangible output and should be opened and closed by deliverable milestones that have been defined by both the deliverer and the receiver.

Even with project support services that usually have no understood deliverables such as Quality Assurance, Contract Administration, and Project Management, it might be possible to partially apply these cost measurement ideas. Since the work package is basically the project management staff, the milestones could be the progress reports that this staff delivers. This package is therefore sometimes managed as if it were a minifunctional organization. In other words, since a very large project might have, say, as many as 10 people on staff (although there are usually less), charges can be controlled through regular reviews, meetings, reports, and so forth. The staff here is actually a small, short-lived function.

A tangible milestone can also be used as an intrapackage deliverable. This is a very useful project management tool to monitor progress *within* a work package. In this case, however, either the project manager or some other functional manager, such as Quality Assurance, must determine the adequacy of the interim deliverable milestone. The receiver of the interim deliverable milestone must be someone who is knowledgeable but *not* involved in the work package itself. It's like an inspector who will accept or reject in-process work. He or she should not be the same person as the receiver of the ending milestone because that receiver is really only interested in the package conclusion. The concluding milestone of the prior work package signifies the opening of the receiver's own work package. If this can't be done and the only alternative is to open the project at the beginning, hoping to close it at the end before expending all the available time, it's like offering an open, unlimited calendar and checkbook to anyone who wishes to charge your project.

8.0 PRICES AND COSTS ARE NOT THE SAME: A QUICK REVIEW

Before going further, let us make a small diversion and clearly differentiate between costs and prices. Determining a price at which to *sell* a particular product or service requires that we include every element of *cost* that the organization expends before adding a profit. The price must include direct costs, a part of the overhead and general and administrative costs, and an appropriate profit. In a free economy, the total price should be set as high as the market will bear. When the unit price is multiplied by the amount sold, the result is called *revenue*. Revenue can come from outside (when the sale is to an outsider) or from inside (when "purchased" by another company division). Whatever is left over after paying for direct costs, overhead, and general and administrative costs is profit. Therefore, total revenue is first distributed for direct costs incurred such as labor, variable overhead, materials, and so forth. The difference between revenue and direct cost is then used to pay for period costs such as overheads and corporate, general, and administrative expenses. If there's anything left over, that's profit. When cost systems do not provide this clear meas-urement of the flow of revenue and costs, the most probable reason is that the organization has adopted conventional (and functionally oriented or absorption) systems by assigning overheads (and/or burden rates) and allocating the costs of the corporate offices to projects. This produces measurements that may not be comparable to the original work package forecasts, thus confusing rather than assisting effective corrective actions taken by project management. Responsibility accounting and direct costing provide clearer measurements upon which to base management.

9.0 SUMMARY

We have now completed our discussion of some of the basic ideas and premises in the two major organizational systems: Forecasting and Measuring. These systems are applicable across the entire organization. I have tried to define them, show how they might be designed, and illustrate some of the ways to implement them. In some ways, these two systems are easier to implement than the two more project-oriented systems of Project Operations and People Management that follow. In some cases, they may be more difficult. They are sometimes easier because the concepts seem to be more widely understood in industry. Bar charts and PERT diagrams are well-established and accepted techniques. Direct costing and responsibility accounting are also familiar to Accounting departments. They are sometimes more difficult just because of the very same reasons. The following attitudes often prevail: "We tried that eight years ago and it didn't work." "That's a great idea in theory, but let's get down to work and stop all this theorizing." (I've actually heard the latter excuse several times during my many years in industry.) Management may understand improved techniques but may not implement them because of historical or cultural reasons. Top level managers often reach their positions *because* they didn't make mistakes, not because they took risks. The ideas we've discussed are still a bit novel and could involve a perception of personal risk. This makes it rough sometimes.

There are no two projects that are exactly alike; therefore, it's impossible to have already "tried it eight years ago," because the situation was different then. Applying useful theory and others' solutions to today's situations prevents reinventing the wheel. But dealing with a total organization in order to design and implement general systems might be similar to stopping a large and organizationally heavy rotating flywheel and then getting it to rotate in the opposite direction. The inertia absorbed is sometimes enormous. However, as we know from physics, the application of a contrasting force (even a small diplomatic one) on a continual basis will eventually stop any flywheel. The decision (and the personal risk of stopping it) is important. Try it. Remember, there are potential profits as well as losses associated with risk taking. Present times seem to reward the risk takers. I think the days of the risk avoiders in management are over. Include these ideas in your project charter. At least you'll be able to start if your suggestions are accepted. If they're not, you are well on the way to becoming a project coordinator, not a project manager. What would you like to be? With inadequate tools, no one can manage well.

Beginning with the next chapter, we start to define and develop two specific systems that are more directly applicable to projects. They are major parts of the management strategy or the predefined solutions to expected variances. They are Project Operations and People Management.

My Suggested Answers to the Case Study

1. If an appropriate milestone had been incorporated into the work package, it would have been impossible for Joyce to report that a task was completed because there would have been no confirming report from the receiver of the milestone. Also, if the receiver had actually reported a delivery, Joyce's work package would have automatically been closed for further charges when the receiver's work package was opened.

    Roger might be doing the same thing, conceptually, but the project has no direct administrative mechanism to highlight this kind of nonfeasance. Of course there's the indirect mechanism of requiring Roger as well as the team managers to produce a total project ETC. This might show the test stand problem. But if Roger really feels that the problem can be solved without violating any of the project's three "golden limits," there's no necessity to re port it as a problem at this time. That's a decision for Roger to make.

2. Most people inflate their forecasts when they either are very unsure of the future or else feel they will be penalized if they don't meet them. It's a very rational response. An unpredictable future can be forecasted more easily if there are provisions for nonpenalty iterations. Obviously, forecasts will be come less lean if the forecaster is not allowed to modify the forecast with time. Inflating forecasts can be minimized by the project manager in the bottom-up meeting when the project ground rules for open management are discussed.

    However, there is always one response if one's forecasts have been made under pressure. Just produce a realistic first estimate to complete report. If that report shows an EAC over the amount in the particular work package, it's an automatic signal that an iteration is needed or contingency resources have to be released.

3. Time should be not charged to a project unless there is work being done on it. This can be prevented through the use of appropriate opening and closing of work packages upon delivery of milestones. Then, the team manager who had expected to work on the project cannot do so and either must assign his or her people to overhead, which the boss will hate, or find another project to work on. In either case, the project manager may be in trouble later on when the milestone is finally delivered and the scheduled work in the following work package is to begin. It's quite possible that the following team manager may now be tied up with other tasks and may cause the project manager to stand in line. It's a two-way street.

4. Roger should immediately initiate a new project forecast incorporating mile- stones, direct costing, and opening and closing of work packages. Since un-

certainty has obviously risen here, Alex should request immediate daily reporting from Roger until he is assured that Roger has the project back under
control. When uncertainty rises, reports must be much more frequent. How
about a daily 10-minute stand-up meeting in Alex's office, during which
Roger will report progress against the forecast he drew up after this meeting?

BIBLIOGRAPHY

HORNGREN, CHARLES T., *Cost Accounting: A Managerial Emphasis* (5th ed.). Englewood
Cliffs, N.J.: Prentice-Hall, 1982.

LAWRENCE, PAUL R., HARVEY F. KOLODNY and STANLEY M. DAVIS, "The Human Side of
the Matrix," in *Organizational Dynamics,* Summer 1977, p. 56–59, New York: American
Management Association.

MEREDITH, JACK R., and SAMUEL J. MANTEL, JR., *Project Management: A Managerial Approach.* New York: John Wiley, 1985.

OTHER READINGS

DUDICK, THOMAS S., *Cost Controls for Industry* (2nd ed.). Englewood Cliffs, N.J.: Prentice-
Hall, 1976.

GALBRAITH, JAY R., "Organization Design: An Information Processing View," in *Organization Planning: Cases and Concepts,* eds. Jay Lorsch and Paul Lawrence. Homewood,
Ill.: Richard D. Irwin, 1972.

MACIARIELLO, JOSEPH A., *Program Management Control Systems.* New York: John Wiley,
1978.

SILVERMAN, MELVIN, *"Common Sense" in Project Cost Systems.* 84 *Mgt.* New York:
American Society of Mechanical Engineers, 1984.

Part II   Project Operations and People Management

PERSONAL
MANAGEMENT
STYLE:
THE INDIVIDUAL
SYSTEM

PEOPLE MANAGEMENT:
THE PROJECT SYSTEM

PROJECT OPERATIONS:
THE PROJECT SYSTEM

MEASURING:
THE ORGANIZATIONAL SYSTEM

FORECASTING:
THE ORGANIZATIONAL SYSTEM

FOUNDATION: PROJECT MANAGER'S AUTHORITY

1.0 INTRODUCTION

Part I dealt with the basics, the organizational systems of Forecasting and Measuring. This part deals with the second two systems: Project Operations and People Management. These systems are less generalized. They apply more directly to the way the specific project is managed. Since no two projects are exactly alike, they have to be modified more than the other two systems. For example, it's relatively straightforward to use PERT charting and the coordinated forecasting process for all projects in an organization as well as measuring systems such as direct costing, estimates to complete, and nonlinear feedback techniques. But the next two systems—Project Operations and People Management—are more specific. Project Operations centers on the project team as a group. People Management centers on individuals.

2.0 PROJECT OPERATIONS: THE TEAM AS A GROUP

Project Operations is involved with the design and operation of the project organization itself, and how it's structured and expected to interact with the rest of the company. One of the outputs of this structure defines the skills needed on the project and the methods that should be used to recruit people with those skills. Usually, there are differences between the viewpoints of project and functional managers. The project manager is interested in completing the project as effectively and quickly as possible and is obviously intent on selecting the best people as team managers or participants for the project's limited life cycle. The functional managers who supply the team managers and participants are interested in optimizing the use of his or her people as scarce resources. Therefore, when internal recruiting begins, there are almost immediate problems if the best people (in the project manager's opinion) are not immediately available for the project. This problem is especially obvious when functional managers attempt to use the project as a training ground for improving or training less-experienced people.

This potential conflict becomes even more complex when the potential team manager's motivations are considered. He or she is naturally concerned about long-term job security. Maybe joining a project would be less attractive than staying with an equivalent functional group. The project goals and procedures must be made clear and the advantages of joining a project carefully pointed out before any responsible individual would even wish to join. This conflict between joining a project or a function is even more apparent if the candidate is recruited from outside the company. What's the advantage of working on this project? The Project Operations and People Management systems should clarify these alternatives and clearly show how joining could be very advantageous in terms of personal growth, promotion, and opportunities that might not be equally available in the more routine operations of a function. Of course, there's also a potential for increased risk in most projects, and this should be made equally clear. It should be a free choice.

Developing a clearly defined project structure is therefore vital to minimize future conflicts due to poor recruiting. In addition to initial recruiting, there could be other potential conflicts betwen project operations and various functions where the project crosses organizational boundaries. These have to be resolved. The situation can be extraordinarily complex. When the team participant is assigned to several projects concurrently, as in a matrix, there may be conflicting inputs from several sets of managers.

In a simplistic sense, the functional manager is generally responsible for the adequacy of the technical inputs that team members contribute to the project (that is, their training) and for administering the various support activities. These typically include benefit plans, vacations, and initiating salary review procedures. The project manager is responsible for integrating the efforts of the various team members into a smoothly operating totality, managing interfaces among the various project tasks and solving unexpected problems that would otherwise "fall in the cracks" (or interfaces). Some conflict is almost endemic; it's built-in. Conflict can often be minimized through the design of the project structure, which is best done throughout the process of recruiting the team members. Sometimes structures are designed beforehand, but this doesn't work as well in my opinion. It's easier to make a list of very general specifications and then complete the detailed design to fit the best people who are available rather than the other way around. People don't change easily; structural designs do. Thus, recruiting is an important part of the Project Operations system.

2.1 COMMUNICATIONS

We have already discovered that communication patterns in projects are different from those in functions because their uncertainty curves are different. Uncertainty is greater in the beginning of the project and lessens toward the end. Thus, communication frequency is especially important when a project is being developed and feasibility studies are being done. The regular feedback that the parent organization is able to produce is inadequate. In one way, the situation during project beginnings is similar to when major disasters or unforeseen events occur in functions. In both cases "uncertainty" is high and the frequency of communication must be increased. Therefore, in the beginning of a project, communication patterns and even the report formats must be defined as an explicit part of Project Operations. Solving problems in these kinds of situations is easier if a system exists.

The two potential problems— (1) project-function conflict involving personnel, responsibilities, and organization structure, and (2) inadequate communications—are the two central issues in the Project Operations system, which is primarily concerned with the management of groups. As we move from Project Operations to People Management, there is more variability. In the latter, we're dealing with individuals. As noted before, it's very difficult to change people to fit into a system. It's easier to change the system to fit the people.

3.0 PEOPLE MANAGEMENT: THE TEAM AS INDIVIDUALS

Motivating and achieving consistent commitment from team participants who are receiving conflicting directions from several different project managers, to whom they are assigned, and from their functional manager is difficult. This is called *role conflict*. Role conflict is built-in because even when a person is assigned only to one project while remaining simultaneously in a functional position, there are a minimum of two major sources of directions that this team participant receives: one from the project manager and one from the functional manager.

Of course, these two sources (like everything else in project management) are interactive, but the general rule is that the functional manager is *primarily* concerned with the adequacy of the technical input (or the "how") and the project manager is *primarily* concerned with the task and when it will be done (the "what" and "when"). There is always some overlap. Since this is often a recurring problem, we'll have to design a system that provides team participants, as well as project and functional managers, with appropriate inputs that are as clear as possible. There should be a coordinated management technique of rewards and penalties directly related to the actvities for which the team participant is responsible. This is called *dual reporting*. It's implemented in the last section of the project charter (Remember item #7 of the Charter: Special Clauses? See Chapter 2).

We are often surprised to discover that the rewards and penalties that we distribute are not valued the same by everyone. There are various theories of motivation that can provide us with different formulas. We will investigate some of these in order to develop an optimum technique for projects. Commitment is in the mind of the team particpant. Motivation is interpretation of observed behavior by the manager. The People Management system is concerned with both motivation and commitment and it should provide a process to improve them.

4.0 SUMMARY

As we move from basic organizational systems to those applying to individuals, the systems become more flexible. There are necessarily fewer constraints and more room for creativity. The relative freedom of the artist's thinking emerges and the linear logic of the natural scientist or engineer recedes. This logic is never eliminated, of course, because our subject is technical projects. Projects are achieved by people organized in teams and effective communications among team members is essential. Therefore, one must know the language. The project manager has to understand what is being communicated. The rate of communications usually decreases as the project ages and uncertainty decreases. This decrease is a guide to the project manager to change his or her way of thinking and the consequent managerial behavior as the project progresses.

Chapter 6 Project Operations: The Third System—Structures and Teams

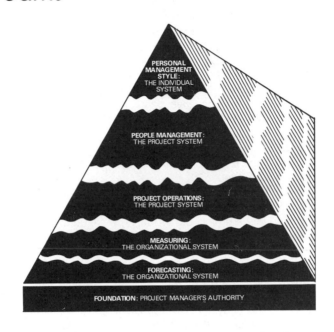

PERSONAL MANAGEMENT STYLE: THE INDIVIDUAL SYSTEM

PEOPLE MANAGEMENT: THE PROJECT SYSTEM

PROJECT OPERATIONS: THE PROJECT SYSTEM

MEASURING: THE ORGANIZATIONAL SYSTEM

FORECASTING: THE ORGANIZATIONAL SYSTEM

FOUNDATION: PROJECT MANAGER'S AUTHORITY

THE CASE OF THE CONFUSED PROJECT MANAGER

Cast

Mary Malone, Engineering Manager
Stan Arion, Design Engineer
Paul Wiston, Purchasing Agent
Pierre Eve, Manufacturing Superintendent

Grozen Engineering and Manufacturing Corporation is a success because of the revolutionary automatic knitting machines that it developed. These machines were more easily programmed than any other existing equipment and produced higher quality fabrics due to a revolutionary, but simplified, knitting mechanism. Over the years, the company had grown as it developed new models with extended capacities, higher reliability, and even shorter turn-around downtime needed to set up a new fabric.

A recent project was the development of the Model 50, which could be reset for a new fabric within the previously unheard of time of five minutes using externally driven, computer-controlled programs. This replaced a one-hour turn-around process that had been more than satisfactory in the past but became less and less so as competition increased. The Model 50 also operated much faster than any previous machines. It was originally designed and developed within the engineering department. The design process had seemed to run very smoothly under the direction of Stan Arion, who was the design engineer.

After the first model was built, it was delivered to one of the company's best customers, NewFab Manufacturing Co., where it was supposed to be field tested before Grozen went into mass production. Trouble developed almost immediately. NewFab reported that the Model 50 was almost a disaster. There were daily occurrences of knitting thread breakage, computer failures, and excessive wear on thread intakes. The machine operators complained that the machine was a jinx. Mary sent Stan out to NewFab to investigate. He wrote the following report which was distributed to Mary, Pierre, and Paul.

To: Mary, Pierre, Paul
From: Stan
Subject: Field Trials of Model 50
Recommendations:

I recommend that a project organization be set up to develop the redesign, re-entry into the market, and field service for this product. It requires a management approach as revolutionary as the equipment itself. The project organization should have total responsibility. It should include Engineering, Purchasing, Manufacturing, Quality, and Field Service.

The Model 50 is unlike any of our previous machines. The thread flow into the machine, the computer systems that control it, and the material output from the machine do not follow our standards for this kind of equipment. It is a major redesign. Since it doesn't resemble any other equipment in the field, due primarily to the appearance of the computer controls, it will need extensive customer and operator training. Many of the field problems have been caused by the operators themselves. We have never done training before, but if we are going to keep technologically current, it appears to be almost a necessity.

There are failures involving the following:

a. *Computer systems:* The computer system needs to be completely redesigned. We were unfamiliar with procurement of this kind of highly complex equipment and had bought it with our standard purchase order forms. We had tested it by installing it on the knitting machine and running it for several hours. While it worked fine in the laboratory, it began to produce erratic results when subjected to the dirt and lint that were part of the environment of a knitting mill.

b. *Thread breakage:* The thread guides within the inlet tubes are the major causes of thread breakage. The threads now move at extremely high speeds and actually wear patterns into the guides. These wear patterns are extra grooves in the inlet tubes. These grooves catch a slight imperfection in the thread when they are large enough. This breaks the thread. Thread breakage increases after the machines operate for a day or so.

c. *Spare parts:* Where spare parts were used in our older machine designs, they were "matched" to each machine by the operator, installed, and then allowed to "wear in." This machine runs so fast that there isn't enough time for slightly mismatched parts to wear in. If they don't fit perfectly, the parts impact upon each other rather than sliding. They don't wear in; they break.

d. *Anthromorphic machines:* The operators said that this machine didn't even look like any other knitting machine. They felt that the computer was "watching" them and recording everything they did. They didn't like that and preferred working with the older, slower models that had no computer attached.

Two weeks after Stan's report was distributed, Mary called a meeting in her office to discuss it. Prior to that, she had requested inputs from Pierre and Paul. When the comments came in, she summarized them, attached her suggestions, and distributed that with the agenda to the other three about a week before the meeting.

Questions

1. If you could put yourself in Paul's position, what would your recommendations be about changes in the procurement organization? What about its procedures?
2. Putting yourself in Pierre's position, what would you suggest for manufacturing?
3. Similarly, what should Mary suggest?
4. Can you control how a customer will use your product? Does it seem unusual to you to have machine operators attribuute living characteristics to machines?
5. If you were Stan, would you have developed a proposed project organization to attach to your report? Why? What would it look like? What would your "reporting authority" be, that is, to whom would you report? Why? Wouldn't this cause you to be split between two bosses? How would you handle this?
6. As Stan, how would you suggest that the project organization be implemented? Would that structure include field training? How would you control that kind of training as project manager after the equipment was delivered to the customer?
7. Are these kinds of problems familiar to you? If so, how did you handle them?

1.0 INTRODUCTION

One important starting task in developing the Project Operations system is the preliminary design of the project structure. Even though this structure may be completely redesigned when the appropriate project team members are recruited, a preliminary structure design process is similar to the first design of the charter. It's an initial top-down design that is later changed in a bottom-up redesign. The amount of redesign may vary depending upon the people recruited. This is typical strategy. The structure is based on a small team design that's different from functional structures.

The classic problem of role conflict (i.e., team members receiving differing directions from two or more different bosses) is a central issue in this system as is building the team. That involves recruiting, which is also covered in this chapter. The next chapter is also concerned with Project Operations. It covers project communications such as meeting procedures and report formats.

2.0 A BRIEF REVIEW

The Project Operations system is not as generic as the Forecasting and Measuring systems. Project Operations provides the general strategy for a *specific* project. It also covers many of the decisions or trade-offs that are specific tactics. For example, the project structure provides general guidance as to who is supposed to do what, when, and with or to whom. This guidance directs daily decision making or tactics. Without strategy or guidance on how to handle expected repetitive problems, tactics may become uncoordinated or undirected. Uncoordinated tactics may help individual heroes but usually do not help team effort. They may not support achievement of the team's project goals.

3.0 STRATEGY DEFINES STRUCTURE

A major task of the Project Operations system is designing the specific project structure or how the team managers are expected to cooperate. This structure follows the general strategy documented in the project manual. The project manual limits and guides daily tactics; it includes recruiting techniques, outlining the general kinds of skills needed. After recruitment, more specific descriptions of individual team participants' authorities and responsibilities can be developed, because the descriptions usually depend on who is recruited. Then the project's general operating processes such as meetings and internal communications links are designed. Those tasks flow in this general sequence as an important part of the basic project strategy.

The structural design is usually begun as part of a response to an external requirement such as a request for Proposal or Quotation, (or an internal requirement such as a specification sheet). It is part of the project charter. The charter has already tied together individual management descriptions in a preliminary organization chart. The structural design has already been outlined in a preliminary top-down form by the project manager, which described the expected repetitive behaviors that should occur. For example, who reports to whom, who is responsible for what and when, and how much it will cost. That preliminary top-down organization will, of course, be modified by the team members themselves when they are eventually recruited.

Since all projects are established to solve a unique and possibly extraordinary problem, the structures are obviously temporary. There just isn't enough time in projects to let the structure evolve formally by top direction and then informally by itself as many do in functional groups. The project structure should be as clearly defined initially as possible. Perhaps the structure of past successful projects could be models, since there is little time to "acculturize" or "acclimatize" project participants. Each project is a "new" organization. This can create difficulties.

When the project organization is large enough to cross the boundary lines of more than one functional group, thereby disturbing these groups' long-term operations, cooperation can suffer among team participants. Clarity in the project structure can help to minimize organizational confusion and disagreement. A stable, well-defined, and accepted organizational structure is also necessary to help decrease the greater uncertainty range with which projects deal.

Let's begin the project structural design simply, by assuming only one functional structure. Let that functional structure resemble a conventional pyramid. That is the standard design in many management manuals. However, it's rarely encountered in just this way. There are the formal structures (i.e., the two dimensions), but there are also a multitude of informal structures that everyone in those formal structures uses to really make the organization work. The number of dimensions can increase rapidly.

However, assuming that this rare, formal, two-dimensional structure really does exist, we add one project structure to it. We now have a formal, three-dimensional organization. Figure 6-1 illustrates multiple organizations. There is the usual two-dimensional structure, for instance, the Engineering department, and there is the project structure as a third dimension, which is perpendicular to the two dimensions of the Engineering department. In this example, John, who is working in both Engineering and the project group, is interacting with both structures almost simultaneously, since they both come together at his own desk. John is therefore involved with two formal organizations at the same time, and the organizational demands (or culture) of these two organizations can be very different.

But, as we know, this three-dimensional structure is really an invalid assumption. Each pyramidal structure itself—Engineering and the project—really has at least three dimensions: a formal, two-dimensional organization and an informal organization. The project organization also has three dimensions. Adding a project organization to the functional organization thereby results in a minimum of six dimensions rather than three. There are three dimensions for the function and three for the project.

This simple model assumes a simple linear interaction between the functional and the project structures. Actually, it gets to be a lot more complex in real situations. These dimensions are not fixed; they vary with time. They also vary depending upon who the participants are. Personalities are involved. Every piece of data doesn't flow through one person at one time. This simplified model of one functional department and only one project interacting at one point is rarely encountered.

MULTIPLE ORGANIZATIONS

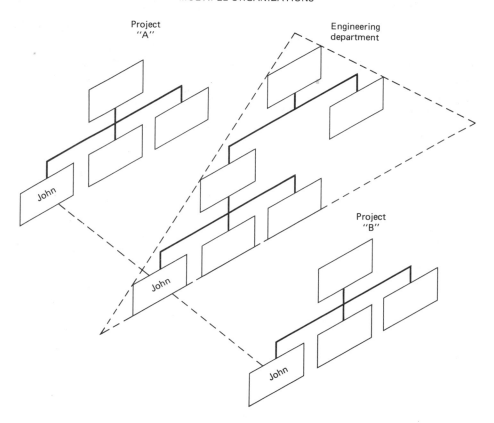

Figure 6-1. Multiple organizations

It often happens that a participant can report to more than one project *and* still report to an originating functional area. That's not unusual in a matrix organization. There is even more complexity when projects are in different divisions of a company. No divisions will have exactly the same culture. Now we have multifunction, multiproject complexity! For example,

> In some firms, a single division may have a matrix structure, while the rest of the company most closely resembles the traditional management pyramid. At the extreme of complexity, the organization chart may be an intricate web of matrices within matrix—or even three-dimensional. Many companies have evolved into the utilization of matrix management sort of unknowingly. (Sheriden, 1979, p.48)

This complexity, with the differing demands placed upon a person in the situation previously described, can be a major block to achieving team cooperation if there is no clear definition of team members' authorities, responsibilities, and working relationships.

There is another level of complexity to consider even after the initial structural definition at the beginning of a project. Project structures are not fixed. They may change as the project itself changes and different team managers move to center stage. The task assignments in the project charter are therefore vital historical foundations for the project structure.

I believe that if the design of the project or matrix structure is not considered carefully initially, the entire project could possibly be crippled later on. Care must also be exercised when using the design process itself, as it can sometimes create a problem by exacerbating potential tensions between functional and project managers. In one case, the design of the operating structure as part of the charter was taken as almost a declaration of war by the functional manager.

> "Don't you trust us to do our part in this most important project? A gentlemen's agreement is not good enough for this guy. He has to have it in writing. Where did this so-called project manager come from anyway? You know that we've always been able to work these things out before without all this formality (which robs me of some of my power)". The songs are familiar to most of us, just the various singers change.

This kind of response can happen when the first top-down charter is presented to management for approval. Even when the response of top management is positive and, in the best of all worlds, the functional manager initially accepts the general idea that the project manager is expected to direct the functional manager's people who are temporarily assigned to the project, the problems may not always end. The situation can become tense again when the bottom-up revision of the charter is completed and the functional manager is informed about the specific responsibilities that some of his people will be assuming. That could mean that he may temporarily lose control over them. This could be difficult for some functional managers to accept.

> "Yes, I know that we have to support the project operations and that some of my engineers, or technicians, or programmers, or whatever, will be assigned to projects, but I don't want any project manager (or my people) telling me what they will or won't be doing. I'll tell them."

Finally, resolving the conflict at this stage may not solve it forever. In one example, the project was released and some of the functional manager's people were assigned to the project with his approval. Some time later, they were reassigned elsewhere by him over the objections of the project manager because of a hot new problem that had to be solved right away. The project suffered. In this case, it required two senior vice presidents to resolve the problem. This kind of situation is usually predictable and, if so, it's much easier to develop solutions to predictable problems before they appear on some organizational battlefield. Sometimes just documenting a solution will even prevent it from appearing at all.

3.1 SMALL TEAM DESIGN: THE OVERVIEW

A major factor in defining the structural design is the numbers of team managers or the size of the project team. Except for major defense or other large governmental efforts, there are very few large projects involving thousands of people or hundreds of subcontractors. Multimillion-dollar technical projects in industry have been easily achieved with a handful of team managers. In a typical technical industrial setting, for example, it is probably possible to operate with five typical kinds of team managers: quality, production, engineering, purchasing/subcontracts and contracts, and administration. These kinds of projects deal with design, development, production, and delivery of tangible products and systems. Other projects may be organized differently, depending upon the project goals themselves. For example, software projects may have teams handling systems, coding, documentation, and testing. These teams develop their own specialized ''project charters'' and replicate their own project structures based on the design in the charter. This charter describes the overall interactions of team managers using task descriptions (see Fig. 6-2).

Developing these charters makes it possible for the team structures to be similar throughout the project. Then anyone who is in contact with any part of the project will be able to more easily understand the team culture and how the project teams interact. The charters won't be exactly alike because each team has different tasks (and people) that make its charter a bit different. But there will be similarities. The beginning or basic structural design to use will be the top-down structure originally outlined by the project manager, perhaps as typically outlined by Fig. 6-2.

TASK DESCRIPTION

Assignee: (Name) _____

Type of task: Design widget

Date: 15 April 19____

Project: Antigravity lifter

Quantity   1

Description: Design main widget

Task Schedule: Attachment #3

Budget: Attachment #4

Reports: Progress

Various attachments

Figure 6-2. Task description

3.2 FUNCTIONS AND PROJECTS: REVISITED

If we compare the effectiveness of a small functional group and an equally sized project,

> We found that the functional organization of technical personnel results in a higher rating of technical performance but usually results in cost/schedule overruns. Complete project teams, both technical and administrative, are associated with lower technical performance but are twice as likely to meet their cost and schedule objectives, and the colocation of administrative personnel with the project team facilitates communication and faster reaction to problems. (Marquis, 1969, p.83)

Therefore, it seems that although the job will get done better *technically* in similarly sized functions, the *time* and *cost* objectives will not be met as well. Projects usually achieve better time and cost goals. What would your strategy be? What do you consider most important? On an ordinal scale of 1 to 10, where 1 is unsatisfactory and 10 is completely satisfactory, can you compare the three overall goals of technical achievement, time, and cost? If you can, you can now begin to decide on whether to use functional or project structures. With tasks that are larger or more complex than functions can handle, there is only one alternative: Projects must be used. In my experience, these project systems do bring technical goal achievement up to a par with the previously cited research findings about higher levels of time and cost achievement. Projects can be restructured to effectively coordinate people's activities and promote effective cooperation within those teams toward those goals. Assuming we decide to use a project structure, the top-down charter would have described each major team manager's authority and responsibilities. Each team manager now redesigns his or her own charter (and structure).

The project manager participates in the development of each team manager's top-down charter, just as each of the team managers previously participated in the development of the project manager's charter in the bottom-up forecast. Each team manager then provides the opportunity for his or her team's bottom-up revision. This supports the development of relatively standard team charters while achieving the project's technical, time, and cost goals. As noted, these charters help to establish a consistent project culture. If everyone can generally understand and participate in developing the ground rules of their own team structure similar to that of the overall project, it becomes easier to develop a common culture within those structures. These goal-directed structures are primary strategic guides for making operating trade-offs. They guide people's repetitive interactions in the project.

The process is even more important when using the ultimate in short-term project structures—that of a matrix format, where the person remains in the functional organization while working part time on several projects. Then, consistent team structures are even more critical in order to provide management continuity, especially when one considers the reversed power and control that individuals in matrices have when compared with those in functions.

The matrix organization is not a simple extension of the traditional pyramidal structure. The pyramidal structure acquires its form from the fact that as one goes up the administrative ladder (1) power and control increase, (2) the availability of information increases, (3) the degree of flexibility to act increases, (4) the scope of the decisions made and the responsibilities involved increase . . . the matrix organization is almost the opposite.

Power and control are given to the individual and/or to groups who have the *technical* skill to accomplish the task, no matter what their organizational level. (Argyris, 1979, p.23 [Italics are mine.])

3.3 AUTONOMY AND DELEGATION FOR UNIFORM PROJECT CULTURE

Projects are usually managed sequentially and at any one time rarely involve more than a few technical specialties of major importance. For example, even though a project might go through a design, development, procurement, manufacturing, and field service cycle, not all of the team is directly in charge at one time. In the beginning, the people in charge of engineering and design teams might be dominant, then those handling procurement, manufacturing, and so on. The project moves sequentially through various phases, and the team manager(s) responsible at any given part of the cycle would be, in effect, managing most of the project at that particular time. This requires a special kind of authority and autonomy definition as a basic element of any structure, one that states that there is no abdication of responsibility at any time by either the other team leaders or by the project manager, even though one team leader is temporarily the *primary* problem solver at a given time in the course of the project.

For example, assuming that the immediate problem of the project is the placement of major subcontracts. The procurement team leader may be the de facto manager in charge, but he or she might be supported by the engineering team (to determine a potential vendor's technical competence), the financial team (to determine the adequacy of the vendor's finances), and, of course, the quality team (to determine if the vendor has adequate internal controls). Therefore, team charters should be able to define authority and responsibility clearly and provide maximum freedom to the team leader.

When the project manual in the charter states,

> . . . . as long as the team manager responsible for a work package provides an estimate at completion, which, when added to the expended time and cost for the particular work package doesn't exceed the original budget for that work package, that team member will have complete autonomy in getting his or her tasks accomplished. . . .

it supports independent problem-solving behavior, within the limits set by the particular work package. However, when the project charter further states,

. . . . if the estimate to complete plus the "expended" exceeds the existing estimate at completion, the team member responsible for that work package is required to provide an immediate notification to the project manager that a change is needed. That notification should be an internal impact statement.

This situation *requires* interaction among team members, because the internal impact statement could involve the technical, time, and cost goals of other project teams later on. Each team manager, therefore, has the explicitly delegated authority to get his or her job done within some overall boundaries and is responsible for the effective completion of the job. How the job is really done within those overall limits is not defined in great detail. It may or may not reflect the details of how the job was originally forecasted. Sometimes changes occur. And the interactions with others when the boundaries are *expected* to be exceeded are also noted. It's freedom within limits. The de facto manager in charge is the one managing the most important work package that is open at that moment. The project manager, of course, is still in charge on an overall basis. Everyone is responsible.

A helpful tool useful in designing the structure has already been introduced— the milestone.

3.4 USING MILESTONES AS DESIGN TOOLS

As previously stated, each work package has been designed to start with a deliverable milestone that must be accepted by the next team manager before his or her particular work package is opened for charges. The team manager receiving the milestone must certify, in effect, to the project manager that he or she is satisfied with the work done by the team manager delivering the milestone. This happens when he or she accepts the milestone. In other words, if the milestone is rejected, the following work package is not ready to be opened for charges. Thus when team manager A has delivered an acceptable milestone that signifies the completion of his or her work package to team manager B, the work package of manager A is closed. The process of working through the project using milestones and determining interim progress using estimates to completion is a concept that minimizes the need for explicitly detailed instructions about the internal team structure, intrateam interactions, and the overall project structure. The definition of acceptability can include all these criteria.

The structure, then, quite clearly allows *authority* to be automatically delegated downward to the respective team manager in charge of a work package. It begins and ends with a milestone. How that authority is used is dependent upon how the person in charge of the particular work package wants to use it. As far as the project manager or the first-level team manager is concerned, the rules (and the consequent "culture") are very clear and straightforward: Just get the technical job done well, in time, and to budget. The overall project responsibility still remains with the project manager. That's why the project manual of standard procedures and controls can be a very slender document outlining various repetitive authority and reporting relationships, since the project manager is minimizing any direction of *how* the work is to be done, just the general interactions among team members. And even that interaction can be limited mostly to administrative concerns such as what is to be reported, how often, and under what conditions. The work package procedures and milestone definitions primarily are the responsibility of the team managers in charge of the work packages.

In summary, the preliminary structure as outlined in the project manual defines the project and team manager's authority levels and provides maximum autonomy for trade-offs or decision-making through documenting the project structure. The structure supports delegation, clarifies relationships, and diminishes the need to coordinate the activities of more than a few team managers at one time for the project manager, since it only deals with overall work packages. The result is intended to be a small, well-coordinated project team with only those team members concerned with the immediate problem directly involved in making trade-offs. At this moment, the other team members are almost like advisors. Thus the overall project management team is almost like a review board that meets often to support the overall project while disseminating information and coordinating project policies and behaviors. However, the structure and the people are always interactive.

4.0 ROLE CONFLICT RESOLUTION

Role conflict occurs when one person receives differing directions about how to behave in a specific job or position simultaneously from two or more people. The structure outlined in the project manual can be a major help in minimizing this kind of conflict. The approval of the preliminary top-down structure by top management and the review and redesign processes of the project manual during the bottom-up process at the project beginning help. Developing those general ground rules and describing structural interfaces in the beginning are useful in decreasing most role conflicts.

In my opinion, without this part of the project manual clearly defined, upper management will too often tend to resolve role conflicts in favor of the functional manager. The pressure for organizational continuity and the usually extensive longevity of functional managers are considerable factors that many project managers cannot overbalance unless there has been a clear direction at the beginning.

4.1 ROLE CONFLICT: STRUCTURAL TOOLS

There are two important tools that we have already covered in the project manual:

1. The reporting level of the project manager
2. The task description for each team leader

Briefly, if the project manager's reporting level is at least one level higher than that level from which the team leaders are drawn, a fairly strong tool for minimizing role conflict exists. If a decision concerning the project goes against the project manager, the next step is to determine the following:

1. Is there any effect on the project's three "golden limits": technical achievement, on time, within budget? If not, the project manager can make an adjustment of the project status as an internal change. No problem.

2. If there is a potential change in overall technical achievement, time, or cost, an impact statement should immediately be issued with the usual trigger clauses. Again, no problem.

The project manager might consider the answers to the second question to be a bit painful, but the alternative is potential failure in the project later on if the impact statement is not issued. (If it's any consolation, "A brave man dies only one time, a coward many times.")

4.2 ROLE CONFLICT: MAKING PERSONAL DECISIONS

The project manual is a valuable tool, but it, like most documents and procedures, often has to be interpreted. (If the law were very, very clear, there wouldn't be any need for judges, would there?) All situations can't be documented, and all bosses are not equal models of constancy and logic. When the structure is inadequate or the two alternatives previously noted are unsatisfactory, this is another technique that might help: Adopt a first-person-singular approach:

> *I* am the "expert" in the work that I have to do, and I am working on an existential (or current) priority list. Therefore, when I receive conflicting inputs, *I* will start the interpretation process with what *I* think is the most important thing to do.

> Can the conflict be defined objectively? Have I tried to develop acceptable alternatives for the boss that will resolve the conflict? If an objective answer isn't obvious or apparently won't work, can I score these alternatives, using an ordinal scale of 1 to 10, thereby giving a subjective value? Instead of requesting a reversal of direction, would these various alternative subjective scores provide "room" for changing the direction, thereby minimizing the problem?

> Is it possible to document the conflict, defining the differences in the various inputs received and request another decision from upper-level management?

This applies most effectively when working on more than one project at the same time. Maybe new data have surfaced or old data have not been correctly presented. When reporting upward to two different bosses who cannot resolve differences instead of to one boss who can resolve problems, role conflicts escalate. It occasionally happens even with clearly written project manuals (that have defined the existential priority of both projects). Then even the most logical presentations may not satisfactorily resolve the conflict. Therefore, the following technique might be considered:

1. Lobby actively with relevant 2-boss counterparts . . . to win support before the event.
2. Understand the other side's position in order to determine where tradeoffs can be negotiated; understand where your objectives overlap.
3. Avoid absolutes.
4. Negotiate to win support on key issues that are critical to accomplishing your goals; try to yield on less critical points.
5. Maintain frequent contact with top leadership to avoid surprises.
6. Assume an active leadership role in all committees and use this to educate other matrix players; share information/help interpret.
7. Prepare more thoroughly before entering any key negotiation than you would in non matrix situations; and use third party experts more than normally.
8. Strike bilateral agreements prior to meetings to disarm potential opponents.
9. Emphasize and play on the supportive role that each of your . . . bosses can provide for the other.
10. If all else fails,
    a. You can consider escalation (going up another level to the boss-in-common).
    b. You can threaten escalation.
    c. You can escalate. Before traveling this road, however, consider your timing. How much testing and negotiating should be done before calling for senior support? Does the top leadership want to be involved? When will they support and encourage your approach? Does escalation represent failure? (Lawrence, Kolodny & Davis, 1977, p.56–57)

5.0 RECRUITING

One obvious source of role conflict can begin when recruiting the team managers. (See the flow diagram depicted in Fig. 6-3.) They are recruited after conferring with the appropriate functional managers. The project manager describes the needs and requests concurrence in the selection of the needed team managers. I'm assuming that those team managers are presently in that functional manager's group.

For example, assuming that the task is to recruit the project engineer and there is mutual agreement between the project manager and the functional manager upon a specific candidate. The project manager can then present the preliminary plan including the task assignment descriptions to the selected candidate (for the job of project engineer) and determine if it is acceptable. If it is, the candidate is invited to the first project initiating meeting, during which the team will develop the revised

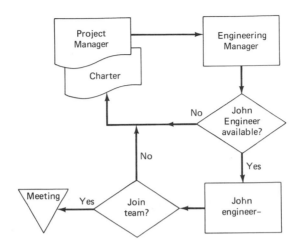

Figure 6-3. Recruiting

details of the project structure and its operating methods. This, of course, is part of the first project iteration—the bottom-up approach noted before. This is the way it's supposed to work; however, sometimes it doesn't.

Consider another scenario. The manager of Engineering won't release the candidate selected, and the project manager is offered an unacceptable replacement. The flow diagram is quite clear: Revise the preliminary plan. It may be that this refusal could modify the potential technical achievement, the time, or the project cost. Now is the time for the project manager to redetermine the cost or the effect of the alternatives presented and to notify the manager of Engineering about any potential change. If this fails to achieve project goals, the project manager escalates, notifying top management because his or her allowable level of uncertainty has been exceeded. There is always a relationship between the expected cost of the project and the presence of the team manager who was selected. Higher costs often get people's attention.

Occasionally there are functional managers who hesitate about assigning one or more of their people to projects. His or her typical response might be, "Don't worry about it, I'll make sure that your engineering requirements are taken care of. I'll deal with my people. You just deal with me."

There are two alternatives the project manager can take:

1. Accept: "That's fine. That means you (the manager of Engineering) will be reporting to me on my project organization chart instead of a project engineer. Of course, I'll also be sending my evaluation of your performance in to your boss as part of my project responsibilities. You understand this, don't you?"

2. Reject: "Sorry about that, but since I know that you're very busy and I can foresee problems, I'll just have to include a time delay and cost factor in my project for those potential problems. That will raise the total time and cost for the project. Of course, these new estimates based on the new time and cost

factors will be put into my project estimates for your approval and then top management's. I'll be back to you as soon as these figures are revised.''

Recruiting is an important step in building a project structure. Whether recruited from inside the organization or from the outside, each team manager has to have a preliminary project task description. That description outlines the proposed major authority and responsibilities. The managers then review that description, negotiate appropriate changes with the project manager and the other team managers, and then work out the descriptions for their subordinates. It's a sequential kind of structural design.

6.0 TYPICAL TEAM MEMBERS: WHO THEY ARE AND WHAT THEY DO

Up to this point, we have implicitly assumed that the project manager would know who was wanted on the project team. This was documented by the organization chart that was initially submitted as part of the first top-down design of the project charter. Prior to the recruitment of the actual team managers, they could have been either specifically named individuals or else specified by general skill levels. For example, ''I want Joe Lam as my project engineer'' or ''I need a Grade 13 as the project engineer.'' The task description, however, can't define everything that must be done. It's impossible to do that.

Most of us are valuable to our organizations because, as humans, we are supposed to be creative and be able to solve new or unusual problems. Accomplishing this requires a great deal of personal freedom. That means we rarely do *exactly* what we are told to do and *nothing else*. (In fact, one of the better aphorisms is, ''If you want to bring this organization to a grinding halt in a hurry, just insist that everyone does only what they are told to do.'') Telling someone exactly *how* to do his or her job is *not* helpful in projects, but outlining the job boundaries within which the particular task is to be done *is* helpful. When projects cross most organizational boundaries, it is also wise to delineate the general, not the detailed, responsibilities of all team managers, not only with respect to their own jobs but also to the jobs of all the other team managers. The task assignment that was mentioned in the project charter partially does this. It has the limitations of any functional job description, since it only covers the bare bones of the job. The ''how to do it exactly'' is still left up to the particular team manager. And if management is an ''art,'' each team manager is an ''artist,'' and that generally means an individualist. But those individualists must also cooperate with one another.

The more specific interacting definition is produced by the responsible team participant during the bottom-up review or project initiating meeting. Some of those definitions might have to be negotiated on a one-on-one basis between the particular team manager and the project manager.

This somewhat closer definition of ''how'' the particular team manager expects to complete project tasks is like managing a type of relay race with each team manager acting as a temporary leader who manages for a certain amount of time, then hands the project over to the next team manager in line. Although each team manager may be dominant only at some point in the project, there is a continuing responsibility that each manager has that is never totally relinquished, even though he or she is no longer nominally in charge. As an example of this relay-race effect, imagine a team with an estimator, project engineer, quality assurance manager, and contracts manager. The first top-down description could be as follows:

6.1 ESTIMATOR

The "estimator" is a central project pivot. . . . All the engineering, planning and managing revolves around the proposed quotation, and he is thus the central character in the project cast. Later, the financial director takes over the center of the stage as fiscal plans are laid out; then it is the systems engineer, the production manager, and so on. Each, in turn, is the "star" of the show, but rather than leaving the stage after his starring role is played, he remains on stage as supporting player. During the design stages, for example, the estimator is still effective, since he is comparing the cost of the final design against his estimated cost of the original design. He has a vested interest in seeing that his estimate is not exceeded due to "minor changes in design.". . . (Silverman, 1984, p.227)

6.2 PROJECT ENGINEER

The project engineer is primarily responsible for the technical subproject within the overall project. Consequently, there is a similar, but more limited, project charter and organization chart for the engineering team. As an example, there could be ongoing responsibilities in a computer company for Hardware, Software, Data Management, Configuration Management, and Logistics. Some of these could include

1. Identifying technical and economic trade-offs.
2. Determining the project systems' functions and sequence.
3. Design requirements imposed by the customer.
4. Developing approaches for production.
5. Ensuring that a usable product is delivered.

6.3 QUALITY ASSURANCE MANAGER

There is one team manager who would prefer *never* to be the center of the project stage and that is the manager in charge of quality assurance. His or her organization chart actually parallels that of the project manager.

This team manager provides guidance to project personnel in eliminating deviations from the forecast. If everyone does his or her job well, the task becomes administrative as he or she documents the satisfactory testing and acceptance from the detail parts to the completed project system. The quality assurance manager will have personnel attached to every part of the project, although the quality assurance organization itself is not expected to originate or produce anything. They monitor and control. They can serve equally well as both the buyer (i.e., your customer) and the seller (i.e., your project team). As the "customer," they're concerned about whether all the materials, designs, final tests, and so forth meet the contractual specifications. As the "seller," they're concerned about developing process controls that must be followed to ensure that the eventual product is produced correctly. "If you do it right, the end result should be right."

6.4 CONTRACTS MANAGER

There is a team manager who, unfortunately, is not often considered to be a major player on the project team. This is the team manager concerned with the project contract (if the project deals with an outsider), the project charter (if the project deals with a company insider), or the project administration (whether it's an inside or outside project). This team manager supervises the satisfaction of both the internal and external administrative obligations of the project: This position is sometimes called the *contracts administrator*.

This manager maintains the central information system that is the clearing point for all project communications. In that role, his task assignment is quite clear: No project documentation can bind the project contractually unless it is sent out by this particular team manager. When the outside "customer" or the inside "user" is made aware of and accepts the need for the contracts administrator, there is an effective control system on all communications. (For more details on this, see Baumgartner, 1963, p.132)

In my opinion, the contracts administrator should also be second in command to the project manager. There may be occasions when the project manager is not available. Illness, other commitments, or field trips, could take him or her away from the project. Whenever this occurs, there should be someone in the project team who is responsible for the project. I believe that the contracts manager is best qualified to act as a second-in-command because his or her job has no deliverables to send to any customer, nor is he or she responsible for completing any project tasks. His or her responsibility is limited to seeing that the contracts are properly documented and sent out. Therefore, the contracts manager can make unbiased decisions in the project manager's absence. Too often projects are held up while a critical decision awaits the return of the project manager. When the decision is critical with respect to time, it is necessary to the success of the entire project to have a second-in-command.

There are probably other team managers who are needed in a project. These descriptions can be a model to use in defining them and their interactions as part of the project structure. Writing these descriptions is only the beginning. The bottom-up, project-initiating or planning meeting provides the way to support intraproject cooperation. We've discussed it generally before, but we'll take another look at it now.

7.0 BOTTOM-UP MEETING: THE PROCESS OF DESIGNING AND IMPLEMENTING STRUCTURE

There are very few meetings that occur in project operations that require the attendance of all the team managers. As noted before, most project-related problems are sequential and not all the team members are usually concerned at the same time, even if they are generally concerned as project team managers. For example, typically, the beginning of most projects deals with technical problems, then production, and finally field service. All of the team managers handling these aspects are interested in the progress of the project, but only in a general way if they are not directly involved at the moment. However, there are three meetings in every project cycle when the whole team has to participate:

1. Project-initiation or bottom-up meeting
2. Design review meeting
3. Project close-down meeting

We discuss the content of the project-initiation or bottom-up meeting here as support for the structural design and implementation process. More detailed procedures for all problem-solving meetings are covered in Chapter 7.

The initiation meeting requires an agenda. A typical agenda should at least include the following items:

1. The project charter—Review of objectives, scheduling, and costs. Connection or importance of this project to the parent organization.
2. The management procedures—Define milestones, work breakdown design, project meeting scheduling, and criteria for measuring performance.
3. The organization chart—developing task assignments, delegation of authority, limits of responsibility, and key areas for each team member (i.e., who those people are and who they will use as alternates). It outlines how team managers are to interact with one another and how they are to report to the project manager.

The primary intent of Item 3 is to help the team to design structure and resolve as many potential role-conflict problems as possible at the project's inception. A formal and extensive agenda might be as follows:

Project Objectives

- Group members begin a dialogue by talking about their expectations and concerns for the team and the project. This inevitably causes differences to surface which need to be recorded and worked through.
- A discussion aimed at developing agreement on the objectives of the group.

Management

- A discussion leading to agreement on how frequently the team will meet, and member's expectations about attending.
- A discussion about leadership in the team—the role of the chairperson, others' responsibilities for initiative in meetings and outside meetings, and the extent to which the group expects the leader to push the group to a decision.
- Roles and responsibilities of group members are discussed, with the aim of recognizing ambiguities and overlaps, not of entirely eliminating them. There are a number of techniques for clarifying roles which have been used. For example, the group can list all major decisions that have to be made and chart who has responsibility for a decision, who participates in it, who is consulted, and who must approve. In this way team members gain clarity about how functional responsibilities relate to those of others.
- The group discusses how decisions will be made. Will unanimity be required? Is consensus sufficient and what does it mean? Will a voting procedure be used? Or will the leader make the final decisions? How will the next level up be involved?
- Ground rules for communication and conflict resolution are developed. For example, does the group want to foster an open airing of differences? What are members to do when they disagree strongly with the team and haven't been able to influence it? How are members to handle information about what the team is doing?
- Understanding is developed about the responsibilities of team members in relating back to their department what is happening on the team.
- Any interpersonal problems are aired so that they do not block team functioning." (Davis & Lawrence, 1977, p.110)

When the team managers participate in this bottom-up problem-solving meeting, the final result must be agreement about the project operational structure. As an indication of this agreement, they sign and date the project organization chart (see Fig. 6-4).

The bottom-up meeting is obviously critical to the future success of the project team. Uncertainty is highest now, and this meeting is intended to minimize a major part of the uncertainty that begins with role conflict. With clear definitions of the various objectives, relationships, and authorities, the structure of the project be-

ORGANIZATION CHART

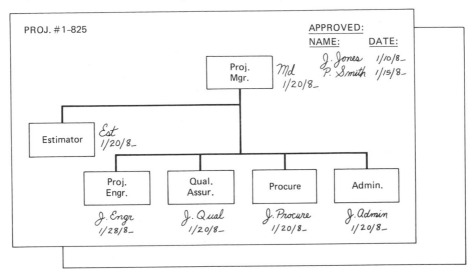

Figure 6-4. Organization chart

comes more apparent. It clearly outlines the responsibilities of the project team managers and the various supporting functional managers. Completing project tasks is obviously the overall responsibility of the project manager, and training and administering the project team managers (and their appropriate teams, of course) are the responsibilities of the appropriate functional managers. When these are well defined, much of the potential role conflict in the "two bosses" problem is minimized. (The bottom-up meeting can also be the first meeting of the coordinated forecasting process [CFP], as noted before, during which the prospective tasks are accepted and the preliminary project flow diagram is drawn.)

Projects can have several different meetings when they are in their initial stage. There might be an announcement meeting, in which top management defines the project, appoints the project manager, and outlines some of the very general limitations under which the project is expected to operate. There could be a meeting between the project manager and top management to review the first iteration of the project charter after the project manager has completed it. Then there can be the first bottom-up review. When the latter is completed, the project structure and operations have been planned in as much detail as necessary. When the customer accepts the project as proposed, there could be a postaward meeting to inform everyone concerned about the final postnegotiation outcome.

[In the postaward conference,] if he can possibly arrange it, the project manager's authority and stature are considerably enhanced by the presence of customer representatives and top-level division, group, or company management. . . . On most projects, however, a "postaward conference" boils down to a series of individual conferences between three or four persons, and most of the people who will work on the project are

never directly oriented towards the functions of the project team as a whole.
(Baumgartner, 1963, p.78)

Unfortunately, many companies don't follow this general meeting sequence
of (1) project announcement, (2) management review of the first charter, (3) project
initiation, and (4) postaward meeting. Also, not all projects survive the initial two
meetings, either because of changes in the project's environment or disapproval by
top management of the first charter iteration. Other meetings may not be thought
necessary by top management.

8.0 SUMMARY

We have developed some techniques and a process for the first designs and implementation of the Project Operations system. This process includes some methods
for defining, recruiting, and integrating project team managers. It was noted that a
project can be managed with a fairly small project team, having from five to eight
members, and that the project manager can develop acceptable task assignments
during the first charter design. The project team can then use the results of that first
design to develop a second bottom-up design in the project-initiation meeting. This
second charter iteration can involve the team managers in the project Forecasting
and Measuring systems and the Project Operations system, specifically emphasizing
structural relationships. This minimizes role conflicts and promotes the personal
commitment of team managers, which is, of course, necessary for project success.
This meeting is part of a necessary process that supports the achievement of project
goals.

Developing improved systems and processes for the effective completion of
projects may not be sufficient if the project participants believe that the cooperation
that is expected or that the personal cost of exhibiting the desired behavior are
greater than the perceived benefits that they would receive as individuals. It always
starts with the individual: In projects, it is always the individual who provides the
creativity necessary for success, assuming, of course, that the project itself won't
violate any scientific laws and is adequately funded. Projects are really small group
structures. Therefore, in the next chapter we deal further with the project team as a
group by setting up communications patterns that help to connect the group to the
individual. Then, in Chapter 8, we'll deal with the individual in the next system—
the People Management system. The individual is really the key to project success.

My Suggested Answers to the Case Study

1. The normal functional procedures that cover the usual materials procured no
   longer apply. Quality Control is no longer adequate to inspect materials as
   they are received. We need a whole new system of quality assurance that
   includes evaluations of vendor processes, controls, and operations to ensure that the products received are representative and meet quality needs.

This means developing an organizational structure that can assist in controlling the processes, not only inspecting the final product. If this is done right, the products will be controlled. More prevention is needed in the field, not correction in-house. Complex subcontracts require closer attention to quality at all levels of procurement.

2. We have to change the way we make our parts. We either have to upgrade engineering and our manufacturing operations to ensure our parts are interchangeable, or else Engineering has to redesign the equipment to minimize the need for exceptionally tight tolerances that cannot be economically produced. If necessary, those parts that have very fine tolerances can be combined into one subassembly that can be replaced as a unit, somewhat similar to electronics manufacturers, who sell a total spare module and no longer replace individual transistors in the field. The manufacturing organization is not up to date, and we've got to change the way that we do things if we want to successfully compete. We need a project set up to develop an entirely new way of manufacturing.

3. We have to develop an engineering project format that will treat these major tasks as if they were total organizations in themselves and not as if they were just parts of the Engineering department. Model 50 is a major change, not only in our products but also in the way we service them in the field. If we continue to split responsibilities as in the past into the Engineering, Purchasing, and Manufacturing departments, we'll never be able to organize the resources in one place, in a timely fashion, and produce the newer, computer-controlled, more revolutionary equipment that the knitting fabric market needs if it is to stay competitive to other kinds of fabrics. If we don't change the way we're organized to improve response to market demands, our competitors will, and we'll eventually be out of business. We've got to get closer to our customers *before* we give them the equipment to test. We've got to improve our links to the outside, and those links can't be split among Engineering, Purchasing, and Manufacturing, as they are now.

4. In my opinion, it's almost impossible to control how customers use a product. Did you ever notice how smudged the walls are next to the "wet paint" sign? Now, why would anybody want to put their fingers in wet paint? But experienced industrial painters know that people have an almost irresistable urge to "test," so they provide a small area for them to do that. That way the rest of the painted areas remain clean. If you can't control your customers, at least minimize any potential damage.

    About attributing human characteristics to inanimate things: They call a ship "she," don't they?

5. Yes, I would have attached a proposed organization chart. The chart would show Stan as the project manager and reporting to him would be representatives from Engineering, Purchasing, Manufacturing, Quality, and Field Service.

    Stan should report to one level above that from which he draws his people. This is necessary in order to be able to resolve conflicts if the people reporting to him get different directions from their functional bosses. It could cause a split, but that wouldn't be Stan's problem; it would be the boss's. If any of the functional bosses' decisions affect (a) technical achievement, (b) overall time, or (c) cost, Stan would issue an impact statement in order to minimize the effect upon himself.

6. The structure should be implemented by Stan as he develops his project charter; recruits his people; defines their jobs on the project; reviews the charter, the structure, and the project forecasts with his team; and designs his internal cost/information systems. Since field training appears to be an important part of this project, it should be managed by project personnel. It will then be no different from any other phase of the project such as engineering, purchasing, and so forth. There will be various work packages and milestones. A milestone in field training could be even the passing of an operator's qualification test by the customer's personnel.

BIBLIOGRAPHY

ARGYRIS, CHRIS, "Today's Problems with Tomorrow's Organizations," in *Matrix Organization and Project Management* (Michigan Business Papers, 64), eds. Raymond E. Hill, and Bernard J. White. 1979.

BAUMGARTNER, JOHN STANLEY, *Project Management*, Homewood, Ill.: Richard D. Irwin, 1963.

DAVIS, STANLEY M., and PAUL R. LAWRENCE, *Matrix*. Reading, Mass.: Addison-Wesley, 1977.

LAWRENCE, PAUL R., HARVEY F. KOLODNY, and STANLEY M. DAVIS, "The Human Side of the Matrix," *Organizational Dynamics*, Summer 1977, p. 56–59, New York: American Management Association.

MARQUIS, DONALD G., "Ways of Organizing Projects," *Innovation*, Project Management Publication, 3, American Institute of Industrial Engineers, 1969.

SHERIDEN, JOHN H., "Matrix Maze: Are Two Bosses Better Than One?" *Industry Week*, June 11, 1979.

SILVERMAN, MELVIN, *Project Management: A Short Course for the Professional*. Cliffside Park, N.J.: Atrium Assoc. Inc., 1984.

OTHER READINGS

BURNS, T., and G. STALKER, *The Management of Innovation*. London, England: Tavistock Publications, 1961.

DAVIS, LOUIS E., and JAMES C. TAYLOR, "Technology, Organization and Job Structure," in *Work, Organization and Society*, ed. Robert Dubin. Skokie, Ill.: Rand McNally, 1976.

FRIESEN, E. N., "The Matrix Organization, Another Dimension," *Mechanical Engineering*, October 1982, pp. 84–87.

GOODMAN, RICHARD A., "Ambiguous Authority Definition in Project Management," *Academy of Management Journal*, December 1976, pp. 301–15.

LAWRENCE, PAUL R., and JAY W. LORSCH, *Organization and Environment: Managing Differentiation and Integration.* Cambridge, Mass.: Harvard University Press, Division of Research, Graduate School of Business Administration, 1967.

LOVE, SYDNEY FRANCIS, *Planning and Creating Successful Engineered Design.* New York: Van Nostrand Reinhold, 1980.

MASLOW, ABRAHAM H., "Behaviors Leading to Self-Actualization," in *Reading Book for Human Relations Training,* ed. Larry Porter. Arlington, Va.: NTL Institute, 1979.

MORTON, DAVID H., "Project Manager, Catalyst to Constant Change: A Behavioral Analysis," *Project Management Quarterly,* 6, no. 1 (1975), 22-33.

OUCHI, WILLIAM G., and RAYMOND L. PRICE, "Hierarchies, Clans and Theory Z: A New Perspective on Organization Development," in *Perspectives on Behavior in Organizations* (2nd ed.), eds. J. Richard Hackman, Edward Lawler III, and Lyman W. Porter. New York: McGraw-Hill, 1983.

PERHAM, JOHN, "Matrix Management: A Tough Game to Play," in *Matrix Organization and Project Management,* (Michigan Business Papers, 64), eds. Raymond E. Hill, and Bernard J. White. Ann Arbor, Mich.: University of Michigan Prress, Division of Research, Graduate School of Business Administration, 1979.

SILVERMAN, MELVIN, *The Technical Manager's Survival Book.* New York: McGraw-Hill, 1984.

WOODMAN, RICHARD W., and JOHN J. SHERWOOD, "The Role of Team Development in Organizational Effectiveness: A Critical Review," *Psychological Bulletin,* 88, no. 1 (1980), 166–86.

WOODWARD, JOAN, *Industrial Organization: Theory and Practice.* London, England: Oxford University Press, 1965.

Chapter 7   Project Operations Continued: Building the Team

PERSONAL
MANAGEMENT
STYLE:
THE INDIVIDUAL
SYSTEM

PEOPLE MANAGEMENT:
THE PROJECT SYSTEM

PROJECT OPERATIONS:
THE PROJECT SYSTEM

MEASURING:
THE ORGANIZATIONAL SYSTEM

FORECASTING:
THE ORGANIZATIONAL SYSTEM

FOUNDATION: PROJECT MANAGER'S AUTHORITY

THE CASE OF THE DISSOLVING MEETING

Cast:

Susan Mitchum, Project Manager, "Sensing" Press Project
Terry Flotum, Contracts Administrator
Warren Blagett, Project Engineer
Floyd Purchase, Buyer
Debra Credits, Accounting Manager

Precision Reproduction, Inc., is a medium-sized company that has manufactured conventional printing presses for many years. Within the past five years, the company began to make a transition from the manually controlled presses that were the mainstay of the industry for years to computer-controlled presses that offer customers many technical advantages. A project management type of organization was recommended by Unctious Consultants, Inc., a consulting group, and a project manager was designated for each of four separate press lines in the company's catalogs. The "Sensing" press line was one of these four. This equipment was intended for major printing contractors, since it was fast and accurate and was able to automatically monitor and dispense the correct ink colors.

Usually the company designed the presses and then depended on major subcontractors to manufacture major subassemblies to specifications supplied to them. Because the subassemblies were so large in size, they were delivered to the customer's site and then assembled, tested, and turned over to the customer by the project management team. Susan Mitchum was the project manager on the latest model E press to be delivered to a major customer (Amalgamated Printing). She had managed the design and procurement effort and called this project meeting for late Friday afternoon in order to finalize all the open items for the press before beginning test runs on the following Monday morning. She had spoken on the phone with her team managers and told them about the meeting earlier that day. The meeting was called for 4:00 p.m. At 4:15 p.m., Floyd Purchase arrived, followed by Debra Credits. Terry Flotum and Warren Blagett showed up a few minutes later.

Susan: Well, let's get started. I suppose that you all have the checklist that we developed when we started this project several months ago, the list that covers all the materials to be delivered and items to be completed before we can start test runs on the E press on Monday. Why don't we go over them to be sure that nothing is missing. Warren, since you're in charge of engineering, is there anything that still is not complete as far as you can see?

Warren: No, everything is just fine. We sent out the installation drawings last week and ordered the extra ink and paper for the on-site test two weeks ago. It looks O.K. to me.

Susan: That's great. That takes care of our end of the test. Terry, do you know of anything that we still need from the customer before we begin Monday?

Terry: Well, as you all know, Amalgamated is a very particular customer. They contracted for both the equipment and the installation so that they can maintain the machinery after we have turned it over to them. I don't share Warren's ideas about being finished, since I haven't received either the drawings from engineering or the operating instructions that we're supposed to deliver to Amalgamated.

Warren: You're always complaining about something. It won't take very long to get those drawings to you, and the operating instructions are the same as for all our other presses. What's the big problem?

Terry: The problem is that we have a contract to meet, and we haven't done it. The customer doesn't really care about your good intentions. All they want is what they contracted for, and they want their drawings and instructions. When will they get them?

Just then, Floyd interrupted.

Floyd: Susan, I don't know why you assumed that we'll be ready for the test run next Monday. According to Apex Ink Co., the company that supplies all our special inks for test runs, they haven't shipped out the three barrels of ink yet because they weren't paid for the last batch that we bought. And I've told you before that Apex is a very independent bunch. They expect to get paid within 10 days after shipment, and our Accounts Payable people rarely can get a check out to them on time. Usually, all that it means is a couple of weeks of delay, but they told me that they're tired of waiting for their money. I just found out that your project is the one that they're going to hold up until they get paid for all the others.

Susan: Floyd, this is the first time that I heard about this problem. Why didn't you tell me this before?

Floyd: Because there wasn't any reason to. We always paid them late, and they always delivered on the next project. I just figured that this would be like all the other times, and I didn't think that it was important.

Susan: Well, what is there to do? Can we get another ink supplier, Floyd? Debra, why can't we pay our bills on time? I know that the project has been charged for those ink supplies. Why wasn't the money sent to Apex Ink?

Floyd: It's going to take at least two months to find another qualified vendor. Just then, Debra interrupted.

Debra: Look, Susan, I don't report to you. I'm just assigned to your project on a part-time basis, so don't complain to me about our bill paying procedures. We're doing the best we can in the Accounting department, and we'll try to get a check out to Apex in two weeks from now. But there's nothing that I can do about it right away.

Susan: Well, what else can be done? At this late date, it looks like this press won't be able to run on Monday, and it doesn't seem as though the problem will be solved here. This meeting is adjourned.

After the others had left, Susan sat at her desk and tried to determine what had gone wrong. She phoned her boss, Jack Johnson, and finding that he had a few minutes, went over to see him. She quickly went over the meeting with him.

Susan: Well, that's about all of it, Jack. Now you know about the delays in getting that press running next Monday.

Jack: Susan, I don't want to appear unsympathetic, but most of this is the responsibility of the project manager, and you're it. By telling me about these problems, you're doing exactly the same thing that your team was doing in your meeting. That's not what is expected. I need answers, not problems. Think about it over the weekend, and let me know first thing Monday morning what you want to do. Meanwhile, I'll call Amalgamated and tell them we'll be delayed for a while.

Questions

1. What was the major error that Susan made in developing this meeting with her team?
2. Was Jack right in his answer to Susan? Would you agree or disagree? Why?
3. How would you handle the problem of the engineering drawings and operating instructions?
4. How could the problem in purchasing have been foreseen? What would you have done?
5. Is a project manager responsible for the accounting procedures of the company? What can the project manager do about them? What about Debra's statement that she's only assigned on a part-time basis? Should Susan do anything about it?
6. Have you ever been confronted by a similar situation? How did you handle it?

1.0 INTRODUCTION: REPORTS AND MEETINGS

The Project Operations system is more than recruiting techniques and job descriptions for the project structure. It also includes developing the repetitive communications patterns that tie the project structure together. These patterns are one of the necessary integrating techniques that help to make the project team an effective working group on a daily basis. Those patterns also provide a means of feedback during the initial, highly critical phases of the project.

For example, the successful development of the basic project structure is obviously dependent upon effective communications. If people do not communicate easily and clearly among themselves, cooperative action is almost an impossibility. These clear communication channels must be maintained throughout the project life. Every member of the team must be able to communicate in an easily understandable (and relatively standardized) way with one another. It's difficult enough solving nonrepetitive technical problems, but when written reports describing those problems appear in unrecognizable formats during poorly designed meetings that seem to go on forever without accomplishing anything, the structure fails. The same applies to oral reports that have no point.

Effectively designed, standardized written and oral communications produce information flows that are clear and to the point, no matter who happens to be producing the data or which manager is in the project's center stage. Effective communications help to support the nonrepetitive, uncertainty-absorbing, cognitive processes of all project participants. Therefore, we'll design next models of reports and meetings that should be adaptable to most projects. These designs should also be part of the project manual. We'll start with written reports, since they should come in only one format. There are several formats for meetings.

1.1 REPORT FORMAT

Every written project report is intended to provide *new* information to the report receiver (see Fig. 7-1). In all cases, it's a "deliverable" and has the same characterisitics of any milestone. Thus, the producer of the report is responsible for delivering a document that will be acceptable to the receiver. (How many times have you received an unintelligible and, therefore, unacceptable written report in a format that was unrecognizable?) Project reports should consist of three easily differentiated sections. The first includes all the addresses, identification data, dates, and so forth. The second is the answer or the recommendation, and the third is "everything else." For example,

1. The *heading* includes all identification data under To, From, Subject, Date, and Identification Number. (The latter serves to identify the report if it's ever needed again. In some cases, it might be a reference to a particular work package, a date, a contract number, or anything else that's appropriate. By looking

---

THE HEADING

To: A. Jones   *From:* M. Silverman   *Date:* x/xx/xx
cc: D. Forsythe
     B. Smith   *Subject:* Pipe Leakage
     C. Davis   *Number:* Project 86-4-2/30
     D. Lee
     F. Luna

THE RECOMMENDATION

All the cast-iron piping in the hydraulic lab should be replaced within two months with high-pressure, stainless steel piping. (See report body for technical specifications.)

EVERYTHING ELSE OR THE REPORT BODY

A. The definition of the problem.

B. Where, why, when, how, who, and anything else.

C. What kinds of investigations were performed. What the results were, including all the test data.

D. The evaluation of alternative solutions to the problem.

E. The reasons for the recommendation.

Figure 7-1. Report format

at the number, one should be able to determine which project and work package it belongs to and where it fits into the report sequence.)

2. The *answer* or *recommendation* is the end result of what the report writer wants to say. Is there an answer? Is there a recommendation of some kind? It belongs here.

3. *Everything else* appears in the body of the report. Here the process that was followed is detailed. All the test descriptions, test results, replications, and relevant data are included in this section.

The report format should be familiar, since it follows the ideas we discussed in the section on milestones in Chapter 6. A report is a milestone, a deliverable. The report's adequacy is determined by the receiver. The receiver can read the first two sections of the report very quickly and then decide if it's necessary to wade through the problem definition, the test descriptions, procedures, data, evaluations, and the rest of it. Even if the report receiver decides to read only the page listing the recommendations, the reports will still be complex enough for the receiver interested in every little detail. The reasons are obvious. This deliverable must be complete in all respects because it is invariably filed somewhere, at least for some time period, and who knows who will read it next? The next person may want the details.

1.2 MEETING FORMATS

Reporting is not always accomplished in writing. Some of the time, especially in the initial, high uncertainty phases of the project, project operations are supported by meetings. These meetings can either produce results or waste time. If they are not designed well, they can become a ritual rather than a management tool. We are all familiar with organizations that regard meetings as a necessary part of all problem solving. For example, consider this author's viewpoint.

> A ritualistic approach to real problems is the ever-ready solution of bringing people together . . . on the naive grounds that the exchange of ideas is bound to produce a solution. There are even fads and fashions to ritualism as in the sudden appearance of favorite words like ''brainstorming'' or ''synergism.''
>
> It is not that bringing people together to discuss problems is bad. Instead, it is the naive faith which accompanies such proposals, ultimately deflecting attention from where it properly belongs.'' (Kets deVries, 1984, p.333)

Many meetings are poorly structured because of the naive faith described above. This faith wastes one of the most valuable assets that the project has—time. Meetings can be divided into two very general categories: problem-solving or information-disseminating. We have briefly discussed some of the content of the first meeting, the problem-solving meeting in Chapter 6, section 7.0. Now we discuss it's overall design.

1.2.1 Problem-Solving Meetings

A properly designed format for a problem-solving meeting requires that most of the individual problem-solving work be done prior to the meeting itself. Meetings are very expensive. They use people's time, and the only reason to have them should be to provide an interaction *among* people that doesn't happen in a one-on-one meeting. Therefore, the contributions of individuals invited to the meeting should be coordinated beforehand. The meeting is then necessary only to achieve group interaction. The problem-solving meeting's format (see Fig. 7-2) includes an agenda, a meeting procedure, and minutes that document what happened and what

MEETINGS: PROBLEM-SOLVING

1. *Agenda*
   a. List and send out problems.
   b. Receive inputs and summarize on "white" paper.
   c. Distribute problems and "white" papers.
2. *Meeting Procedure*
   a. Discuss each problem.
   b. Junior members vote first.
   c. Summarize the decision.
3. *Action Minutes*
   a. Summarize who, what, and where.
   b. Set follow-up date.
   c. Include preliminary agenda for next meeting.

Figure 7-2. Meetings: Problem-solving

is to happen in the future. It's a very logical way to proceed, but it assumes that the problem-solving meeting is *not* a crisis meeting. (Most of the time, a crisis meeting becomes an information-disseminating meeting, with the meeting leader telling everybody what has to be done.)

The Agenda. The agenda is designed as two interdependent sections—an initial list of problems and then a summary of possible solutions produced by the meeting attendees for each problem. The first section—the list of the potential problems with which the meeting is expected to deal—is originally formulated by the meeting chairperson. The list is distributed to meeting attendees in sufficient time *before* the meeting for recommended answers to these problems to be written by the meeting invitees and received by the meeting chairperson. In other words, any attendee who has a particular contribution to any of the problems is supposed to respond in writing to the meeting chairperson. These comments are summarized by the meeting chairperson into a summary for each problem. The summary lists the problem and any comments, suggestions, solutions, and so forth that have been received, about the problem. Some people may not comment on all the problems; therefore, the summary for a specific problem may not be very long. The summary is then distributed back to the meeting attendees before the meeting as the *agenda*. In other words, the meeting agenda is a list of problems and summaries of answers. When the meeting starts, everyone knows the sequence of the problems to be solved and what everyone else (who took the time to reply) thinks about the particular problem.

The Meeting Procedure.   The discussion about any particular item on the agenda should be started with the question, Does anyone have anything to contribute about problem 1 (or whatever) that has occurred *since* you received the original problem list? Anyone who attempts to use the meeting as a speech-making platform or hasn't taken the time to reply, usually considers his or her time more valuable than all the other meeting attendees. When this happens, the meeting chairperson should have no hesitancy about saying that. After all, the meeting was called to get interaction of the group. Individuals with independent inputs should provide them beforehand.

Assuming that the team members are capable of independent decision making and have accepted the responsibility for themselves, the following rules are applicable as part of the meeting procedure:

1. *Everyone is present.* If you have been invited to the meeting, you have agreed to be there. Unless you are the star of the show, the rest of us will not wait for you to appear. The meeting will start on time.
2. *Silence means assent.* If you have been invited and you don't say anything, you have agreed with everything that went on at that meeting. Of course, if you don't attend (or don't send a proxy), rule 1 takes over automatically.
3. *Junior members vote first.* Senior members then won't be influencing the juniors' votes.

The meeting chairperson summarizes the final position on each point immediately after everyone votes to be sure that there are no errors in the minutes. If there are no objections to the oral summary, rule 2 applies.

Action Minutes.   The following are guidelines for the oral summary and minutes:

1. The oral summary should be immediately documented. Action items should include the usual who, what, where, when, and how.
2. There must always be a follow-up mechanism listed, unless the item closes a problem.
3. The minutes are documented and then distributed as soon as possible after the meeting ends.
4. If any items have been concluded, the minutes should say so. If any have been delayed, the reasons and the date for the next problem-solving meeting in which a solution will be presented should be included.

Running a meeting this way is sometimes difficult, but it can be done when it is organized beforehand. It requires that all the attendees and the meeting chairperson think out possible contributions *before* they come to the meeting. As previously noted, meetings are expensive and there's no reason to allow one person to take the time of everyone else while he or she does a mental "memory dump." Problem-solving meetings are called to gain the group's interactions in solving problems that have been defined beforehand. Everyone at the meeting is "there" according to meeting rules number 1 and 2; otherwise they shouldn't be invited. Why waste people's time? Therefore, the chairperson gets all the individual work done before the meeting starts, so that the meeting can then deal with the cooperative efforts of the group.

(These kinds of problem-solving meetings are not the usual kinds of project meetings. They are more applicable to functions, since functions are "eternal," that is, "We will have staff review meetings on the first Monday of every month, *forever.*")

Operating problems in a project generally appear sequentially. Thus, not all project team managers are immediately concerned with the "crisis *du jour*" while projects are in operation. Why attend a problem-solving meeting if one can't contribute to solutions? For example, a problem in engineering design could be of direct concern perhaps to only three people: the project engineer, the project manager, and possibly the next manager in the project sequence, e.g., the manufacturing manager. The other project team managers, such as the purchasing manager or the service manager, might not be immediately concerned, although they probably would like to know what happened. They can read the minutes for that.

These mini-problem-solving meetings dealing with the "crisis *du jour*" should also follow a definite format. They should have all the elements except a two-step agenda. They should have a short agenda (the "crisis *du jour*"), a procedure (prompt starting time), everyone should be present, and silence means assent. Minutes should be taken in a report format, which are then distributed so that everyone is informed about what was decided. As noted before in Chapter 6, section 7.0, there are just three occasions when a *formal* problem-solving meeting involving all project team managers should occur in a project life cycle. These are the following meetings:

1. Initial forecasting or bottom-up meeting
2. Design review (see Appendix 3) meeting
3. Project close-down meeting

Since we've already covered the first bottom-up meeting in Section 7, Chapter 6, and the Design Review is available in Appendix 3, let's discuss the project close-down meeting now, even though this could logically be covered at the end of the book. The close-down meeting could also be used *during* the project life cycle if the project is to be turned over to another project manager or else is being subjected to a complete project iteration. This type of meeting results in a summary or closing document. Of course, the next step might be reopening the project again using the by now familiar top-down and bottom-up sequence.

The close down of the project life cycle rarely has much of an effect on the technical success of a project, because by this time the technical objectives have been achieved or else they haven't been achieved. The close-down meeting, then, does not help much to achieve success of technical goal achievements. However, it may have a lot to do with the success of the project manager. Residual attitudes towards the project after close down by the client, the project team, and senior management can either be positive or negative, and the completeness of the close-down process is often one of the deciding variables.

1.2.2 Closing the Project: The Last Problem-Solving Meeting

The meeting to close a project is done *before* a project is completed. It's almost like the last design review. There are three parts to this type of meeting:

A. *The "outside," or customer's or user's requirements,* are placed on an "open-item" or "punch" list. Looking at the project's cumulative cost curve in Fig. 7-3, we know how much time we still have (i.e., remaining time or △ time) and how many hours or dollars are left (i.e., remaining cost or △ cost).

If these resources are sufficient to satisfy all the items on the "punch" list (see Fig. 7-4), there is no problem. If not, now is the time to issue another impact statement. This is just like defining a new project—determining the close-out process. After all the customer's items are listed, we deal with the project itself.

B. *The "inside" or the internal or administrative close-down* sequence that I

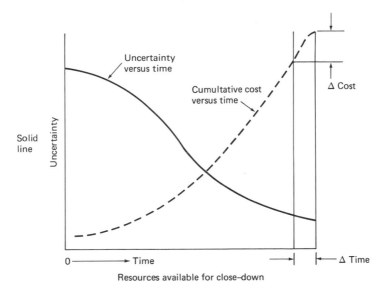

Figure 7-3.   Resources available for close down

A. THE OUTSIDE

1. *Open Items*—"Punch List"
   versus
   contracted items
2. *Resources*—Required time and cost
   versus
   ▷ time and ▷ cost

Impact Statement?

B. THE INSIDE

1. People
2. Excess inventory
3. Excess capital
4. Data, drawings, and documentation
5. What else?

C. THE SUMMARY

Figure 7-4. The Outside, The Inside, and Summary

suggest or even the items that are included on this "inside" list are not important. Yours may be different. I always try to deal first with

1. The people: Who will stay until the end, and who will be released for other projects? Those who are released will be provided with an appraisal of the work that they have done on the project since their last appraisal. Releasing your team managers is somewhat like breaking up a family, and it can be stressful unless you plan this aspect very well. Some of the areas that you might consider are:
   • Manufacturing: Are materials and services completely documented and controlled?
   • Field Service: Has training been completed? Will there be a continuing need? What about service contracts?
   • Accounting: Are all projects accounts closed? Any nonresolved open items?
   • Engineering: Are all operating manuals and service instructions completed?
   • Marketing: Is there any use for the project achievements in the future?

- Sales: Are all contract items completed: purchasing, legal, distribution, and so forth. (For other details see Meredith and Mantel, 1985, pp.407–421).

2. Excess inventory: If there is any, who owns it and what should be done with it?

3. Excess capital: Did we buy something specifically for this project? What should be done with it?

4. Data, drawings, and documentation: Are all prints and documentation to match the final product or describe the final service that the project produced?

C. *The project summary* is probably one of the more important documents about your project. Here you can outline the history of the project. For example, what did you start to do? What actually happened? This doesn't have to be very long, but it should be complete. If the project is being terminated because it has accomplished its goals within the three "golden limits" (technical tasks, time, and cost), this is a reasonable place to state that. Conversely, if the termination happens because those goals have not been reached, there are obviously excellent reasons why, and since you have very carefully documented those changes through impact statements, the summary document can point out the positive aspects in closing this project down now and the positive decision not to continue. The basic purpose of this document is to help future projects, and every project contributes something to that. Many summary documents are no more than two or three pages long. If further details are required, supporting documentation can be found in the various reports that the project has generated.

Report and Meeting Timing.   Project reports and meetings are very important in project operations since they deal with solutions to unique problems. Consequently, as noted several times before, the frequency of project reports (and meetings) is high at the beginning of a project and decreases with time and uncertainty.

The differences between functions and projects are not handled well in many organizational measurement systems. When functions and projects are then required to use similar control systems in "order to get a standard reporting format," project measurements suffer. In many project beginnings, while there isn't much money spent in the beginning, there's a lot of work going on in small groups trying to drive uncertainty down. Accounting systems that are based on monthly (or any other standard time periods) are, therefore, inadequate. And when the organization tries to achieve overall standardization (of course, using the familiar functional model), project management isn't helped, since project participants should be able to design their own operations including having the necessary amount of personal freedom to manage and control the project. When organizational structures, operating systems, and controls developed for relatively standardized functions are imposed upon projects, project goals are often not optimally achieved.

In other words, recalling the uncertainty curve and the idea of nonlinear feedback reporting because of the need to absorb uncertainty, it quickly becomes appar-

MEETINGS:
INFORMATION-DISSEMINATING

1. Agenda
   a. List of speakers, subjects, and length of time to speak
2. Meeting Procedure
   a. Silence means assent
   b. Everyone is present
   c. The past is over
   d. Summarize
3. Action Minutes (Same as problem-solving meetings)

Figure 7-5. Meetings: Information-Disseminating

ent that there has to be a lot of feedback at the beginning of a project. As noted before, since most reporting systems won't show this need, since not much time or money has been spent, we will use the information-disseminating meeting to keep track of things after we have organized the project in the first problem-solving or bottom-up meeting.

1.2.3. Information-Disseminating Meetings

The information-disseminating meeting has the same general format as the problem-solving meeting, since it also has an agenda, a procedure. and minutes (see Fig. 7-5). It's a lot different in operation, however. It's purpose is to inform and advise the whole project team concerning present status and predicted problems. It is also based upon reports received or expected to be received before the meeting. It's a partial extension of those reports but for information only.

All team managers are aware of what the topics are *supposed* to be at an information-disseminating meeting because those topics have been developed in the project forecast. Those team members who will be disseminating the information should have distributed a written report (in the standard format) to all attendees before the meeting. The information-disseminating meeting, therefore, is used to clear up any minor points not understood and to be sure that everyone is informed. The meetings are based upon the written reports that all the team members have received prior to the meeting.

Agenda. The agenda is just a list of items reflecting the written reports that were distributed to meeting participants several days before the meeting is scheduled. This meeting is like any other since it is to provide interaction within a group, not promote speeches by individuals. The project manager calls each speaker on the

phone several days before the meeting and asks if he will be ready to talk at the meeting and how much time he will need. Since the person being called already knows that he is due to report (that was part of the project forecast), there should be no surprises. Everyone has received a *preliminary* copy of the agenda before the phone call is made. It has some very basic data on it—the usual To, From, Subject, Identification Number, Meeting Place, and Meeting Start Time.

If an expected meeting speaker will not be ready and has not informed the project manager of this beforehand, a line is drawn through his or her name, which indicates that the speaker was aware of his or her scheduled participation and neglected to inform the project manager prior to the phone call that something unforeseen had come up. This is a definite management failure. However, if the speaker, knowing that the project manager will call beforehand, calls the project manager first to inform him or her of unforeseen delays, the project manager requests a new estimate and notes the new date on the agenda next to the specific topic. There are only three alternatives that appear as a result of this phone call:

A. *"I'm ready."* Therefore the project manager asks how long he or she will speak and notes that on the agenda.

B. *"I'm not ready, and I didn't tell you beforehand. Therefore, I have failed."* A line is drawn through the name on the agenda.

C. *"I'm not ready but I called you last week to tell you about it. This is when I now expect to be ready."* A date is placed in the agenda showing when the report will occur.

The *final* agenda is sent out just as drawn: with the times, the lines and/or the new dates. Lines drawn through a name indicate an obvious management failure with that person.

Procedure.   As I've previously said, a project is sequential. Therefore, everyone might not want to attend the meeting, but they should be informed. The procedure includes four general rules:

a. *Silence means assent.* It's almost impossible to determine what someone is thinking. Therefore, if there is no objection during the meeting, it is assumed that everyone has agreed.

b. *Everyone is there.* Team managers are responsible people who are as capable of absorbing uncertainty as they are in managing their tasks. Therefore, whether they attend the meeting or not is not important since they have been informed and have received a written report. They are responsible for that report's contents. If they wish to read it in complete detail, they can do so. When they receive the final agenda showing that someone will speak and the amount of time scheduled, if they wish to attend, they can at that time. Being informed by the final agenda of an information-disseminating meeting that an oral summary will be given, automatically makes the decision to learn more about it dependent upon the particular manager. If the manager cannot appear (or decides not to appear), he or she has automatically agreed to whatever

happens at the meeting since they were "silent." In other words, if they could not (or decided not to) attend, according to rule 1, they have agreed with everything that happened. Remember, they could have always sent a proxy, if they couldn't be there, or else they could have notified the project manager. Why hold up a meeting of all the other team members just because one cannot make it?

c. *The past is over.* Sometimes new data appear that indicate an error has been made or a problem occurs that should have been solved in the past. We cannot relive the past, just learn from it. We'll now set up a new task to handle it. If we knew then what we know now, would we have done it then? Obviously not, therefore we just go on and solve the "new" problem.

d. *Summarize.* When an oral report has been delivered and other people have had their questions answered and comments noted, the meeting chairperson should summarize what happened. This is the time to eliminate any possible errors that could be written into the minutes. As in the problem-solving meeting, action minutes should be immediately written and distributed. Who, what, where, when, and how should be included. If a task is concluded, the minutes should say so.

There is another kind of meeting that is very difficult to classify in either of the two general formats—problem-solving and information-disseminating. It concerns people who may not be as familiar with project operations as the project team is. But the purpose of this meeting is basically the same as the others—to support cooperation. This kind of meeting is with upper management or the customer.

1.3 UPPER MANAGEMENT OR CUSTOMER REVIEW MEETINGS

Review meetings with upper management or the customer may not be able to follow the project's meeting format. Depending upon the nature of the subjects covered, the carefully detailed communications design of meetings supporting the project structure sometimes goes wrong.

Obviously some of the agendas and procedures must be modified. If it's a problem-solving meeting, you can rarely count on any return of a preliminary agenda. Therefore, present all the background data that you can when you send the agenda out the first time. (They may be so surprised at getting an agenda that they may even follow it.) If it's a problem-solving meeting intended to cover design review, the problem-solving format should be unchanged. Participation in a design review meeting is the same, no matter who is there. The issues to be dealt with more directly determine the life of the project. If it's an information-disseminating meeting, the meeting chairperson should be sure that appropriate backup reports are provided to all attendees. The agenda for this meeting should show no failures or any delays in presenting information.

The procedures outlining the "silence means assent" and "everyone is pres-

ent'' rules don't work very well for these meetings. (Top management or the customer may not agree.) Even though the project team is providing information, these major participants are really not integral members of that team. Therefore, these rules will probably have to be suspended. However, these meetings can still have a well-defined structure if the project manager is the meeting chairperson.

Sometimes the logic of these processes gets lost because the further away from or the further up the organizational ladder someone is from the immediate work responsibility, the faster, easier, and cheaper the job appears to be and the less knowledge of (or empathy for) project problems. Some meetings of this type can be less than satisfying (or even orderly). But here are some suggestions. They are not as logical as those concerning the meetings managed within the project but, then, it may not be a very logical situation.

- Be aware of or else learn very carefully the specific interests of the people who must approve your proposal or your actions and direct your attention to those interests.

- Rehearse your presentation. It will build up your confidence and give you a chance to make adjustments that would never have turned up otherwise. Then review and revise. Rehearse—the best impromptu speeches are those that have been well rehearsed. . . . When you think you've finally got your act together, why not try a dress rehearsal? Gather a few of your associates for a dry run. See how your timing works out.

- Nobody knows as much about your system as you do. Here is your chance to tell your story, to reveal your planning skills, and to make management or the customer an offer it can't refuse.

- Try to make your delivery sound like a conversation rather than a sermon.

- Let's see some enthusiasm on your part. It's catching. And you'll be surprised at how good a salesman you are!

- Don't reject any suggestions or comments quickly. They may help and sometimes someone who isn't as close to the problems can see an ''obvious'' answer.

- Don't ''defend'' an answer that is being questioned by becoming emotional. Just present the supporting data.

- And finally, remember the KISS principle (Keep It Simple, Stupid). If you must get complex, do so only after you have carefully covered the basic data and you are sure that your audience is following you. Ask questions such as, ''Am I being clear?''

- Visual aids—don't waste your time trying to describe a system if you can show a picture of it, and don't use a photo if you can lay your hands on a film clip. (Abstracted from: Kannewurf, 1980)

2.0 CONTROL OF COMMUNICATIONS

The project's primary documentation such as charters, requests for proposal (RFP), functional specifications, drawings, standards, and various contractual paperwork is

normally controlled by procedures that are fairly well established in the parent organization. These procedures, to be satisfactory, should include a numbering or date control system indicating when a document was produced, who wrote it, and who received it, and where it is filed. As part of the Project Operations system, there should also be a subsystem that also controls *who* is authorized to write the various kinds of project documentation. This subsystem is part of the administrative section of the project structure. It should be noted in the project manual. The following rules are generally acceptable:

1. Any team manager can send letters, memos, or documents within the company. There is an implied delegation of authority here. It means that a team manager originating a document is best qualified to determine what, when, and to whom documentation should be sent within the company. Each manager understands the limits of his or her position and can handle his or her job well in accordance with those predefined limits.

2. Purchasing is solely responsible for written materials sent to vendors. Procurement can be a delicate matter, and it should be handled by the appropriate team manager responsible for purchasing. This doesn't prevent anyone from contacting vendors or receiving catalogs from them. But no one other than the purchasing manager should be able to contractually obligate the project to buy anything or change anything after it was purchased. This could be a difficult rule to maintain unless all the team managers are explicitly informed what they may and what they may not do when dealing with vendors.

3. The contracts administrator will process all correspondence external to the company except that concerned with purchasing. This includes documentation received from or sent to the customer, legal agencies, or anyone else external to the project (and the company). It should flow through a single channel. For example, the results of an engineering test may have to be sent to the customer. This should be sent out only by the contracts administrator, who will transmit it with an appropriate letter of transmittal.

4. Copies of correspondence should go to the project manager only if the project's three limits—technical achievement, time, or cost—are affected. If team managers are controlling their own tasks and they produce a document that refers, in their opinion, only to their own operations, it isn't necessary to send a copy to the project manager. In fact, the only time that a project manager is concerned is when something will *not* happen according to the forecast or if something of major importance *will* happen that was unforeseen or will affect other team managers. And that concern is important only when the responsible team manager can't handle the problem alone. The correspondence, however, should be appropriately coded with some kind of numerical code on the document so that anyone can track it down, if needed.

3.0 SUMMARY

The third system, Project Operations, contains techniques that are less well defined than the first two—Forecasting and Measuring—because it has to be more closely tailored to specific projects. As we move closer to specific projects and people, the systems become less concrete because we're more dependent upon the unique "art" of the project team managers as they manage their project teams. Because the project crosses conventional organizational boundaries, the project structure is often in endemic conflict with the parent functional organization. When the functional managers are suppliers of the resources (both human and physical), recruiting project participants can be difficult, and welding them into a coherent, cooperative team requires more flexible (and more pragmatic) tools in order to quickly build a project "culture." The techniques have to be more adaptable to the particular situation.

Resolving role conflict in the project beginnings helps to develop the project "culture." Standardization in project reporting and communicating also helps. In addition, the increased amount and higher frequency of information flows during those initial, critical parts of the project require the development of different meeting formats for prompt, effective information dissemination. Regular organizational information systems are usually unable to supply project data quickly. Fortunately, all of these problems can be dealt with during the second project iteration—the problem-solving or bottom-up meeting—and the perceptive project manager will take as much time as necessary in the beginning of a project to resolve them. Even meetings with upper management or the customer can be managed with some pragmatic rehearsal and presentation techniques.

But the project team consists of individuals, and it is the personal commitment of these individuals that ensures project success. As we get closer to individuals, the techniques in the system become even more subtle and variable, just as the individuals themselves are. The next system, People Management, follows this trend toward individuality. We'll be dealing with individuals, and, therefore, we'll be very concerned about motivation, appraisal, and the control of individual behavior.

My Suggested Answers to the Case Study

1. Susan did not correctly structure this meeting beforehand. This was a *problem-solving* meeting. Therefore, there should have been an agenda distributed with all the open items listed and specific written answers requested from meeting participants before the meeting. Then these answers should have been discussed on a one-on-one basis before the meeting. The meeting itself is needed primarily for making decisions based upon the interactions among people. There should have been no surprises.

2. I agree with Jack. Assuming that Susan was the "expert" on her project, she should have gone to Jack with both the problems *and recommendations* for solutions to those problems, and those recommendations should include areas she couldn't control. Otherwise, Jack couldn't help her.

3. First determine exactly what was contracted for and the estimate to complete to actually deliver the drawings and instructions. Then, let Terry contact the customer and determine how much time the customer would allow. Try to match the customer's demands with the elapsed time in the estimate to complete and make a recommendation including additional resources to Jack Johnson if the project didn't have the resources to meet the customer's needs. Those recommendations could be in the form of an impact statement including, for example, having work done on the drawings by others (thereby increasing the cost), or adding overtime to Warren's estimate to shorten the delivery schedule.

4. The purchasing problem is similar to the engineering problem, since both could have been eliminated by using tangible milestones. A checklist showing when tasks were to have been started (i.e., the purchase order placed) and completed (i.e., delivery of the ink before the test was to be scheduled) would have pointed out the problem. This would be part of a usual project control system.

5. Not usually, but that is no reason to be limited by them. Most accounting systems are set up for general corporate purposes, and the project manager must determine whether they are adequate for project purposes. If developing a milestone control showed that the company was receiving materials late because of late payments, then the project could adapt by issuing purchase orders sufficiently early to account for this delay. Failing that, advise management through an impact statement what the cost of the delay would be.

   Debra's statement cannot stand unchallenged. Specific guidelines for Debra should be established by Susan and reviewed with Debra. Then they should be reviewed and approved by both Susan's and Debra's bosses.

BIBLIOGRAPHY

KETS DEVRIES, MANFRED F. R., *The Irrational Executive*, Psychoanalytic Studies in Management. New York: International Universities Press, 1984.

KANNEWURF, ADOLF S., "How to Present Your Proposals to Management," in *Achieving Success in Manufacturing Management*, Society of Manufacturing Engineers, Charles F. Hoitash (Ed.) Used by permission. 1980.

OTHER READINGS

ARGYRIS, CHRIS, "Today's Problems with Tomorrow's Organizations," in *Matrix Organization and Project Management*. Dearborn, Mich.: University of Michigan Press, 1979.

HON, DAVID, "How to Hold Productive Meetings with Your Peers," in *Training/HRD*, May 1980, pp. 56–63.

MACIARIELLO, JOSEPH A., *Program Management Control Systems*. New York: John Wiley, 1978.

MARTIN, CHARLES C., *Project Management: How to Make It Work*. New York: AMACOM, American Management Associations, 1976.

MEREDITH, JACK R., and SAMUEL J. MANTEL, JR., *Project Management: A Managerial Approach*. New York: John Wiley, 1985.

NORTHCRAFT, GREGORY B., and GERRIT WOLF, "Dollars, Sense, and Sunk Costs: A Life Cycle Model of Resource Allocation Decisions," *Academy of Management Review*, 9, 2(1984), 225–34.

ROSENBLUM, R., and F. WOLEK, *Technology, Information and Organization: Information Transfer in Industrial R & D*. Cambridge, Mass.: Harvard University Press, Graduate School of Business Administration, 1967.

Chapter 8   People Management: The Fourth System

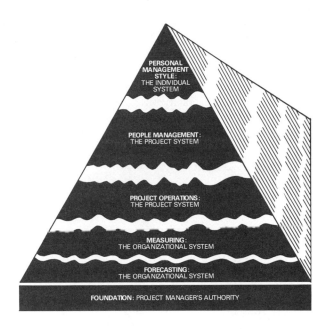

PERSONAL MANAGEMENT STYLE: THE INDIVIDUAL SYSTEM

PEOPLE MANAGEMENT: THE PROJECT SYSTEM

PROJECT OPERATIONS: THE PROJECT SYSTEM

MEASURING: THE ORGANIZATIONAL SYSTEM

FORECASTING: THE ORGANIZATIONAL SYSTEM

FOUNDATION: PROJECT MANAGER'S AUTHORITY

THE CASE OF THE MISSING ENGINEERS

Cast

John Donahue, Project Engineer
Peter O'Reilly, Chief Engineer
Shiela Russo, Designer
Alan Low, Internal Consultant

Oil Cities Industries is a major supplier of special computer equipment to the petroleum exploration industry. For many years, it had been at the forefront of technology with the application of computer controls to the exploration for new deposits of petroleum. It produced high-quality systems that were rugged enough to stand the rigors of extended use in unfriendly field environments. Within the recent past, competition had been able to carve some markets away from Oil Cities by producing very serviceable equipment that was not quite as advanced technologically but that could be purchased at a much lower cost and that could be received on a faster delivery schedule.

Top management of Oil Cities was concerned. They had called a meeting of all the company executives, and Peter attended as the Engineering representative. He received a directive at that meeting to cut down on the "excessive" re-

search and development costs that every new project seemed to undergo. Management implied that the Engineering department was "gold plating" the products and never getting them out in time at a low enough cost to meet competition. Peter had protested, but had no defense when the company accountant showed that these costs had been steadily increasing for each project for the last few years. And product complexity did not seem to be a major factor. Peter asked the company's internal consultant and trouble-shooter, Alan, to meet with him the next day to get some ideas. They met the next day at 9:00 a.m. in Peter's office.

Peter: Just take a look at these figures. They show that every new product development project has exceeded it's original technical, time, and cost estimates. Oh sure, the project manager always issued an impact statement, but when we investigated, we always found out that the requirements for additional resources were justified.

Alan took a moment to review the data that Peter had provided. All of the original estimates had been thoroughly analyzed by top management and had been accepted with minor modifications. When impact statements were later issued, they also had been thoroughly investigated. All of them were justified since it always appeared that unforeseen problems had appeared.

Alan: Look, this is a wild guess, but is it possible that we don't really know how to estimate correctly? Is it possible that some of these unforeseen problems could have been forecasted by somebody on the project?

Peter: No, our project managers have lots of experience with our products and our customers' needs. Every estimate is thoroughly analyzed during the last coordinated forecasting process meeting, and management always brings in the Accounting people to test those estimates. There were sometimes a few changes then but nothing of real importance.

Another thing is that even though we have about the same Engineering staff that we've always had, their productivity has decreased tremendously over the past few years. I've checked all the time sheets and there's no absenteeism there. In fact, not only has everybody been working on their jobs, but they've even been spending a lot more overtime lately. It's just as if our Engineering department had half the staff missing.

Alan: Let me do a little investigating of my own. I think I'll look into it by getting together with some of your people, John Donahue and Shiela Russo. I've found that it's always a good idea to start any problem solving with people who are closest to the actual work being done.

Later that week, he met with both John and Shiela, explained what was going on, and gave them the project documentation that Peter had given to him. About two weeks later, he had lunch with both of them. It was a very enlightening lunch.

John: Alan, I'm sure glad that we were able to get the concept of the project charter in place here about ten years ago. The part that was particularly applicable was the impact statement, because as the project engineer reporting to the project manager, I was responsible for all the engineering work on the last three major computer systems, and I had to write a lot of impact statements.

Alan: Why, what happened? Didn't we have enough data in the project beginning? Why didn't we do some feasibility studies if there were that many unknowns?

John: Oh, it wasn't the project unknowns that caused all the trouble. Most of the problems were caused by top management. When Engineering originally had outlined a functional specification that seemed to satisfy the customer's basic needs, it only took several weeks of work before some vice president would get a bright idea that it would be nice if this computer system could do the company payroll in addition to the calculations that the geologists wanted.

That meant a lot of changes that we hadn't anticipated at the time of the original forecast. So I wrote an impact statement, which was, of course, approved. Then, a few weeks later we'd get another request like, "Say, I know we decided to make the new equipment dust proof because it is to be used in deserts, but wouldn't it be wonderful to have a universal unit that could be used anywhere? So why not also make it waterproof?" It's a wonder we can get anything out since our productivity is measured only by the end product and not by the "changes" that have been included.

Shiela: Of course, since I'm only a designer, I don't get direct inputs from top management like John does, but the same general kind of thing happens to me. Just last month, I was working on the design of some equipment structure when one of the sales people met with the project manager. Before I knew it, my functional boss called me into his office and said that the project manager had changed my work package and gave me another set of specifications that was supposed to cut the weight of the equipment by 10 kilograms because they wanted it to be airborne too.

That redesign only cost me a few days, but it increased the cost of the unit tremendously. The lost time of a few days cut into my output. Productivity goes down when you have to do the work over. Of course, the final design cost was way over budget for the project and we got blamed for it.

Alan: Look, I can appreciate what you're telling me, but changes are a way of life around here. We all have to cope with them because that's the way our customers are; they want everything to be the latest technology.

John: You may be right, but I'm not sure that our customers are really the problem; our company is. The competition's products are not very sophisticated but they do the basic job required of them, at much less cost. I know, because our sales guys occasionally come crying in here with some sad story about competition and bring us some of those products to show us "how well those other guys are doing."

Later, Alan met with Peter. He reviewed some of the comments he had received and asked for some feedback.

Peter: Alan, we can't stand still technologically. If there are a few changes as the design progresses, that's the nature of our business.

Alan: Our customers don't seem to be as concerned about highly advanced technology as we are, since they are buying the less costly competition's products. Those products may not be as fancy as ours but they do work, even if in more limited circumstances. Maybe the market has changed and the state of the art is not what's needed anymore. I've learned something else from checking the sales records. We have been selling fewer pieces of equipment but at a higher price than our competition. So our situation has declined even more drastically than expected, since the higher unit prices covered up the decline in total quantity. In other words, the sales revenue has changed from relatively high volume and moderate prices to low volume and high prices. We're selling to fewer customers than before. This can't go on.

Questions:

1. Is it possible to minimize external changes in rapidly changing technology? How would you do it? Is it desirable? Why?
2. How would you expect this situation to affect the motivation of the engineering personnel? Why? How would that affect productivity?
3. Can you develop a system to solve this repetitive problem? What is it?
4. Is this typical of any situation in which you have worked? What happened?

1.0 INTRODUCTION: TEAMS THEN PEOPLE

The third system, Project Operations, describes several ways to manage the project as a group of people. The project structure describing jobs and interactions is documented in the project manual. For example, written interactions among team managers follow standard report formats. These formats treat a report as a deliverable milestone. They allow the receiver to define the acceptability of any report received by determining how much to read after reading the "recommendation" section. Meeting designs help people save time by concentrating on human interactions rather than wasting it by having people just speak without a plan. Role conflict, caused by differences in directions received from project management and functional management is diminished by a clear definition of reporting relationships for each team manager and participant outlined in the structure described in the project manual. The impact statement translates temporary setbacks into recommendations for upper management action, when those setbacks seemed large enough to affect overall project technical goals, time, or cost. The impact statement notes the reasons for any setback, recommendations are explicitly defined, and approval is requested within a specified response time.

The Project Operations system also provides information feedback in a nonlinear way that matches the drop in the project uncertainty curve. As the project team solves major problems and becomes well coordinated, formal information-disseminating meetings are displaced by quick, informal, small group reviews. The frequency of these meetings drops with time as uncertainty decreased. Increased coordination among the project teams decreases the need for formality. The next system, People Management, is even more adaptable, because it deals with the indi-

2.0 GROUPS AND PEOPLE IN THEM: CONFORMITY
AND SELF-SELECTION

The overall behavior of groups or project teams is easier to predict than that of individuals. In other words, as groups increase in size, there are fewer behavioral extremes because people acting in the middle of an acceptable behavioral range will reinforce one another and those at the ends will tend to cancel out. It's a fairly common occurrence based on statistics of a normal distribution. And there is usually a normal or expected pattern of behavior (perhaps another word is *culture*) that defines the limits of the acceptable range. For example, assume that the following statement describes a typical situation: "The design engineers generally won't give an opinion on the useability of the new motor line until they have completed the life and endurance tests, but Marketing feels right now that the new motors will be great because of the novel features that these motors have." Group behavioral norms in these departments are apparent. Now consider an individual who has been very successful in Marketing and is transferred from the Marketing department into the Design Engineering department. Personal enthusiasm is still accepted, but it is limited or tempered by a cultural requirement of a complete, logical, extensive series of operating and life tests. If the transferee is to survive and succeed in this new environment, his or her behavior must conform to that of the accepted norm. Therefore, those behaviors that are in a middle range (for each group) reinforce one another. "Deviant" behaviors in specific individuals on either end of the "normal" curve for the group either cancel one another out or, if the behavior is very deviant (for that group), the individual is eventually expelled or is uncomfortable enough to elect to leave. Overall group responses are simpler to predict.

Then there is another factor of self-selection that affects the original entry of individuals into the group. In most cases, project participants have been selected (or have volunteered) because they seemed to understand and respond to the expectations of the group's behavioral environment. The project charter partially describes this environment, but there's more to it than that.

Self-selection begins very early in life. It's the process that occurs as one is associated with organizations and groups, each of which has its own series of "filters" that will not accept an individual with unacceptable behavior.

For example, all project engineers have probably already been initially "filtered" by their secondary education. Technical training usually emphasizes predictable, deterministic answers to questions. If they provided "creative" answers (or behaviors) in college during classes in basic physics, chemistry, or mathematics, it was probably neither appreciated nor rewarded. In fact, the end result could have been failure in basic sciences at that time. Unless they accepted the basic premises of a university technical education, they didn't get through. Then applying for a job required acting in certain acceptable ways, and finally joining the project team required behaviors that were defined by the project manual. Those behaviors were "organizationally acceptable" behaviors, that is, task assignments, and so forth.

When people know what is expected in a situation, the potential that they will exhibit behavior that matches this expectation is increased, provided they will gain something. Unfortunately, this same self-selection "filtering" process can often screen out the more creative, risk-taking behaviors that we need in many projects. It's almost a contradiction. As managers, we'd like to have predictable behavior but that predictable behavior should not stop creative or nonpredictable thinking! If projects are intended to solve novel problems, it's reasonable that novel answers might be needed. These novel answers can only spring from individual creativity that the self-selection process might have eliminated. It's almost a tautology. (As an example, you can't get a job without experience but the only way to get experience is to have a job. Or, you can't be creative until you have become a conformist.)

We have to design a People Management system that stimulates creativity and risk-taking behavior, even though this same behavior might not have been rewarded in prior training, jobs, and even tasks performed in the same organization. One way to change the organizational environment and encourage risk taking is by delegating maximum freedom and authority. We've partially done that in the other systems discussed by emphasizing the responsibility of the individual in the bottom-up forecast, responsibility accounting, estimate to complete, milestones as measurable and deliverable.

Creativity does not live well in an environment that rewards individual conformity and predictability. Our systems provide an integrative framework intended to support a team effort, but that team effort is also supposed to support individual risk taking. The individual team manager has almost complete freedom to act within the work package limits, and if the risk fails, as it sometimes does, the impact statement immediately requests assistance. That impact statement is *not* intended to be an admission of failure. Some organizations might penalize creativity by transferring or firing people who have not achieved predetermined (but not exactly predictable) goals. If failure to achieve for legitimate reasons or if risk taking is punished, sensible people will avoid these actions. But we must have creative risk takers in projects, since projects, by definition, are created to solve novel problems and often require novel solutions. Risk taking may itself be risky since sometimes the risk doesn't pan out. If the individual feels that failure is punished, he or she won't take risks. Thus there will be little individual creativity.

Conversely, if the individual feels that failure is regarded as a learning process, the little stress that may be incurred stimulates without paralyzing creativity. Some stress helps. Too much hurts. Humans react to excessive stress by preparing to fight or flee. When too much stress is perceived, the brain receives less oxygen because the blood supply is diverted to the muscles for either fighting or running away. But the amount of oxygen diverted varies with the person. Different people have different perceptions of the same situation and consequently feel different levels of stress. One popular way to decrease stress to acceptable levels and still get novel solutions is spreading it across a group. In other words, some managers appoint a committee. When committees are managed well, research shows that they can be very effective in developing creative solutions when the problems are unstructured or not well understood. This requires effective leadership. Conversely, when the problem is structured relatively well, individuals do better at problem solving than groups do. (Kolasa, 1969)

We have encouraged group creativity by using a bottom-up (more creative) approach to solutions to the poorly structured problems presented by a project. Now we have arrived at the point in the project systems design life cycle where the problem(s) are relatively well understood by the group. It is now the individuals who must be creative to solve them. That's logical, especially in projects since problems are sequential. For example, the overall technical specifications of the ''new'' turbine have been developed by the customer's committee and modified by the project team. The project forecast and estimate have been delivered and accepted by everyone. Now the particular configuration design problem that will exactly determine the size of various pumps, piping, and so forth is no longer handled by a committee. It moves into the special domain of the ''expert'' at the moment—the project engineer. Later, when the problem is solved technically and creatively, it may be approved by the team in the design review meeting.

Therefore the project manager must support individual creativity not only when it succeeds (obviously), but also if unpredictable failure occurs, especially when there has been a direct connection between creativity and personal commitment to the particular project goals. Creativity without individual commitment to accepted goals is relatively useless. Children are creative but they may not be able to design an innovative cooling system for a new turbine. Therefore, internal project goals are assigned to various responsible team members. These team members are unique individuals, and they require management systems that deal flexibly with individuals, their needs, their commitments, and their consequent creative inputs—in other words, their motivation. There is an increase in the number of variables because we're considering individuals. This makes this system more complex than the other project systems. This system must be carefully and delicately designed to support, protect, and encourage individual creative contributions within both the overall goals of the project and those of the parent organization. If success in the project doesn't reflect eventual success in the parent organization, it's reasonable to expect that there will be very little success in the project.

The People Management system is not an easy system to develop when you consider the different pressures on project participants. Their behaviors are affected both by the history that they have experienced and by their expectations of the future. Working on a project can induce stress such as role conflict from differences between functional managers and project managers (as in matrix operations) or from opposing demands of long-term goals versus short-term goals (e.g., ''Keep your technical competence updated by going off to be trained but also stay on the job and get that project done *now!*''). Fortunately, however, most People Management problems are also somewhat repetitive, and even though we are concerned with the wide variability in the responses and needs of individuals, the problems can be handled predictably in a manner that is acceptable to most project participants. The basic premise of individual differences means the system dealing with them must be applied equitably, but it doesn't have to treat everyone alike. The results can be different. We are not all alike. We have individual goals and can be rewarded differentially as we achieve them.

3.0 INTRODUCTION TO PEOPLE MANAGEMENT

We'll start our People Management system design as usual with definitions, then go on to explore some of the relevant theories about motivation. Then we'll connect motivation with commitment and productivity on the project. As before, we'll use a "first-person-singular" concept. Each project manager is unique, as are all project team managers. Therefore, this system must be more situationally dependent than the others. Since motivation partially depends on personal feedback and measurement, we'll also deal with performance appraisals and finally conclude with a further application of the dual reporting subsystem that is intended specifically for People Management systems in projects.

Motivation has been a fascinating subject for thousands of years. After all, what could be more interesting to any observer of human behavior than to think about why people do the things the way they do? Since scientists and philosophers have been considering this subject for quite some time, it's obvious that preliminary thinking about human motivation covers a much broader area than the one we are limited to: our work in projects. Originally, motivation was a major part of the way a person was supposed to live. It extended far beyond the work environment when first discussed in ancient Greece.

4.0 MOTIVATION: THE BEGINNINGS

Aristotle, who was perhaps the initial developer of the scientific method, defined *motivation* as the very basis of anyone's life. He believed that the aim (or motivation) of life was not goodness for its own sake, but happiness as defined by people themselves. (We have echoes of this in our present political documents, for example, life, liberty, and the *pursuit* of happiness.) The chief condition of that happiness is a life of reason that values clear judgment, self-control, and above all, logic. This condition is achieved through a "right" or middle way that fits a particular situation. It's the golden mean.

> "Right," then in ethics or conduct, is not different from "right" in mathematics or engineering; it means correct fit, what works best to the best result.
>
> The golden mean, however, is not, like the mathematical mean, an exact average of two precisely calculable extremes; it fluctuates with the collateral circumstances of each situation, and discovers itself only to mature and flexible reason. Excellence is an art won by training and habituation: we do not act rightly because we have virtue or

excellence, but we rather have these because we have acted rightly; these virtues are formed in man by his doing the actions; we are what we repeatedly do.'' (Durant, 1926, p.76)

In other words, our behavior guides our thinking, and it is correct when it is based in, or motivated by, a correct fit with the particular situation. Correct behavior trains us to have excellence, which, in turn, enables us to again act correctly in accordance with our evaluation of another situation. The behavior and the needs of the situation must match. Our evaluations are, however, generally guided by some mental golden mean that prevents the ''correctly motivated'' person from interpreting the environment in an extreme way and behaving in any excessive fashion.

Applying Aristotle's concept to projects, the guidelines produced in the project manual are supposed to be similar to a project-oriented ''golden mean.'' It is assumed by the team member that these guidelines were produced rationally and that they will work to everyone's self-interest. As applied to projects, the Forecasting, Measuring, and Project Operations systems will operate well because they define group behaviors within which the team member has agreed to work. We, therefore, have an overall ''behavioral perimeter'' somewhat similar to the ''cost perimeter'' developed for the project charter. Now we move toward a different view of the individual for comparison.

A more recent theorist, Sigmund Freud, gives us a different explanation of behavior. He assumes that a rational person is motivated by his or her own best interests, whatever they might be. Those interests are based on instinct and related to unconscious drives as primary motivators of behavior. *Instincts* are defined as inherited or innate psychophysical dispositions. Their related unconscious drives may not even reach consciousness. Of course, the problem with this idea is that completely rational people (if any actually exist) might have instincts, since they were born with them, but they would never have any unconscious drives. Therefore, complete rationality is almost a contradiction in terms. For Freudians, emotions are an important basis for motivation. Freud connected the individual's emotional state with the motivation to work, by viewing work as a way an individual develops and confirms his or her sense of self-esteem. (Kets deVries, 1984, p. 67) Although Freud made major contributions to psychology (and motivation theory), they are difficult to use because the search for instincts resulted in lists of hundreds, which were difficult to easily attach to any person. Some psychologists even began to question whether the unconscious motivations themselves were really unknown or whether they were somehow *learned* by the person. (Steers and Porter, 1979, p.10)

Frederick W. Taylor (1911) and his associates in the Scientific Management school defined motivation in accordance with several less complex assumptions about people. For example, Taylor assumed that work was inherently distasteful to most people and what they did at work was less important than what they earned doing it. Few people wanted creativity, self-direction or self-control in the work place. Therefore, this school of thinking maintained that management was responsible for the success of the organization through development of clearly established procedures and close supervision of people to ensure that those procedures were followed. If people at work did well, they were supposed to get paid well, using an incentive system. As a practical matter, however, incentives did not work constantly, because workers quickly perceived that as production rose, management changed the conditions of the work and decreased the incentives. To many of the workers, it was a "no-win" proposition. The decline in work output, then, became a self-fulfilling prophecy as far as management was concerned.

Scientific Management and its management philosophy was important because it was one of the first disciplined and systematic approaches that used motivation theory based on both aspects of human behavior and economics. It promised that there was the "one best way" to manage. If management followed the precepts of the theory, success would surely follow. If success didn't occur, it was because of the manager's failures, not the workers'. Therefore, the manager was really a key figure in this theory.

4.1 "NEED" THEORISTS

Abraham Maslow extended motivation theory by providing us with a basic "need" hierarchy (see Fig. 8-1). He proposed that people have a succession of needs, which must be satisfied in a predetermined manner. He suggested that there are five needs and that they occur in this sequence: physiology, safety (security), love (social), esteem, and self-actualization. (Maslow, 1943) A lower need had to be satisfied before the next higher need would emerge. That is, the need for food, water, air, precede the need for personal security. The need for personal security would then be dominant until it was satisfied, and the need for love (or some other being like you) would then occur. After this need was satisfied, then group esteem was desired. Finally, the self-actualization need would surface. Research findings that purported to test this theory obtained little support (Wahba and Bridwell, 1973) because the needs described are not uniformly applicable to everyone.

However, since management is an art, it can use part of a theory. It doesn't have to throw a whole theory out, just because part of it is invalid. Art can use pieces of theories very well. There are several concepts in this need theory that are

MASLOW'S NEED HIERARCHY

Figure 8-1. Maslow's Need Hierarchy

very useful. One is that a satisfied need (however defined) is no longer a motivator of behavior, and another is that there seems to be a succession of needs or accomplishments that people want. This succession may not be defined similarly for everyone, nor may it ever be completely satisfied for a particular individual. This theory, like Freud's, was based on clinical experience with individuals.

There are other "need" theories, such as McClellan's (1961) that proposes that each of us has needs for achievement, love, and affiliation. This theory was even expanded to include whole cultures and civilizations. Although the data supporting this theory were extensive, the theory had the same problem as Maslow's— individuality may not fit the theoretical mold.

Frederick Herzberg and associates developed a motivation theory that is also based in need theory, but it segregates those needs into two separate categories— those external to the person and those that are internal. They feel that there is a level of zero motivation that is attained through the satisfaction of externally caused needs. To reach that level, the individual must feel that typical working conditions such as the compensation levels, fringe benefits, management, and working facilities are adequate. Once those needs (called *hygiene needs*) are satisfied, the internal needs satisfied only by the work itself become important. These could be satisfaction, personal challenge, etc. (see Fig. 8-2).

In other words, according to Herzberg, who took much of his data from questionnaires administered to engineers and accountants, there is no way that any person will be motivated by the work unless all the other hygiene needs have been satisfied. Now if we put the theories of Maslow and Herzberg together, which probably neither researcher would agree with, maybe we can use them in our People Management system. By overlaying Herzberg's theory on Maslow's theory at the "esteem" level, we have a person who is physiologically satisfied, psychologically secure, and has at least one person who likes him or her. That would satisfy the hygiene needs. Therefore, we would now be concerned with how he or she regards the group and if there are personal needs that can be satisfied by the group (see Fig.

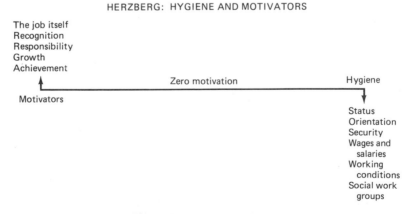

Figure 8-2. Herzberg's Theory

8-3). Group fit is defined by comparing the more general Project Operations system with the more specific People Management system. Therefore, we can move on to the individual and the attempt at self-actualization promised by the People Management system.

4.2 BEHAVIOR MODIFICATION

If we assume that the individual is goal-seeking and responds to rewards or penalties, a straightforward approach involving Behavior Modification theories could be a starting point. This approach involves presenting a predetermined stimulus to the team member (a potential reward or penalty), which is expected to result in some predetermined behavior. This is called *operant conditioning*. It requires the manager to do the following:

A. Clearly define both the behaviors that are desired and those that are not.

B. Provide continuous feedback or measurement to the individual against work goals.

C. Match the rewards or consequences offered for good performance against the behaviors exhibited by people.

D. Deliver different rewards (or penalties) according to different performance levels. (Steers & Porter, 1979)

As you can see, some of the ideas in Behavior Modification theories parallel the concepts in Scientific Management. That is, work should be measured continuously, with the results fed back to employees so that they can correct their errors and improve and with an incentive (or reinforcement) attached to better performance. The manager is the determiner of goals and the deliverer of the reinforcement or reward, and the workers determine the resources needed and respond accordingly.

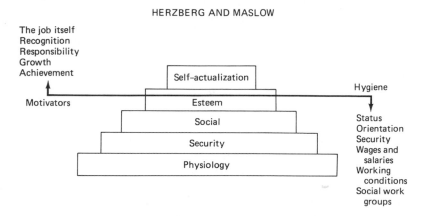

Figure 8-3. Theories of Herzberg and Maslow

But everybody does not respond linearly or automatically to rewards or penalties, especially when those rewards or penalties are interpreted through their own unique thinking processes. We all perceive and interpret the world differently, and those perceptions greatly affect our responses. In other words, reinforcement does not automatically condition behavior but affects it through and in conjunction with the individual's mental contents and processes. (Locke, 1977) Therefore, it's not as we might expect it to be. There are other theories that might help us further.

4.3 EQUITY THEORY OR TO WHOM SHALL I COMPARE MYSELF?

Let us assume that the individual's mental processes include a comparison of the rewards or penalties obtained versus those delivered to someone else whom the individual has selected as a criterion. When there is a perception of inequity according to this theory (something that would never occur in a Behavior Modification theory because a reward or penalty is considered the same for everyone), tension is created that is proportional to the amount of perceived inequity. That tension motivates or drives the person to reduce the inequity to acceptable levels. (There is always *some* tension necessary, otherwise the worker would suffer from a lack of stimulation.) The strength of the drive is directly related to the level of perceived inequity.

> Since theories or models of social processes are ways of making sense out of our environment by simplifying relationships between variables, it should not be surprising that any given theory fails to capture the complexity we know to exist in the real world. . . . It is possible that individuals who believe events that happen to them are under their control (internals) would have a greater propensity to attempt to reduce perceived inequity than individuals who believe events are largely beyond their control (externals). (Mowday, 1979, p. 135,140)

As the originator of the Equity theory, Adams (1965) suggested that some people can also cognitively restructure the situation as one way to reduce excessive tension. There could also be actions taken by the person to change the ratio of his or her work inputs to the outputs (i.e., value received). For example, "I didn't want that promotion anyhow." Other alternatives could be leaving the situation, quitting the job, or mentally changing the person who is being used as a criterion.

Equity theory is useful because it seems to predict the actual ways in which people are supposed to react, not only the ways in which we are supposed to view the world, and also how we interact with it. This theory provides several alternative methods of dealing with individual differences. Although we really can't tell how any specific individual values differential rewards when he or she compares them with the inputs required to get them, the theory predicts the several courses of action that the individual might be willing to take. But even here, we still have complexity because of the unknown motivational *sequence* of these various alternatives. In other words, which alternative will be taken first, or second, or third? This theory is very unusual because it is one of the few theories that predicts that the individual will not try to maximize values received, just optimize them based on a comparison with others.

People are complex, and no one of these theories can really entirely predict a specific individual's responses. Even though the project team is relatively small, we still have to build a composite theory that uses elements of many other theories to allow for individual differences. We will start with simplicity and then build complexity as hypotheses are tested and we learn more.

Therefore, a pragmatic beginning might be to start with an obvious fact: We can't do anything about the past. People's instincts and past experiences are beyond our scope. A motivation system has to be concerned with the present and with the individual's present expectation of the future. It shouldn't assume that everyone basically responds the same as described by Maslow, Herzberg, Freud, McClellan, or the others. It is malleable and applies to specific individuals, those in a small project group. It is based on Expectancy theory.

4.4 EXPECTANCY THEORY

Expectancy theory describes movement or a process rather than some kind of end goal or static condition. The majority of the other theories, in essence, are saying, "If such and such happens, then this is the probable result." Expectancy theory attempts to explain the relationships among variables as they change dynamically rather than as they might be at some end point. There are no fixed needs to satisfy as in the need theories, nor is there any golden "middle" that is attainable as in Aristotle's theory. There are no rewards or penalties to be valued and achieved either independently or in comparison with another as in Behavior Modification theory. The Expectancy theory (also called *Valence theory* because people are supposed to "value" a changing series of expected future rewards) assumes that individuals are largely autonomous beings who independently determine expectancies of the future, but they do it in the present. They also determine the instrumentalities or the methods by which they will achieve goals and valences (or values to themselves) for various types of behavior that they can exhibit.

This can be contrasted with Behavior Modification theory, which assumes individuals to be passive responders to forces that are recognized in the environment. Expectancy theory is primarily, but not entirely, cognitive. It deals with thinking beings who reason and anticipate connections between their actions in the present and some future event. The anticipation includes elements of emotion. Accordingly, it proposes that the closer the person sees a connection between some defined behavior *today* and some expected personal outcome related to the job in the *future* that results from that present activity, the higher the job performance. (Vroom, 1964) And the closer the connection between that job performance and some valued reward to that person (or the avoidance of some penalty), the higher that person's motivational force will be to perform that present behavior.

For example, an engineer completes the exhaustive life tests on her design of the new brake pads. The design resulted when she was able to overcome many design problems. The tests prove that she has a design breakthrough for the new brake product lines. Therefore, she might expect a very high performance rating for that job. She knows, of course, that she can expect to receive that rating in the *future* for the actions she completed *today*. And when there is a bonus, a special award, or some other kind of recognition that she values highly which will result from this higher performance rating, her motivational force is assumed to be high. That motivational force, then, is dependent on two sequentially multiplicative, dependent calculations. Assuming for example, (on an ordinal scale of 0 to 1) that the life test has subjective value to her which is very high, almost 1 (say, 0.9), she values the high performance rating also very highly (say, 0.8), and the bonus has the highest value of all to her (say, 1.0). Then, she has a final motivational force of

$$0.9 \times 0.8 \times 1.0 = 0.72$$

These figures, of course, are all subjective, but so is motivation. Now let's take the example where the life test still has the same intrinsic value and so does the high performance rating, but she doesn't think that there will be any relationship between job performance and a bonus, (that is, she believes there *might* be a bonus but the potential achievement, she thinks, is very low, not quite 0, say, 0.1). In other words, the tests themselves and the success of the performance are still highly valued by that engineer, but she doesn't believe that management cares or will connect successful performance with some kind of reward (it happens sometimes). The motivational force can then be calculated as follows:

$$0.9 \times 0.8 \times 0.1 = 0.072$$

and that's quite a difference. In this example, she wants or values the bonus, but doesn't think that anything will happen to gain her one even with the outstanding job she did on the brake pads. One interesting consequence in this theory is the linking of the person's valences (or internal values) with expectations of future consequences.

Let's consider an alternative scenario where she doesn't care one way or the other about any rewards that management can provide but is primarily concerned about her own satisfaction, which also sometimes happens. (You know, "work" is not the same thing for everybody and different people view it differently. One definition that I really like is . . . "Work is something that you do for somebody else's benefit. Otherwise it's really not, it's fun, and you even get paid to do it!") The figure for the bonus value is irrelevant here. We then might have

$$0.9 \times 0.8 = 0.72$$

This might apply for some people where the major reward is like a pin-ball game. If you win, you get a chance to get free-plays on the machine and you can play the game again. Putting it in project management terms, ". . . if you're good on this project, you get a chance to work on the next great project that we have coming down next month." Here her motivation might be to be able to work on the next project, rather than receive a monetary bonus.

4.4.1 Using Expectancy Theory

Applying Expectancy theory within our People Management system is fairly straightforward, since now we can use others' cognitive processes without having to delve into them. We'll let them handle those internal processes themselves as long as the expected behaviors occur. As managers, we can define motivation behaviorally and then develop procedures that provide goals that are valued by the person, according to Expectancy theory. Our definition of motivation is an inferred process that takes place in the person's mind but is observed only through behavior. That behavior, in turn, is interpreted by the observer. This assumes that all behavior is motivated by some mental process. Therefore, taking the definition a step further, motivation is the inferred selection of the behavior observed from those behaviors considered possible by the person being observed.

MOTIVATION DEFINITION

The inferred selection of the behavior observed from those behaviors considered possible.

In the People Management system, motivation is behavior that has been evaluated as such by the beholder, as Shakespeare also said about beauty. Another way of putting it is that the project manager observes some kind of behavior (or output as a result of that motivation) and interprets that behavior in accordance with his or her own thinking processes. By defining motivation this way, we have eliminated the requirement to enter into the thinking processes of the person being observed. We know it's there, but we're only concerned with our interpretation of its output. And we interpret one kind of motivated behavior as happening when an expected result is attained. In many ways, this is easy to work with because motivation is valued according to the behavior that the team or project manager thinks is the outward result. That also places the responsibility for goal setting where it belongs—on the team or project manager. There now has to be a definition of the end result that the desired behavior (i.e., beginning with motivation) should produce. Since we deal with independent individuals, those definitions must fit a thinking and responding person who is determining the appropriate behavior required to achieve project goals. This relates achievement to some standard of job performance. Then, there is another mental calculation of the probability of a connection between job performance and some personal gain (or avoidance of some penalty). In other words,

1. Behavior is determined by a combination of forces in the individual and in the environment (as perceived by the individual).
2. People make decisions about their own behavior in organizations.
3. Different people have different types of needs, desires and goals (values).

4. People make decisions among alternative plans of behavior based on their perceptions (expectancies) of the degree to which a given behavior will lead to desired outcomes.'' (Adapted from Nadler & Lawler, 1977)

In summary, a manager must clearly define what is wanted. A valued reward should be associated with a well-defined, measureable goal. Designing a People Management system using Expectancy theory as a basis is then relatively easy to conceptualize. There are three simple steps:

1. Outline the goals and how they will be objectively measured.
2. Determine the value of achieving that goal and the value to the goal setter, since that is the value that will be paid. Provide objective feedback to the other person during the goal-achievement process.
3. Pay off positively or negatively, dependent upon the result.

However, conceptualization and application are not the same. A successful application requires the following:

A. A one-on-one approach because each person is unique.
B. Completely open goal-setting and measuring processes.
C. The manager must have the authority and independence to reward or penalize.
D. Complete commitment on the worker's part toward achieving the goal and on the manager's part to support the process in any way possible (see Fig. 8-4).

The manager determines the specific goals that are wanted. Those goals have to be measureable some way. For example, lifting morale in the design group is *not* a measureable goal unless you can provide a way to test morale or connect output with morale or some other measurement. The measurement may be one such as meeting the three golden ''limits'' of the project—technical achievement, time, or cost. The person being motivated under this type of system then determines ''how'' to meet the goal, that is, what will be needed, when, and the cost, and the methods

---

SIMPLE MOTIVATION SYSTEM

---

1. Set up goals
2. Measure achievement against these goals
3. Pay off

Figure 8-4.   Simple motivation system

that are expected to be used to meet the goals. It's also partially an educational process for both parties. It educates the other person because it requires thinking through how a problem might be solved (that is, creativity), and it educates the manager who is being shown a new way to do it.

As a manager, the job is then to "test" the proposed solution, set up measurements, measure actual performance against goals, and finally pay off. The other systems are used to perform the test:

1. Using the Forecasting system, do a bottom-up forecasting against goals.
2. Using the Measuring system, compare against that forecast using milestones.
3. Using the Project Operations system, provide feedback (the estimate to complete).
4. Using a part of the People Management system, pay off after achievement.

It's a deceptively simple sequence to understand, but it can be difficult to implement since it requires a definition of the relationship between measureable performance and either reward or penalty. (I'm not suggesting that negative consequences be used equally with positive ones because the prospect of penalties seems to induce stress. Creativity suffers under too much stress.)

As project or team manager, the alternatives available for rewarding people may be very limited. Be sure they are known before setting up this aspect of the People Management system. Generally, these are some of the advantages and disadvantages in using this system. Some of the advantages are:

1. Goals are set, and they are measureable.
2. The manager is not directly involved in the unknown thinking processes of the team member.
3. Both positive and negative payoffs may be used (but I don't like negatives).
4. If it becomes clear that a goal cannot be reached or a target will be missed, the team member must bring that up immediately and show why an unforeseen factor is causing a problem, in order not to endanger the total reward for goal achievement. This means that the team or project manager may have to provide a partial payoff, even if no goal is achieved as forecasted.
5. It pressures both project manager and team member to agree on things that can be done and those that cannot.
6. It attaches achievement to behavior. The measurement of that achievement is interpreted the same way by the project manager and the team member.

The following are among the disadvantages

1. If the project manager is unable to provide either a reward or a penalty that is recognized by the team member, this won't work.
2. If the team member cannot provide or affect a measureable outcome for a

task, goals cannot be established. This may occur when there are many other factors that affect the goals that are beyond the control of the team member, such as when the task is to improve profits by 10 percent and the team member controls sales but not costs.

4.4.2 The Point Technique: Cafeteria Plans and Pin Ball Games

One answer is to set up a point system for projects. The points can be established for each predetermined goal. The manager then pays off in these points, according to achievement of a predetermined standard or goal. If you just change the points to money, you have a basic form of individually valued and control incentives. It's possible that points can be used for nonmonetary purposes. For example, 500 points might be worth an extra week's vacation, a private parking space, or even a bigger office in which to work for a month.

Very few appraisal systems are absolutely digital, that is, either the worker has made it or hasn't. The Points system is analog. It is analog because even if the specific goal is worth, say, 50 points, and the person was able to achieve only half the goal because of unexpected difficulties beyond his or her control, there is still a reward of 25 points. The person can ''spend'' those points however he or she wants, since each of them has different needs. This, of course, assumes that the valued reward for which the points are spent is available within the organization and redeemable for a given number of points.

It's almost like a cafeteria in which the buyer can pick and choose from a variety of foods, paying for them with his or her money. Points might be used to receive a money bonus or to receive other things, such as preferred special parking spots outside the office, extended vacations, extra time off, or even the home use of company equipment such as a computer. They could be used for that. Another nonexclusive alternative is that those with the highest amounts get another chance to ''play the game again.'' It is made clear that when recruiting team managers on another project, a high point winner is always a first choice. In any event, the end value of the points is in the participant's mind.

This idea has been used successfully in organizations where the project manager was not formally permitted to reward by recommendations for salary increases. One project manager ''required'' a particularly successful engineer to accompany a field trip of a vendor evaluation task force. Of course, the vendor was located in a particularly attractive vacation spot. Another manager distributed a semiannual memo outlining the superior achievements of the team participants who met their project-related goals. The memo, naturally, became part of the participant's personnel file.

5.0 APPRAISAL: PERFORMANCE MEASUREMENT— ''CONSTANT ERROR'' AND ''HALO'' EFFECTS

When the goals are absolutely clear and achievements equally apparent, appraisals become very objective. But clarity is a goal that is not achieved as often as we would like. There is often difficulty in defining a measureable objective itself or else one that fits into the project goals. Unforeseen problems may interfere in meeting goals, or human preferences may change. Managers have to make judgments on performance, a process called *management appraisal*. Management appraisal can be learned. The steps to follow are straightforward:

1. Focus on the problem, not the person (i.e., minimize personalizing).
2. Ask for the person's help and discuss ideas on how to solve the problem. Try to have both of you cooperate concerning the external problem.
3. Come to agreement on steps to be taken by each of you and how progress will be measured.
4. Plan specific follow-up dates.

No matter how straightforward or obvious expected behaviors are supposed to be, appraisals may be difficult even when carried out under the best of circumstances. Some of us carry mental "baggage" that relates appraisal processes to the grading processes of schools. In my opinion, they are not the same as schoolroom grading, since appraisals are much more interactive and helpful. But our appraisal processes can be affected by two noncognitive psychological processes that we may not be aware of. One is the "constant error" effect and the other is the "halo" effect.

The *"constant error"* effect involves applying a predetermined set or mold, irrespective of the situation. It is a predisposition of the appraiser to appraise everything in the same way. For example, consider an academic situation in which a perfect examination paper is given 99 because the instructor believes that no one is perfect. Therefore, there are no perfect scores. Or the appraiser may never give someone a zero, because no one is that bad.

The *"halo"* *effect* involves applying a particular characteristic in a specific situation to the entire appraisal process. It is a predisposition of the appraiser to allow one element to affect everything. For example, if the appraiser is a male with a beard and, unknowingly, has the idea that "all people with beards have a better job performance," assuming that all other things are equal, it is unlikely that female team members will ever receive as high scores as their bearded male coworkers.

Now these are unconscious psychological processes. If they were cognitive, they would be classified as prejudices. We know how to deal with prejudices, but dealing with things of which we are unaware can be difficult. The project manager can try to determine if these processes are happening by using a *forced rating system*. Although such a system has never been tested for validity or internal consistency, it is often helpful to expose one's own thinking processes. List the names of the project team vertically and list various characteristics horizontally. The characteristics themselves can be anything that the appraiser considers to be related to job performance. In Fig. 8-5, I have listed them as A, B, C, and D. Then, using an ordinal scale from 1 to 10, rate each team member on the basis of a particular characteristic. Repeat the process *for each characteristic* until all of them have been

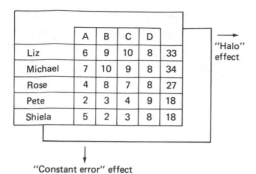

	A	B	C	D	
Liz	6	9	10	8	33
Michael	7	10	9	8	34
Rose	4	8	7	8	27
Pete	2	3	4	9	18
Shiela	5	2	3	8	18

→ "Halo" effect

↓
"Constant error" effect

Figure 8-5. "Halo" effect and "constant error" effect

completed. Then put the rating sheet with all the scores away for several weeks. After that time period, without looking at the scores for the prior period, repeat the process using a clean rating sheet. When this has been done several times, assemble the rating sheets and determine if any patterns emerge.

In my opinion, it's easier to determine if a "halo" effect exists than a "constant error" effect. When a particular team member has a very high or a very low score, a "halo" effect may be indicated. Is it possible that one characteristic of that team member is causing you to always give the individual a high rating? Or, conversely, you may just have a very good or a very poor (depending upon the score) team member. The "halo" effect may become apparent by comparing scores of individuals over time. The "constant error" effect is a bit more difficult to detect. If no one gets a score close to either 10 or a 1 on any characteristic, you may either have a "constant error" tendency or a group of mediocre people. Unfortunately, there's no way that I know to definitely determine if either a "halo" effect or a "constant error" effect is operating in one's appraisals. But using this method of forced rating on artificial characteristics and then comparing your scores over time may provide clues that can sensitize you to any potential problem.

6.0 DUAL REPORTING: MULTIPLE APPRAISALS

The complex design problems in developing an effective People Management system become very apparent when the system has to operate in a matrix environment in which people are assigned to many projects simultaneously and spend limited time with any one project. The complexity is due to increased potential for role conflict when many project managers interact with the functional manager of one person. A clearly defined project structure and a correct reporting level will often tend to minimize role conflicts, but often not entirely. Therefore, the People Management system that is concerned with individual goal setting, appraisals, and consequent rewards must interact with functional systems that are not very similar. *Dual reporting* is one answer to solving the problems in this type of situation.

To briefly review, there are at least two management axes interacting in a pro-

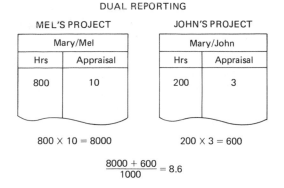

DUAL REPORTING

MEL'S PROJECT   JOHN'S PROJECT

Mary/Mel	
Hrs	Appraisal
800	10

Mary/John	
Hrs	Appraisal
200	3

800 × 10 = 8000   200 × 3 = 600

$$\frac{8000 + 600}{1000} = 8.6$$

Figure 8-6. Dual reporting

ject (or matrix) organization: the *functional* or *vertical axis,* which is concerned with ''supplying'' competent personnel to the project and handling administration and training, and the *project* or *horizontal axis,* which is concerned with ''using'' these personnel optimally and dealing with goal setting, measurement, and job performance appraisals. In order to simplify matters, I use the term *dual reporting* to describe this type of situation. There might even be more than two axes, such as when there are multiple matrices using the same personnel on a part-time basis. (This was very briefly covered in Section 7 of the internal charter when there was a provision for performance reviews to be done by the team or project manager. In effect, each team or project manager who has team members reporting to him or her was given the responsibility to appraise the job performance of that team member.)

Therefore, let's use the simple motivation system as I have previously outlined it:

1. Set up goals.
2. Measure achievements against those goals.
3. Pay off when achievements occur.

and develop a dual reporting format that allows us to use it in a matrix situation where the team members work only part time.

We have already developed the goals during the bottom-up session, and we are now ready to use those goals as a measuring tool. We have even set up the point system. For example, let's assume that I am appraising Mary Engineer. I have defined her project-oriented goals, and she has come back with a forecast of how she will reach those goals. Then both of us have determined how it will be measured. She then works on the project. Now there are several alternative appraisal tools that I can use.

1. *Ordinal scale:* Assuming a straightforward ordinal appraisal scale, at the end of the task I provide Mary Engineer with my evaluation. She has done a fan-

tastic job and has been rated as a 10. She also has worked on another project, but only got a 3 there. Both scores are forwarded to her functional boss for consequent rewards. Her total rating is then 13 out of a possible 20.

2. *Ordinal scale and time:* She received her 10 and that was multiplied by the amount of time spent on the project. For example, out of the last six months or 1,000 hours, she spent 800 hours on my project and 200 on John's. Her 10 on my project is multiplied by 800. Her 3 on the other project was multiplied by the 200 hours spent there. Her total rating is now $10 \times 800 + 3 \times 200 = 8,600$ divided by the total of 1,000 hours worked. Mary receives a weighted score of 8.6.

3. *Points:* Mary received 85 points from me and 45 points from her other project manager. She can therefore use these "points" anyway she likes, provided the rewards are in the functional manager's authority to deliver.

4. *Points and time:* This measuring system is the same as alternative 3, *weighted* or multiplied by the proportionate amount of time spent, as in alternative 2.

I think the last alternative is probably the best because it accounts both for the complexity of the job performance (using points) and the time spent on a project. Any of these alternatives (and there are probably others) can be used in dual reporting. The advantages are that the criteria for successful job performance are predetermined and both the project participant and the project manager are required to make a clear connection between achievement on the project and job performance and then between job performance and rewards.

Dual reporting can be easily implemented by the project manager. Just incorporate the concept into the project charter. Whether the first evaluations sent out are positive or not, both the person being appraised and his or her functional manager will find it difficult to ignore them. But even if they are disregarded by the appropriate functional manager, the team or project manager can always issue an impact statement if the person's performance is unsatisfactory. Since impact statements involve higher costs when there is insufficient cooperation from the functional organization, they will certainly get upper management's attention. It's better to get it in the beginning before a complete disaster hits. Sometimes just the appearance of dual reporting is sufficient to gain your ends.

In many ways, evidence that something is working is probably less important than the belief that it is working. For it is the belief that sustains the activity, not the evidence. Evidence may serve to strengthen the conviction that something is working; still it is the belief that sustains the pursuit. . . . The important thing is to do something and to have everyone affected believe that what is being done is the right thing to do. As Max Weber suggested in his analysis of bureaucracy, the appearance that something is right and proper to do legitimates doing it and sustains the efforts. (Salancik, 1982, p. 216)

7.0 SUMMARY

This chapter covered concepts and techniques for the fourth system—People Management—and the techniques to build personal motivation and commitment. The conventional motivational theories that we have briefly reviewed provided insufficient flexibility to account for individuality. It is the individual who provides the creative spark needed to solve the specific problems that a project faces. The group structure provides overall guidance, and the individual's original technical training can provide the logical, consistent cognitive process. But the situation needs more than that. It needs originality and commitment. Therefore, any motivational system must contain techniques that support risk taking. A future-oriented, Expectancy theory–based system is very appropriate. The end goals are determined by the manager, and the team member determines the resources needed and how to get it done. There must be a direct connection between achievement (or even "positive" failure if a feasibility study is being completed) and a valued reward.

In some ways, the technique requires both the manager and the team member to cooperate in defining and measuring performance. It also requires the team manager to define personal values and determine if they are related to the job performance. It can be difficult or very gratifying. It's probably one of the better ways to train one's self in human relations. And that training filters down.

> One of the most frequent requests that come to psychologists from top management is that psychologists train first-level supervisors in human relations. Rarely does top management understand that the way in which these supervisors handle people is a reflection of the way in which they and their subordinates behave. (Yankelovich, 1978, p.83)

The People Management system that I've outlined in this chapter influences both the manager and the team member to interact openly. When that interaction results in positive behavior for both of them, it not only improves the performance of the team members but also the managerial skills of the team manager. It's practically a no-lose situation for everyone.

In the next chapter, we begin to develop the most difficult yet the most flexible, but surely the most personally profitable system of all—your own personal style as a manager.

My Suggested Answers to the Case Study

1. Determination of the project objectives is definitely a management prerogative but those objectives should be firmly defined during the first top-down estimate. One way to minimize changes is to treat all changes as if they had been imposed by a customer. That means that *all* requests should be costed

by the project team and then *funded* by the requester before implementation. While that may be the policy for external customers, it is not usually done for inside customers. With this new policy, every change that is imposed will cost someone. The amount of engineering productivity can then be compared against the original task plus the changes.

It is definitely desirable because the requester of the change has to provide the time or funds *before* any work is done. When there is a "trigger clause" on the change, the project isn't held up while someone makes up his mind.

2. This present situation should decrease motivation if it is unchanged. Multiple design modifications usually slow up productivity. This affects the appraisal process since the "target" keeps moving. How can achievement be measured this way? Without improvement, the situation causes lessened opportunity for personal growth, which would probably decrease the motivation to produce.

3. One technique might be to assign a "value" to each work package. Each person working on that package is entitled to receive the specific reward if the package is completed either on time or ahead of schedule. The "value" may be anything, for example, 60 points for task A, which is to be completed in six months. If the task is completed a month sooner, you get 70 points; a month later, 50 points, and so forth.

The number of points achieved may be dependent upon achievement, time, or some other criterion. The engineer will get part of those points even if the work package is not completed because of changes imposed by others. Then, management will have to recognize the cost of any change, because they will be paying for partially completed tasks.

Those points can then be translated into a raise, extra vacation, benefits, a private parking space, or whatever, dependent upon the desires of the person "spending" those points. It's almost like getting a bonus.

BIBLIOGRAPHY

DURANT, WILL, *The Story of Philosophy*, (35th printing). New York: Pocket Books, 1926.

MOWDAY, RICHARD T., "Equity Theory Predictions of Behavior in Organizations," in *Motivation and Work Behavior* (2nd ed.), eds. Richard M. Steers, and Lyman W. Porter. New York: McGraw-Hill, 1979.

NADLER, DAVID A., and EDWARD E. LAWLER III, *"Motivation: A Diagnostic Approach,"* in *Perspectives on Behavior in Organizations*, eds. J. Richard Hackman, Edward E. Lawler III, and Lyman W. Porter. New York: McGraw-Hill, 1977.

SALANCIK, GERALD R., "Commitment Is Too Easy!" in *Readings in Management of Innovation*, eds. Michael L. Tushman, and William L. Moore. Marshfield, Mass.: Pitman Pub., & Ballinger Pub. Co. 1982.

YANKELOVICH, DANIEL, "The New Psychological Contracts at Work," *Psychology Today*, May 1978, pp. 47–83.

OTHER READINGS

ADAMS, J. S., "Inequity in Social Exchange," in *Advances in Experimental Social Psychology* (vol. 2), ed. L. Berkowitz. New York: Academic Press, 1965.

BRUNER, JEROME S., "The Conditions of Creativity," in *On Knowing: Essays for the Left Hand,* Cambridge, Mass.: Belknap Press of Harvard University, 1962.

CORNISH, EDWARD, *The Study of the Future.* Washington, D.C.: World Future Society, 1977.

COTTLE, THOMAS J., "The Mosaic of Creativity," *Confrontation: Psychology and the Problems of Today,* ed. Michael Wertheimer. Glenview, Ill.: Scott, Foresman, 1970.

KETS DEVRIES, MANFRED F. R., ED., *The Irrational Executive—Psychoanalytic Studies in Management.* New York: International Universities Press, 1984.

GALBRAITH, JAY R., "Organization Design: An Information Processing View." *Organization Planning: Cases and Concepts,* eds. Jay Lorsch, and Paul Lawrence and others. Homewood, Ill.: Richard D. Irwin, 1972.

GREEN, THAD B., AND OTHERS, "A Survey of the Applications of Quantitative Techniques to Production/Operations Management in Large Corporations," in *Proceedings of the 36th Annual Meeting of the Academy of Management,* Kansas City, Mo., 1976, pp. 202–206.

HARRISON, F. L., Advanced Project Management. New York: John Wiley, 1981.

HERZBERG, FREDERICK, BERNARD MAUSNER, and BARBARA BLOCH SNYDERMAN, *The Motivation to Work.* New York: John Wiley, 1959.

HUNT, PEARSON, ED. "The Fallacy of the One Big Brain," in *Harvard Business Review: On Human Relations.* New York: Harper & Row, Pub., 1979.

KOLASA, B. J., *Introduction to Behavioral Sciences for Business.* New York: John Wiley, 1969.

LAWRENCE, PAUL R., HARVEY F. KOLODNY, and STANLEY M. DAVIS, "The Human Side of the Matrix," in *Organizational Dynamics.* New York: AMACOM, American Management Associations, Summer 1977.

LOCKE, EDWIN A., "The Myths of Behavior Mod in Organizations," *Academy of Management Review,* vol. 2 (1977), 543–53.

LOYE, DAVID, "The Forecasting Mind," *The Futurist,* June 1979, pp. 173–77.

MCCLELLAN, DAVID CLARENCE, *The Achieving Society.* New York: Van Nostrand, 1961.

MARCH, JAMES G., and HERBERT A. SIMON, *Organizations.* New York: John Wiley, 1958.

MARQUIS, DONALD G., "Ways of Organizing Projects," *Innovation,* Project Management Publication, 3, American Institute of Industrial Engineers, 1969.

MARTIN, CHARLES C., *Project Management: How to Make It Work.* New York: American Management Associations, 1976.

MASLOW, ABRAHAM, "A Theory of Human Motivation," *Psychological Review,* vol. 50 (1943), 370–96.

MILLER, DAVID W., and MARTIN K. STARR, *Executive Decisions and Operations Research* (2nd ed.). Englewood Cliffs, N.J.: Prentice-Hall, 1969.

ROBBINS, STEPHEN B., "Reconciling Management Theory with Management Practice," *Business Horizons,* February 1977, pp. 38–47.

SCANLON, BURT K., "Philosophy and Climate of the Organization," *Vectors Magazine,* 4, no. 5, (September-October 1969), S.M.E. Dearborn, Mich.

STEERS, RICHARD M., and LYMAN W. PORTER, *Motivation and Work Behavior* (2nd ed.) New York: McGraw-Hill, 1979.

TAYLOR, FREDERICK WINSLOW, *Scientific Management.* New York: Harper & Row, Pub., 1911.

VROOM, VICTOR, *Work and Motivation.* New York: John Wiley, 1964.

WAHBA, MAHMOUD A., and LAWRENCE G. BRIDWELL, in "Maslow Reconsidered: A Review of Research on the Need Hierarchy Theory," *Proceedings of 33rd Annual Meeting of Academy of Management,* 1973, pp. 514–20.

WOODS, DONALD H., "Improving Estimates That Involve Uncertainty," *Harvard Business Review,* July-August 1966, pp. 91–98.

Part III   The Fifth System: Personal Style

PERSONAL MANAGEMENT STYLE: THE INDIVIDUAL SYSTEM

PEOPLE MANAGEMENT: THE PROJECT SYSTEM

PROJECT OPERATIONS: THE PROJECT SYSTEM

MEASURING: THE ORGANIZATIONAL SYSTEM

FORECASTING: THE ORGANIZATIONAL SYSTEM

FOUNDATION: PROJECT MANAGER'S AUTHORITY

1.0 PERSONAL STYLE AND CHANGING YOURSELF

The last and, in many ways, most difficult system to develop is one's own management style. It is difficult because we are redesigning our own behavior. There is less predictability in solving any potential problems and usually less objectivity applied in solving them. It requires a lot of personal discipline to optimize one's own behavior. That personal behavioral "style" that will be covered now typically includes delegation, leadership, conflict resolution, implementing change, handling stress, and ethics.

2.0 DELEGATION AND CONFUSION

By definition, every manager must delegate. And unless the delegation is carried on logically and consistently, it can quickly deteriorate into confusion. There has to be a rational technique used for the process of delegation.

But just flatly stating that delegation is necessary doesn't explain the reasons behind this process. Those reasons are based primarily in our limited mental capacity to process and act on information. As an example, let's assume that the manager has only one person reporting to him and that individual is processing approximatey the same quantity of information as the manager is. It is impossible for the manager to understand everything that the individual is doing *in addition* to understanding his or her own job. There isn't enough information processing capacity in that manager's organic computer that he carries on his shoulders. Our brain's ability to process data cannot handle the load. The best that we can do is to "sample" the flow, and then allow someone else to act. We can't be everywhere at once. But allowing someone else to act means that limits must set upon those actions. The limits define the areas within which the other person has the freedom to act. Simplistically put, defining limits means delegation. The example grows as we add more subordinates.

Now that we know the reasons "why", the next step is to define the "how". Delegation cannot be a random process. No two delegations are exactly alike, because no two situations are exactly alike; therefore each delegation requires a careful delineation of the actions that the person being delegated to can or cannot take. That delineation comes from an analysis of the relationship of the work or potential task that is to be done and the resources available to do it. Each of us will have a slightly different process of delegation that reflects his or her own personal "style". Developing that "style" is one topic of the next system.

Delegation and leadership are closely related. Both are interactive exchanges between the leader and project team participants. Those exchanges are affected by three typical ongoing *sources* of confusion and conflict.

1. Those within the project caused by opposing role requirements imposed on individual team members by both their project and their functional managers. (Some of that has been resolved in the third system, Project Operations.)

2. Those between the project manager and functional managers caused by the differences in the authority level of the project manager vs. the functional manager. (Some was resolved in the development of the initial project charter, when the project size determines the reporting level of the project manager.)

3. Those occurring because of changes in levels of uncertainty over time. (Some of this was resolved through more frequent reporting at the project beginning and design of the various meeting formats during the "bottom-up" project initiation meeting.)

Projects are "new" organizations and these endemic sources of confusion and conflict raise an obvious need for effective delegation, conflict resolution and leadership criteria. These change with the project life cycle.

3.0 CULTURE AND POLITICS

Projects are small "cultures" and they require fast and effective management techniques for causing acceptable behaviors that fit the culture or the needs of the situation. Some are formal, such as Management Training and Organization Development and are managed by the functional organization. Some are not so formal. These include learning the "politics" of the way things are done around here. They can be important if the project is to succeed within a parent company. Therefore, they are part of the management "style." But basic thinking patterns are difficult to change. They have been functioning for the person for years. However, when there are obvious misalignments between behavior and the needs of the situation, behavior can be adjusted with various motivational techniques. That adjustment process improves management "style."

4.0 STRESS, OPPORTUNITIES, AND ETHICS

Starting and managing projects involves absorbing varying levels of uncertainty and solving unforeseen problems. This can cause stress. Recognizing the signs of personal stress and learning how to deal with them is as important as any other of the elements of personal management style.

Finally, at some time in the project life cycle, all of us have to make decisions that are based in our own personal ethics. Those decisions are really the most crucial ones. Ethical decisions must be made because of the nature of the projects themselves. Just like physicians who "practice" medicine, engineers and scientists are "practicing" engineering when working on projects. There are always tradeoffs to be made and those tradeoffs are often based on opinion without complete and objective supporting data. When uncertainty is high, decisions are primarily based on informed experience. That experience or opinion depends, to a great extent, on how we view the world. For example, what *safety factor* do you use in calculating the loads in a structure? If the data for the materials strength is noted in your design handbook, why use a safety factor at all? On a more applicable note, even when the project ends, uncertainty never drops to "zero" because there are always a few things that we're not sure of. Professional opinion is always eventually rooted in ethics. That's what is meant by "practicing" a profession.

Increasingly, society holds us *personally* responsible for our decisions. That is similar to a "medical" model of competency to practice, e.g., "Have you made that decision in accordance with the best practices of your co-professionals?" That personal responsibility applies to professional employees even though we have made those decisions for the benefit of our employers. If some regulation or law has been violated, lack of personal gain is no longer a factor in your defense; only personal competency is. We will conclude this section of the book with some processes and techniques that will support improvements in our "practicing" decisions.

5.0 SUMMARY

Personal "style" is the most variable project system since it deals with the most numerous, least tractable, but yet very familiar kinds of variables, those that affect ourselves. It is probably the most rewarding system to develop since improving one's management style is often both personally gratifying and a prerequisite to eventual success in an organization. And success is quite portable. While it may not fit as well next time, achieving it the first time shows everyone that you can do it. It gets easier the next time.

Chapter 9 Personal Management Style: The Fifth System

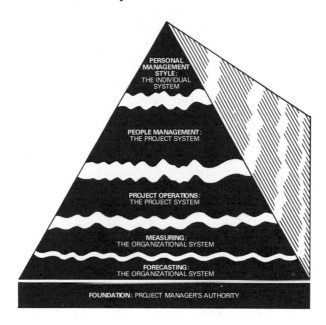

THE CASE OF THE DIFFICULT PEOPLE

Cast

May Brenner, Project Manager, Hydropack Project
Bill Weston, Chief of Project Management
Ira Zimmerman, Chief Engineer
Carol Horstmann, Manufacturing Supervisor

May Brenner was beginning to regret her decision to accept the job of project manager on the Hydropack Project. She sat in her office and mentally reviewed the situation. Several weeks ago, her former boss, Ira, mentioned to her that she was doing very well as an engineer but there was a possible promotion to project manager available. Did she think that she could handle it? She said, yes, and with Ira's support, she put in her application. She was interviewed by Bill Weston, accepted, and assigned as the project manager for the Hydropack Project. The project was to design a revolutionary kind of road-building machine that controlled the thickness of hot asphalt as it is laid down on city streets. For the first few days, she was busy becoming familiar with the customer's specifications. The preliminary feasibility study had just been completed. The results of a tensile test of the steel used in the machine mainframes were missing, so May scheduled a meeting with Carol, in manufacturing, to get new specimens made. The meeting was not satisfactory.

She had asked Carol for a delivery schedule to machine different specimens that were needed to confirm the tensile test. When Carol asked May when the specimens were needed, May said, "As soon as possible." Carol promised that they would be ready in a week. May really didn't believe Carol, but decided not to press the matter and the meeting ended. Carol did not deliver the specimens on time, and May had to phone her and remind her to finish them in the following week.

There was another problem that bothered May. She had been through several training sessions on managing projects so she felt quite confident as she developed her preliminary charter and selected her people. She then called her new boss, Bill Weston, and requested that he review it. During that phone conversation, he told her that he was too busy at the moment but to go ahead anyhow. May then recruited her team, had them revise the charter, and then brought the final revisions to Bill. He had to approve them before submitting them to top management. Several days later, Bill asked her to come into his office. Promptly on time, May showed up at Bill's office.

Bill: May, I really don't know what this stuff is that you brought me. It doesn't make any sense; it doesn't seem to have any relationship among the various parts; and I don't understand how you're going to get this project accomplished because those plans are just not satisfactory. I can't submit them to top management; they'd laugh me right out of the room. It looks just like some kind of classroom exercise. Let's be practical. You can't do that here.

May: Well, I followed the procedures outlined in the company project manual. Did I leave something out? All the people on my team seemed to have agreed with them. What's wrong?

Bill: Look, you came here highly recommended; you've done a good job for the company for several years now; and your former boss, Ira, says that you can solve problems. I took his recommendation and you seemed to be a logical person during the interview. I'm sure that you can figure out what has to be done to fix up this project. By the way, you'd better get it fixed by the end of the week, because I just don't have the time to do your job and mine. This meeting is over.

May quickly walked back to her office with the project charter and the other files that Bill had shoved at her. After spending most of the day fruitlessly reviewing them in an attempt to find out what was wrong with them, she walked over to Ira's office and was relieved to find that he was alone.

May: Hi, can I come in? I need a lot of help but what I really think I need is a transfer back here to my nice, secure, engineering job.

Ira invited her to come in and sit down. He shut his door to show that he was not to be disturbed; then he slowly walked back to his chair, sat in it, and regarded her very gravely. May then reviewed what had happened that day. After a few moments, he spoke.

Ira: May, designing is much easier than managing projects. In Engineering, when you follow Mother Nature's rules, you usually get a reasonable answer. Human beings often lack the consistent logic of Mother Nature.

Our management world is only partly a logical place. It is also an emotional one. There is even a greater emotional demand placed on project managers than there is on functional managers, because the level of uncertainty on projects varies much more than it does for functions. To be successful as a project manager, you've got to be a lot more concerned with the differences between and the needs of the people with whom you work. Now, I'm not saying that functional managers don't have similar problems. They do, but the magnitude of the uncertainty in functions is usually more predictable. Therefore, functions can usually provide a firmer, supportive training ground for people. Many projects don't have the time or the resources to do that. The familiar environment of the Engineering department has been replaced with the novel and less forgiving environment of the project team. You're not only the engineering manager now, you're also a general problem solver and everything else. It's not an easy job.

May: Ira, I know that what you're saying is right, and I've had a few problems dealing with some of my project people, but I'm dealing with a boss who criticizes and rejects everything that I do. He doesn't even tell me what's wrong; he just rejects the whole thing. I agree with you about training because he doesn't seem to be providing any.

Additionally, I'm dealing with Carol Horstmann, who is turning into another problem. When I was in the Engineering department, I would deliver the drawings to her using the correct engineering drawing release forms, she would then complete the production routing sheets, and the parts would be made on the schedule that she had forecasted. That schedule was usually O.K. because we seemed to have more time when I was in the Engineering department. We don't have as much leeway in our projects.

For the last several days as project manager, I was trying to understand my project and get control of it. I couldn't take the time to complete all the forms, and so I asked her informally for a delivery schedule. She would always answer by asking me what delivery I wanted, and after I told her what we needed, she would agree with it, but then she was invariably late. So I started to call her on the phone, just before they were due and ask if the delivery would be made on time. She would always say, yes, but it wouldn't happen, and she always had some kind of a reason for it. I put up with it, but I feel sure that when Bill finds out about it, he will blame me for the late deliveries. I'm at my wit's end.

Ira: May, remember that when you asked for delivery schedules before as a member of my department, Carol would give you a schedule that often took eight to ten weeks. What did you do then?

May: Well, nothing, because I wasn't in a hurry for the parts, but now I am. Why can't she give me a straight answer now? If I don't like her delivery schedule, I could always go to an outside contractor. Of course, I've been told that it's best to have things made inside and that's why I've gone along with her. With the situation the way it is, Carol's broken promises and Bill's screams are driving me up the wall. What should I do?

Questions

1. Are these problems typical and how would you define them?
2. How does the present situation really differ from when May was in the Engineering department?

3. How should she handle the problem with Carol? Why?
4. How should May attempt to resolve the problem with Bill? Why?
5. Is this case study a familiar one to you? What happened in your situation?

1.0 INTRODUCTION: STYLE

At this point, it might be reasonable to review the descriptions and applications of the five project management systems in order to see more clearly how these systems apply progressively from the "general" area of the company through the "more specific" area of the project to the "most specific" area of the project manager.

The first two systems, Forecasting and Measuring, can apply to the whole company.

The third system, Project Operations, applies to the project group.

The fourth system, People Management encourages personal commitment and positive motivation for people on the project.

The fifth system, Personal Style, is the day-to-day behaviors of the project manager that are intended to deal with unusual, least predictable situations.

In other words, it is first the company and then the project group, followed by others as individuals, and finally yourself. This last system of Personal Style provides guidelines for the project manager to use in solving the problems that fall "between the cracks" or into the interfaces among the various team managers. No repetitive management system is prepared to handle these completely, since it requires a personal flexibility and responsiveness to nonpredictable problems. Therefore, the fifth system is the most interactive of them all. It can encourage or discourage human creativity and motivation and build or destroy the integration of a project team. One part of this fifth system, Personal Style, is the application of daily tactics within an overall strategy. Strategy is the general or accepted ways of dealing with variances, those variances that are the important differences between forecast and measurement. Another part is the specific methods that one chooses within an overall behavioral scheme to solve nonforecasted problems or make nonprogrammed decisions. Style depends upon how the individual perceives and interprets the situation. It relates primarily to human interactions such as behavioral norms and expectations. It doesn't operate in an organizational vacuum. Getting along with the people in the company and getting what one wants are directly related. It reflects the politics of the situation and those politics are facts of organizational life. Understanding strategy and tactics, and perceiving and correctly interpreting the organizational situations is basic to developing the best personal management style.

1.1 STYLE: MORE DEFINITION

Basically, a useful but simplified definition of *style* is behavior. It's the observable actions or outward results of how we relate to our external reality and to our internal dispositions. We interpret both, that external reality and our internal dispositions too. Others interpret our behavior according to their criteria, not ours. Our external reality is interpreted through conscious (or cognitive) and unconscious processes. Our internal dispositions begin in our biology. External reality, therefore, is always partially subjective. When our behavior is able to be generalized over many different kinds of situations and is interpreted by others as consistent and predictable over time, it is sometimes called *personality,* rather than *style.* This is a misnomer, of course, since personality involves other factors such as the unobservable thinking processes that affect our internal disposition. Our minds do interact with our bodies. These processes cannot be seen; they can only be inferred since our unconscious processes are not accessible. Inferences are interpretations. By confining our discussions to the visible behavioral part of the personality called *style,* we'll try to minimize interpreting. But we understand that it can never be entirely eliminated since observable behavior is always interpretable.

Style starts with biology as modified by time. In other words, it starts with our mental and physical attributes, shaped by past experiences. Since we can't modify our biology nor the past, we must deal with ourselves in the present situation and our expectations of the future. In management, those perceived present situations and expectations of the future are primarily modified by our relationships with other people in the work environment. Somewhat arbitrarily, we can divide those relationships into those concerned with others

1. in the external parent organization
2. in the project

It's arbitrary because some people can be part of both situations, but this classification makes explanations slightly easier. When present and future relationships are relatively predictable, they are a major source of organizational "politics." In this case, predictability means that specific organizational signals will probably result in some expected human behavior.

1.2 STYLE: THE EXTERNAL POLITICS

Every organizational environment provides different signals to the people in them. These often define the limits of acceptable management behavior. Being able to read, understand, and respond appropriately to these signals is obviously very important. Not all signals are the same. There are differences among organizational levels and between functional and project managers, even when they are at the same level. For example, the vice president of Engineering responds to different inputs than the project engineer. That is a difference in organizational level. When the project engineer responds to different inputs than the standards engineer, that is a difference in jobs at the same level.

As a general rule, the higher a person rises organizationally, the more his or her success depends on the ability to read and understand the company's political signals. Usually entry or beginning levels require technical skills, and the newcomer's style only needs to indicate promise. At middle levels, interpersonal skills (getting things done with peers, at meetings, and in relating effectively to superiors and subordinates) become increasingly dominant. At the top levels, neither technical nor interpersonal skills are sufficient by themselves. There must be a display of consistency with the total political set. That set is sometimes called the *organizational culture*. (Nichols, 1971, p.134). When promotion occurs, the signals that initially were responded to are no longer applicable. In fact, responding to those past signals can be a major fault in the newly promoted manager. It is now a different job. Promotions usually depend upon how well your personal style matches the signals and requirements of the prior organizational level. Success in a new job depends upon how quickly one stops responding to the previous signals and starts responding to new ones.

Fortunately, the signals are often obvious. Many existing political relationships among different functional management jobs are ingrained and survive a change in managers very well. For example, the (functional) manager of Purchasing who always insists on approving all purchase orders no matter who was the originator of the original requisition for materials will make it quite clear to the new chief engineer what are accepted behaviors. The new chief engineer will be informed about it and therefore will be expected to place all purchasing requirements through the manager of Purchasing, even if this was not part of his or her old style at the previous job (or company).

In comparison to the relative stability of functional signals, the project manager has a more difficult task modifying personal style to respond to the new political signals. There are fewer standard inputs. Projects are usually new entities that establish a somewhat different organizational structure each time. Therefore, there is less past history and fewer clearly articulated organizational signals. And when projects cross functional organizational boundaries (as many do), there is an inherent political conflict that the parent organization has difficulty resolving. Nobody knows *exactly* what to do even if there have been other projects, since it's a novel situation with a new management team in place. The signals are not very clear, and even when they are clearly established (by using the charter and the project systems) and accepted initially, the project itself may change. The level of uncertainty changes, the project tasks may change, different team managers may be at center stage, and even the project manager may be changed. Obviously, the politics will also change.

1.3 STYLE: THE DIFFICULTY IN ADAPTING TO THE ORGANIZATION

Project managers, therefore, have to be able to use a more flexible management style than equivalent functional managers use. For example, a yearly appraisal and performance review may be acceptable for the slower functional groups, but entirely inappropriate for projects, especially if the projects last less than a year. The organization's signals to the various functions may dictate yearly reviews, but the effective project manager performs dual reporting on a shorter time cycle. There is a possible clash of styles. The organization prefers standardization, but the project requires individuality. It's a very delicate situation, and, as in this example, the optimum project management style for the problem at hand might seem to occasionally be in direct conflict with the parent organization's signals to the rest of the company.

Facing these differences in signals in the beginning of the project through clarifying discussions with top management and following through with the design of project systems helps. One part of an appropriate personal management style is a correct interpretation of the signals received and an attempt to minimize the differences in signals as much as possible. In any event, however, the primary signals should be those of the project. If this is *not* so, the project will not be primarily directed according to the three "golden" goals—maximum technical achievement within minimal time and cost. It will have other goals that are probably not under the control of the project manager. If you recall Chapter 1, this type of situation requires a project coordinator, not a project manager. Someone else is really managing the project. It's wise to bring this out in the beginning.

1.4 STYLE: INTERACTING WITH INDIVIDUALS

However, even after differences in signals between the project and the parent organization are well defined, there are still those less than predictable individuals providing signals within the project itself. They have to be understood at the source.

The project manager does not actually produce many of the project outputs. Project managers are more likely to be concerned with the people who really must make and carry out the development decisions to be sure that there are no problems that are overlooked. These people are the sources of signals within the project. In order for the project manager to receive these important sources of signals, he or she must have

> personal energy, powers of influence, and quickness that will be crucial in keeping things moving, avoiding holdups, and resolving seemingly unresolvable problems.
>
> As one observes these managers, they seem to be engaged in a ceaseless round of "political" give-and-take. (Sayles and Chandler, 1971, p.495).

That political give-and-take obviously involves all kinds of data coming from individuals in the project. The many subjective inputs are now as important as those objective inputs concerning technical competency or business acumen. For example, product knowledge and familiarity with the legal language of contracts are necessary but incomplete. Style responds to some inputs that are not even directly related to the work being done. It also is based on how the manager interprets those signals. For example,

> "Many executives believe that developing a style which encompassed compassion and empathy would bring them into conflict with corporate goals. . . . One was flabbergasted," he wrote, "by the very idea of sensing his subordinates' feelings and developing an ear that listens. If I let myself feel their problems," he said, "I'd never get anything done. It would be impossible to deal with people." (Maccoby, 1976, p.100)

This example seems to treat people as an impediment because of their extreme variability. This manager's style will rarely respond to many apparently irrelevant inputs. There's a lot that can be missed. Conversely, a manager who responds solely to others' signals may not have the capacity to steer the project in a relatively straight, forecasted path. A middle road is needed that includes an adaptation of one's behavior to fit the signals in the situation.

One acceptable behavioral style could be the one that is sometimes adopted when discussing design alternatives. A calm and thoughtful comparison of different technical solutions could be optimum. This style fails when a shipment has not been made as scheduled. A highly charged, emotional style of behavior requiring intense physical activity intended to minimize the problems caused by the late shipment is now more appropriate. Therefore, the style or the behavior that you exhibit should be fitted first to the organization's general expectations (that's why we developed those systems) and then those of the particular present situation or person with whom you are interacting.

That style includes the whole person, not only the part of you that happens to be an engineer, a chemist, a physicist, etc. The logic of systems is only the basis or strategy. When tactics are needed, the style must fit the existing, present situation. The whole person responding in any situation uses both the logic of original training, as an engineer, physicist, chemist, or whatever, coupled to specific emotional creativity. The resulting personal style is observable behavior, interpreted by others providing an acceptable output to perceived inputs. These inputs may vary from those of the organization (such as company culture or standard policies), the individual project team members (such as internal motivation), and your own inner thinking processes and tendencies (such as compassion and sympathy). When the management style follows both internal (your own thinking) and external (others' behavioral) signals, it can be appropriate. When it matches the apparent needs of the situation, it is appropriate.

1.5 HOW TO LEARN STYLE: TO ADAPT

The adaptation of one's own style is a process that can begin cognitively. For example, using a modified version of the familiar, Scientific method, we can

1. *Define the problem:* What is the difference between my behaviors and those expected by the participants in this particular situation? ''The Purchasing department has presented me with three alternative vendors, and none of them can meet the delivery requirements that my customer insists upon. This is a problem that interfaces among the project engineer (who has to approve new vendors), the Contracts team manager (who might have to contact the customer and ask for a delay), and the Purchasing team manager (who provides recommendations as to which one to choose). These managers expect me to make a decision, and I expect them to supply one to me. I define the problem as unwillingness to absorb uncertainty.''

2. *Develop a personal theory:* What general explanation or predicting framework will fit best? What are the general solutions that are available? Using a problem-solving meeting format selected from the Project Operations system, I might list the alternatives in an agenda, ask for written recommendations from the three team managers, and after receiving them either in writing or orally, sum them up as the final meeting agenda and have the meeting. Using the modification of the Expectancy theory motivational technique, I might just delegate the whole problem back to Purchasing, outlining that their suggested solution should not affect any technical considerations, allow for a minimal delivery slip of two weeks but increase the amount of funds available to speed up delivery as much as possible. (We'll cover this process of delegation in greater detail a little later.)

3. *Determine what the appropriate style would be:* What style would best fit the situation? Then select a trial behavior pattern. What is my most comfortable problem-solving mode—risk avoidance or risk taking? What would be the various potential gains or losses involved in any alternative? (I might even want to score them on a scale from 1 to 10.) Based upon my calculations, I'll start by turning the problem over to Purchasing and ask them to come back with a forecast in three days. Then, if the solutions presented look acceptable, we'll implement them. If not, we'll try the problem-solving meeting. If that comes up with no answer, maybe we'll just have to call a review meeting with the customer and ask for relief. If no positive answer develops after that, we may have to consider canceling the project. We now have a style or sequence of problem solving that reflects a calm, take-charge project manager to the outside world.

4. *Test that style:* As the response data come in, I might find it necessary to modify the style. I might even want to go back to Step 1 again.

It's really a never-ending personal development program that requires one to respond appropriately to different kinds of signals. By the way, your answers may be entirely different from mine. It's to be expected. Style is unique to the person. But there are parts of it that are useful in many project situations. As an illustration, we can use the same kinds of bricks to build a house, but my house will be quite different from yours. In other words, we've all had to pass calculus, strength of materials, geometry, and basic psychology, but the way that each of us uses this information and combines it in our minds is very different.

Since we have handled the inputs of organizational signals through the design of the prior four systems, most of the predictable, conflicting inputs for the project should have been resolved. Those were relatively standardized, strategic processes. On a more limited scale but equally important, the fifth system of Personal Style can use some relatively standard solutions to predicted problems. These are the tactical "building blocks" for the project manager. Three of the more important tactics or building blocks of personal style are the methods for handling delegation, project leadership, and conflict resolution. These building blocks are responses to the following typical problems:

1. What should be done to be sure that there are never too many, too few, or the wrong kind of tasks handed out? Delegation.
2. How can one assure that the project moves in the direction that you have laid out as the project manager? Leadership.
3. Which techniques can be used to handle conflicts or disagreements that are based in human differences? Conflict resolution methods.

1.6 DELEGATION: DEFINITION

Delegation is a process through which authority and responsibility are assigned to another person. In a very simplified form, the elements of this process are

1. Define the task to be done.
2. Assign it to an individual.
3. Agree upon the parameters, for example, resources needed, completion criteria, completion date, and appropriate authority.
4. Determine the in-process or follow-up procedures.
5. Measure achievement against forecast.
6. Reward or penalize accomplishment.

It's really a simple process to understand, but difficult to follow. For example, it's obvious that every delegation requires the assignment of *both* the responsibility for getting the job done and the authority to do it. Responsibility and authority have to travel together. But responsibility is not linear; giving it away doesn't get rid of it for the delegator. The delegator still has as much responsibility (or accountability) for the task after it has been delegated as before. That's unusual. Things are clearer with authority. When that is given away, it cannot be retained. You either have it or you don't. In other words, authority is almost digital but responsibility is partially analog. Too many times we try to assign the responsibility and keep the authority. That *never* works. Unless a person has the authority, he or she can never accept the responsibility. They will just pass *all* decisions upward. And even when we do assign the authority, we must recognize that in the end, we are still responsible.

1.6.1 Delegation: First Delegating Downward

We defined the general process of delegation, but how does one implement or use it in his or her own personal style? For example, how does one select the right person to whom to delegate? How does one define the task size, complexity, time limit for feedback reporting, and so forth? How does one measure delegation? If we can measure it, we can begin to use it.

In the process of assigning tasks to others, delegation is

1. Measured by the amount of uncertainty, that is, by the amount of trouble that you can get into.
2. Delegating what you know.
3. Eventually having all your delegations equal (see Fig. 9-1).

Expanding Rule 1 above, a definition for *trouble* could be the potential loss that one would sustain if the task is not completed well. In other words, the mana-

---

DELEGATION

The process of determining the
a. task to be done
b. within a time limit
c. given the authority to do it
d. measured by the amount of trouble you can get into.

Delegate what you know!

Figure 9-1. Delegation

ger's evaluation of the potential damage to himself or herself if the delegation is not completed well. To illustrate, assume that there are two tasks to delegate, one in electronics design and one in mechanical equipment procurement. If we were discussing my unique style, I would be the only one who can determine which of these two tasks is more critical to me as the project manager. But how would I determine which is more critical?

Rule 2 says to delegate what you know. Assuming that my personal expertise is in mechanical equipment, it would be easier for me to determine errors when dealing with equipment than it would be in dealing with the electronics design. Therefore, I would probably delegate the equipment procurement first and concentrate on the electronics design, because I would be exposing myself to less trouble with the mechanical equipment. This also applies when I evaluate to whom I wish to delegate those tasks. Knowing an individual better should lessen the amount of trouble that I can get into, assuming I believe that individual to be competent. Conversely, knowing the individual may also cause me to delegate less if I believe that the individual is less qualified.

Rule 3 requires having all your delegations be equal. As I concentrate on the things that I don't know (and I am responsible for), I eventually begin to learn more about them. This means that I will be able to delegate more of them next time. Eventually as the "unknowns" begin to decrease, the potential trouble becomes equal on all tasks delegated. Then I'll be delegating equally.

To briefly review and expand these ideas, the three rules work for people as well as things. Continuing the example of having less knowledge about electronics, even though the electronics designer assigned to the task is supposedly the best designer in the company, I still would require more follow-ups, with a smaller time interval, than I would with the equipment procurement. Typically, progress would be reviewed every week in electronics and every month in mechanical equipment procurement. My having less knowledge about electronics than mechanics is the reason behind this difference. The things one doesn't know can get one into trouble quicker than the things that one does know. As my familiarity with electronics (or the confidence in the electronics designer) improves, I lengthen the time between feedback sessions.

Delegating requires self-analysis. One has to determine what one knows (either about things or people) and how much trouble can happen by delegating. Trouble is measured by uncertainty, that unmeasurable, powerful determinant of management decision making. It follows that using these three steps means that I will always be delegating according to my perceptions of uncertainty. Different tasks will have different follow-up periods, but they'll eventually be equal in the amount of trouble that can personally be sustained. It's a self-training program, because I will always be following up those areas that are known least well. But that's reasonable, because a manager is supposed to be lessening uncertainty.

Another point to consider when delegating to any person is the amount of time that is allowed for that person to reach a goal *without* general supervision or feedback. By delegating appropriately, the manager is helping that person increase the amount of time that he or she can work unsupervised. This improves the person's ability and then his or her organizational stature. That's not bad either.

These rules apply when one is in charge of the delegation process. But when one is not the delegator but the delegatee and a task has been assigned improperly, what then? For example, when a task has been delegated with no authority to accomplish it, the delegation process should be reversed.

1.6.2 Delegation: Then Upward

Delegating upward is legitimate when authority doesn't match the responsibility. Let us assume that delegation is partially complete. The simpler tasks associated with the job have been completed, i.e., the forecasts, the completion criteria, and date have been set by the delegator, and the person to whom the job has been delegated has outlined the resources needed, including the necessary authority. If that authority is not granted, the task will not get done, since the formal organization will not respond to requests for resources and assistance.

The lack of authority requires that the person to whom a task has been delegated inform the delegator what will happen due to the limited authority. This usually means a reduced scope of the work.

Then it's up to the delegator. If the delegator wants the original task accomplished, adequate authority must be supplied. The word has been passed upward that a lesser task will probably be done without the appropriate authority needed to complete the whole task. When this happens, it is a *delegation upward:* "Give me the tools to do it all, or based on what I have, this is all that I can do." Defining the appropriate authority and getting it, of course, is as important as the definition of the task itself and the time allowed to complete it.

Obviously, authority is not linear. After it is gained, one has to know how to use it and to be able to exhibit the equally appropriate behaviors. That's leadership, which is the next topic in the Management Style system.

2.0 LEADERSHIP: AN IMPORTANT PART OF STYLE

Leadership has been a subject of great interest for thousands of years. But it's only within the last few years that there have been significant breakthroughs that have disgarded the universalist ideas that implicitly say, "This is the way every leader should act," "Leaders are born, not made," or, conversely, "Leaders are made, not born." The central questions have always been "What is a good leader?" and "How can I become a good leader?" Much research intended to answer these questions was based in the assumption that leadership was something that happened when authority was delegated formally by the organization. We know that's only part of it, but few theories discussed the informal authorities or the differences in the situation. While it's true that you cannot lead or manage anybody if they don't recognize your authority to do so, it's also true that authority is more than the formally delegated organizational "blessing." Leadership is often defined as multidimensional. For example,

> Managerial authority refers to that . . . which has been delegated to him by the institution he works in. Leadership authority refers to that aspect of his authority derived from the recognition of his followers of his capacity to carry out the task. Managerial and leadership authority reinforce each other; both are, in turn, dependent upon other sources of authority, such as the leader's technical knowledge, his personal characteristics, his human skills, and the social tasks and responsibilities he assumes outside the institution. (Kets de Vries, 1984, p.48)

But I believe that this type of multidimensional distinction between managers and leaders is not particularly relevant in technical projects. The project manager and each team manager must be both manager and leader in order to accomplish the project's goals (managerial) using the commitment of individuals (leadership). Therefore, it is important to obtain both of these sources. One of these sources— that of managerial authority—has already been covered in the Chapter 1. To simplify matters, we'll define *leadership* as a totality.

The *totality of leadership* is the behavior exhibited by one individual (the leader) to induce another (the subordinate) to act in a manner intended by the leader. It's a two-way process between the leader and the subordinate. This is very important in technical organizations, because the subordinate, being an information worker, controls both the quality and quantity of work done, and can be unwilling to grant the leader the desired level of authority. The subordinate, in this case, has the freedom to withhold cooperation (and you may never know it), or can move to another job or firm.

Now that we have a definition, we can start an overall approach (or theory) to developing appropriate leadership behaviors and then determine how well they fit the situations (i.e., the testable hypotheses). Traditional management theory has defined leadership behavior as the nonquantifiable tasks of "planning, organizing, and controlling." If we could operationalize these tasks and if leadership criteria did not change, learning how to lead technical organizations would be relatively straightforward. But that's not what happens; situations change.

In Homer's *Iliad*, we find that leaders were always supposed to exhibit behaviors motivated by great shrewdness and cunning. But recommending that there is only "one way" to plan, organize, and control or that good leaders must be motivated by "shrewdness and cunning" implicitly assumes that the resulting behaviors are supposedly universal and fit every situation. This kind of universalist theory doesn't seem appropriate, because it doesn't allow for differences in organizational cultures, the nature of the project task, or even the backgrounds of the people assigned to a project. For example, the leadership behaviors required to lead a laboratory team developing a unique, creative solution to some previously unsolved problem are not the same as those for getting a disciplined work force to produce a short term increase in output of high-quality products in a factory setting. There are major differences in the situation, specifically in time and culture.

Not only do situations differ in time and culture, they differ from company to company, and, of course, even within divisions of the same organization. Projects are prime examples of changing situations. There are always iterations and situational changes, which means we need fewer universalist generalities and more specifically defined leadership behaviors. And even in the project life cycle, when the project situation changes, leadership behaviors must also change.

This is quite different from the relatively fixed environments of equivalent functional managers. In many cases, the usually slower pace of technical functions emphasizes consensus: "That's the way we always solve our problems around here." Such a functional environment is rarely as applicable to projects, since creativity and consensus may not mix well.

Consensus seems to flourish well in a relatively calm environment with a minimum of strong emotional currents. Consensus happens when the behaviors are interpreted as a series of small individual compromises in decision making. This works very well when goals are well defined and measurement of outcomes is explicit. The following describes an extreme of this:

> The Japanese system of consensus decision making. . . . Policy proposals are debated at lower levels . . . when they finally reach the Chief Executive, his role is partly symbolic, attesting to the consensus (Kets de Vries, 1984, p.123).

This kind of consensus reaching might be appropriate in the middle of most project life cycles. However, it is not appropriate during other parts of the project, when policy is developed top-down before being modified in the bottom-up meeting. In the beginning, the chief executive (or the project manager) is an active participant, not at all symbolic. In the beginning of most projects, there are also open discussions of opportunities and changes and a tolerance of an endemic underlying level of conflict as differences in creative approaches are proposed. This is similar to the situation that arises when a project faces an unforeseen problem. The situation changes later on when decisions are made with less uncertainty and with people further down in the project structure making them. To summarize, universally recommended and constant leadership behaviors are not applicable to every situation. There is no "one best way" to lead a group. It is multidimensional because it changes with the situation.

2.1 LEADERSHIP: THE BUILDING BLOCKS

But even if there is no universal or fixed model, we can select some standardized building blocks from relevant theory and then create our own individualized, and more complex, appropriate leadership style that specifically fits what we have observed in many project situations. The leadership structure is then unique and changeable, even if the blocks from which it is built are not. Some of the building

LEADERSHIP THEORY

1. *Trait theory:* The combination of various traits that constitute a leader.
2. *Gestalt* or Great Person theory: A great leader that succeeds irrespective of the situation.
3. *Gestalt-Situational theory:* The interaction of the leader in the supportive, but unchanging, situation.
4. *Contingency* or *Situational theory:* Interaction of the leader and the situation

Figure 9-2. Leadership theory

blocks for personal leadership theory are quite old. Trait theory, which is one of the oldest theories or building blocks, has been around for literally thousands of years. One of intermediate age is the Great Person theory, and then we have the relatively modern Situational or Contingency theory (see Fig. 9-2).

1. *Trait theory:* The leader is characterized by a series of internal traits. Therefore, if we can define and measure these traits, we can then measure anyone against the desired trait scales and determine if he or she will succeed as a leader.
2. *Gestalt or Great Person theory:* A leader is not characterized by a series of traits, but he or she is a total person. Therefore, a leader is born or trained to lead well and will always be a leader, irrespective of where or when.
3. *Gestalt-Situational theory:* The leader is a great person who exhibits the right leadership behaviors. The environment is not important.
4. *Contingency or Situational theory:* There is a changing interaction between the leader and the situation. Therefore, different leadership behaviors (some of which the leader is born with and some of which he or she learns) may be required in different situations or are contingent upon certain circumstances. Consequently, a leader may be successful or not, depending on how well his or her behavior and the situation or circumstances fit together.

2.1.1 Trait Theory: What Is it and Can We Use It?

Trait theory seems obvious and logical. Just determine who are good leaders and measure their traits with some kind of test. Then, by using the same test, measure a potential leader for the required traits. If desired, train potential leaders by compensating for leadership deficiencies or weaknesses. In effect, a good leader-manager is supposed to be able to manage anything, anywhere. This approach is used by the armed services in junior officer training. When a situation is well defined, as it usually is in the military or in any large functional organization, the application of Trait theory can be very useful. For example, under certain conditions, leaders managing people performing intellectual tasks are more successful when they, themselves, are intellectual. (Fiedler, 1985).

But Trait theory has some important flaws. The first flaw is that the basic definition of the particular traits may depend upon the one who is doing the measuring, since traits cannot be directly observed. They are inferred. For example, the usual way to define the well-known trait of intelligence is by an intelligence test. This is an operational definition. It means that the operation itself defines the comparison of the end results. The problem, of course, is that the test may not indicate intelligence at all, but something else. Perhaps, it is a better indicator of test-taking ability. Thus, the person doing the measuring may incorrectly infer that a person does or does not have a trait called *intelligence* through some correlative measurements.

Additionally, the trait selected as the independent variable may covary with some other traits. (Wiggins, 1973, p. 368) Since much of Trait theory is based on correlational analysis (e.g., the ''cunning and shrewd'' behaviors of Greek heroes that correlated with great leadership in Homer's epic poems), if there are other variables besides being cunning or shrewd that affected the leadership outcome, the statement about the relationship of ''cunning and shrewd'' with leadership success may not be true at all. Another unknown variable, for instance, a high pain threshold, may be affecting the correlation.

Another problem is that the more recent scientific work in Trait theory is based on controlled research done under laboratory conditions. The laboratory is obviously not the same environment as that of an organization, and the subjects in the experiment may, therefore, not react in the same way as they would under real-life conditions. Finally, there is an assumption that trait measurements taken at one time are stable over an extended time period. In other words, people don't change the important parts of their leadership styles, their traits. They may not change their basic personalities totally, but we have many examples of leaders who have changed their public behaviors as situations changed. Basically, then, Trait theory is inadequate.

2.1.2 Gestalt or Great Person Theory

Great Person theory states that people are not just assemblies of traits, they're a *Gestalt*, a totality or a whole person. The Gestalt or whole person has certain properties that are different from just the sum of a group of traits. Therefore, this approach proposes that people cannot be analyzed into parts but must be considered as complete personalities. The Great Person theory is similar to the Trait theory when it comes to leadership measurements and leadership training. In Trait theory, there is an implication that a person can be measured against some optimum series of traits. Then, by training, the traits that scored low can be strengthened and the traits that scored too high can be deemphasized. A similar concept might apply in Great Person theory. We could measure ourselves against some overall Gestalt score obtained from the "great person" and attempt to match our overall score to the "great person's" score. Combining the Trait theory and the Great Person theory, one could say that if a person has the right mix of traits or else has the right Gestalt, he or she will be a good leader-manager. And if he or she is a good manager, any situation can easily be managed.

It's an attractive idea since it says that leadership success travels with the person irrespective of the organization or the situation in which he or she operates. McGregor's theory X and theory Y is an example of an integration of Trait and Great Person theories (McGregor, 1960). In brief, a theory X manager feels generally that all people have these traits in the work environment—they are lazy, they don't want to work, and they have to be constantly watched to be sure that work gets done. This is Trait theory in others. Therefore, the best Gestalt leadership behaviors are always directive. Conversely, the theory Y manager feels generally that all people have these characteristics—they work as naturally as they do other things such as play; if they are supplied with adequate resources, they will willingly produce without being closely watched. Therefore, the best Gestalt leadership behaviors are always supportive. The thinking patterns of the leaders in both these examples are holistic (Gestalt).

Great Person theory also relies on something called *personal charisma*. Charisma is another way to describe a *total* combination of unusual qualities in an individual. This combination of qualities is attractive to others and results in special attachments, if not devotion, to the leader. A charismatic leader, therefore, leads primarily through the influence or authority granted by followers.

But the Great Person theory also has damaging flaws. The great leaders against whom we're supposed to measure ourselves were interacting with a specific situation that may not apply to us. For example, consider an obviously great person, Alexander the Great. Some of the conditions that affected this great person and probably supported his drive to conquer the world are

1. He was born in Macedonia at a time shortly after his father had come into possession of valuable silver mines that were to finance the Macedonian army.
2. He was the son of a conquering king.
3. He was taught logic and thinking as a student when Aristotle became his tutor.
4. He was to inherit a superbly equipped and trained army. That army itself was motivated by the spoils that came from active warfare and was anxious for more.
5. He was initially up against a series of quarreling Greek city-states and then a decaying Persian empire.

A great person (or leader) must have an appropriate situation in which to lead. If Alexander's army wasn't well trained and in place, his tasks would have been much greater. Possibly he wouldn't have been able to win those early and very necessary battles. Going a bit further, Alexander had to have motivated soldiers who felt that it was to their self-interest to follow him. (By the way, Alexander always led the way. When he died and his body was being prepared for burial, it was noticed that he was heavily scarred all over the front of his body, but there were no battle wounds on his back.) He was appropriately matched to his environment. He also had charisma. It was a very positive situation that included the combination of formal authority, informal influence, and great personal charisma.

2.1.3 Gestalt-Situational Theory

The combination of the great leader and the appropriate situation is therefore an obvious next step in leadership theory. This assumes that when there is an optimum match between leadership behaviors and the environment, the leader will succeed. It also assumes constancy, that once a great leader has been established appropriately in the situation, there is no longer need for behavioral change. This is an important flaw. Alexander did not have a fixed leadership (management) style. It changed to fit the situation. Therefore this model would not apply to him. He fought the Persians in battle and after winning, insisted that his Greeks marry Persian women, because the situation then required integration, not combat. However, Great Person theory gives us more to go on than Trait theory, because it deals with the whole person. Great Person–Situational theory is even better, because it deals with the situation in which the leader acts. But, it doesn't provide guidance when the situation changes relatively rapidly, as in project life cycles. Therefore, all these theories are only partially applicable. Rapid change is a central issue in our next theory—Contingency or Situational Theory.

3.0 CONTINGENCY OR SITUATIONAL THEORY

Contingency or *Situational* theory proposes that there is a predictable interaction between the leader's behaviors and the organizational environment in which his or her behaviors occur. When the behaviors and the environment fit together well, an optimum group output or performance results. So it's not the leader alone, nor is it the group (or the organizational environment), it's the combination. This can be both an asset and a major difficulty. Most people are fairly stable in their behaviors and can learn to be successful as a leader in a stable environment. But when the situation changes, the previously successful leadership patterns may no longer be completely appropriate.

For example, the hero in the war movies who trains the usual group of nondescript draftees into a compact and deadly fighting force that wins the battle is a very directive (and obviously successful) leader. Now place that same leader an advanced research and development laboratory. He has to deal with creative scientists and engineers. I can just imagine the first directive: "Beginning this week, we're going to have more creative work done around here. If we don't produce at least one patent a month, I'll impose some tough restrictions!'' Clearly, this situation requires different leadership behaviors. Situational or Contingency theory helps us to define and understand the situation and suggests which leadership behaviors are best for different situations. In effect, it says, "If this is the situation, then behave in this way." It provides many different building blocks that are appropriate for the leader to use in different situations. It has two parts: the leader and the situation.

3.1 THE LEADER: GENERALLY

First we will discuss leadership behavior and then the interactions between that behavior and the situation. We can simplistically classify leaders into two generalized and opposing groups: those that might be called the *no-nonsense, bottom-line-oriented leaders,* who are primarily interested in the end result or goal, and the *supportive, how-can-we-cooperate-better-oriented* leaders, who are primarily interested in helping others to achieve agreed-upon goals. The first group of leaders includes those who keep their distance psychologically from their subordinates (Blau and Scott, 1962). Some are more cold and aloof; these leaders are task-oriented (Fiedler, 1967). Others have a need for achievement (McClellan, 1961) and exhibit production-oriented actions (Katz et al., 1957), and concern with production (Blake and Mouton, 1964). When these behaviors are attached to an inability to trust subordinates (McGregor, 1960), the leader will probably exhibit behaviors typical of a theory X leader; that is, he will have the general belief that subordinates are fundamentally lazy and they have to be closely supervised to keep them working. The contrasting leadership behaviors describe the nondirective, supportive, instructional patterns typical of a theory Y leader, who believes that people like to work as much as they like to play and the best leader is one who provides the training and resources, and help people achieve goals.

Much of the research has this kind of descriptive flavor to it, that is, dividing leaders into two different groups according to their inferred beliefs. These beliefs seem to result in behaviors showing either a concern for people or for production. The leaders are either task- or relation-oriented. They understand the world as primarily theory X or theory Y. But are they really opposite? (McGregor later stated that his theory X and theory Y managers are not on opposite ends of a spectrum but are two *different* ways of looking at the world. McGregor, 1967, p.79). In effect, the same person may change his or her view of the world under different conditions.

The theories we discussed assumed that the situation didn't affect the outcome if the "right" leader was managing. In other words, they predicted that a "good" leader will succeed anywhere. From our own experience, we know that this doesn't always happen. There are situations (especially in the beginning of projects) when a semidirective leader is needed to outline the charter (an end result) and (in the middle of the life cycle) when a supportive leader is needed.

Contingency theory not only describes this; it offers prescriptions for optimum behaviors. In my opinion, one of the best models in this general body of theory is that proposed by Fred Fiedler. His hypothesis seems to provide a better "fit" between the typical project situation and the leader's behaviors than other theories do (See Blake and Mouton, 1964 or Hersey and Blanchard, 1969). It starts, as do most others, by inferring how the leader totally views the world based on interpretations of the behaviors he or she exhibits. Then it defines the organizational situation. It finally suggests an optimum "fit" that depends upon a leader behavior–situation interaction. In other words, one can determine how one "is," what the situation "is," and how best to fit them together. Since people don't change quickly, the theory logically provides guidelines to change the situation, if required.

3.2 THE LEADER: SPECIFICALLY
IN THE SITUATION

The *contingency* or *situational theory* also proposes to classify leaders into two very broad categories: those who regard the world in an overall task orientation and those who have a relations orientation. It is almost a total, Gestalt-like kind of orientation. Task-oriented people are more comfortable when getting the job done, and relations-oriented people are more comfortable when getting cooperation from the group. (This, of course, is a simplification. For the full flavor of the theory, I refer you to original text: Fiedler, 1967.) Now comes the interesting part. The leaders who are task-oriented are more successful in situations that are either very favorable or very unfavorable for the leader, whereas the leaders who are relations-oriented are better in situations that are intermediate. (The terms *favorable, unfavorable,* and *intermediate* are defined in the next section.) It's a bit complex, but then so are the situations in projects.

The leader's degree of overall task or relationship orientation can be measured by a pencil and paper test called the *Least-Preferred Coworker* (LPC). In summary,

> a person who describes the least-preferred coworkers in a relatively favorable manner tends to be permissive, human relations oriented and considerate of the feelings of his men. But a person who describes his least preferred coworker in an unfavorable manner—who has what we have come to call a low LPC rating—tends to be managing, task-controlling, and less concerned with the human relations aspects of the job. (Fiedler, 1967a, p. 499)

The LPC test measures people's feelings concerning a person with whom they can work least effectively. Low LPC people (task-oriented) gain self-esteem and satisfaction from successful task completion. High LPC persons (relations-oriented) gain self-esteem and satisfaction from successful interpersonal interactions. The complete tests for both LPC and the "situation," with the method of scoring, may be found in Fiedler, 1967, Appendix A.)

In other research, it was found that there were some correlations between LPC scores and age, experience, and managerial level. Low LPC scores, which indicate high task orientation, varied according to increasing age and experience, whereas high LPC scores, which indicate relations orientation, were positively correlated with managerial level (Bass, 1981, p.343). So we now have some general measurements of how leaders view the world and the relationship of those views to age, experience, and organizational level.

3.3 THE SITUATION

The *situation,* according to Fiedler's theory, is defined by a combination of three measurements obtained from questionnaires completed by the people in the situation. They are

1. *The Leader's Position Power* which is measured by the degree with which the power of the leader is perceived by the people being led. This includes the authority granted by the position itself. Power enables the leader to get the group members to comply with and accept his or her direction and leadership.

2. *The structure of the task* is measured by how well the task to be done is understood. In other words, a well-structured task is enforceable while the unstructured, ambiguous task is difficult or impossible to enforce.

3. *The personal relationship between the leader and the group members* is measured by sociometric preference ratings completed by the group members. In effect, it's how well the leader is liked by the group members. While there is no reason to believe that a well-liked leader will give better or wiser orders, it seems clear that his or her orders are more likely to be followed than orders from a leader who is less well liked.

A word of caution before we go further. This is research, and it has all the advantages and disadvantages of every research program. Under laboratory-controlled conditions, it's always possible that the situation may not reflect all the variables that happen in real organizations. For example, the experimental controls that are intended to minimize extraneous variables or factors may be eliminating variables that are really important. Conversely, with fewer or no controls, we would be unable to distinguish among those factors that do seem to affect the outcome and those that don't seem to. Fortunately, the predictions of this theory have been tested a number of times in different environments and under different conditions and the findings are quite consistently supportive. On a personal note, I have tried to use this theory in managing real projects (not with any rigid controls, of course) and find that it does work, at least better than some of the others I have tried. But since human behavior, represented by leadership, is based in many, many variables, this and other research recommendations about behavior should always be evaluated against your interpretation of your particular situation. Remember that you have to build your own unique management systems. Remember the first person singular approach.

Now let's put this description into a typical project framework and compare each of these measurements with the situation at various stages in the project life cycle. For example, when a project is beginning the situation is not immediately "favorable" for the project manager, in my opinion. Position power, the first measurement, would be fairly low.

Then we have the second measurement: task structure. The project task is not very well structured in the beginning, since, by definition, a project is created to solve a novel or unusually complex problem that the regular functional organization can't handle. Consequently, this second measure of Fiedler's situation seems to also be low when the project is started.

Finally, we come to the last measurement: leader-member relationships. In many cases, the project manager and the team are "new" to one another. Consequently, on an overall basis, we can assume that the leader-member relations measurement would also be fairly low. They just don't know one another very well yet.

Now if we can relate these measurements of the situation to our basic idea of "uncertainty," we might find them useful. In my opinion, uncertainty and situational unfavorableness are directly related.

The theory predicts that a task-oriented leader's behaviors would be appropriate when the situation is either very unfavorable or very favorable for the leader. The situation at the beginning is very unfavorable; uncertainty is high. The team leader's behaviors should be result-oriented or goal-directed, and he or she should be good at organizing the way the project will operate (what the repetitive systems will be.) However, this result-oriented behavior should *not* be expected to limit the creativity of individual team members. It should just provide the overall boundaries of the end results, the time, and the cost (the three "golden limits"), within which individual team members must determine *how they operate* their parts of the project. Team members make these definitions when they do the bottom-up forecast and later when they report on their respective estimates to complete. It also fits our other discussions about delegation, dual reporting, and so forth. In other words, at the beginning the leader exhibits task-oriented behaviors by outlining the end result and the overall boundaries. The individual team members determine their own methods for contributing to achieving project goals.

Now let's move to the middle of a project. Assuming that the first design review has determined that critical end results look like they will be achieved, the situation has changed. The leader's position power is now fairly well established. By this time, everyone understands what the leader can and cannot do or will and will not do. Second, the project tasks are structured fairly well. "We know that the project is feasible; all we have to do now is to get it done on time." Finally, the relations between the leader and the team members are very clear and (hopefully) mutually supportive. Thus, the situation in the middle of the project is "intermediate" and more relaxed. Uncertainty (as my subjective description of that same situation) has lessened a fair amount, but it still partially exists in the project's middle.

A more relations-oriented leadership style would now be optimal. This style of leadership is more supportive and is intended to help team members achieve their individual goals. The outlines of the eventual solution to the project's problem are visible, and the team members can concentrate on the immediate tasks at hand rather than the need to design the total amount of work to be done. They have worked out interactions among groups and are working well. There is less need for overall direction now. In this kind of situation, the project manager can be concerned with things dropping "between the cracks" and can be less concerned with overall direction of the entire project team. Now directive leadership behavior may even be dysfunctional, because it might interfere with an individual team member's tasks. The leader should now show a more relaxed behavior intended to support team members if needed.

Finally as the project begins to wind down, the situation changes again. There is a reawakened need to directly reset the limits for the whole project, because the project manager is getting ready to hold the last problem-solving meeting—the close down meeting. The overall limits must again be developed, this time for close down. Taking the three situational measurements again, you will now probably find that the leader's position power is fairly strong, since recommendations for the next assignment of the project team members are coming up. When one has the capacity to control another's future, that's power. As far as task structure is concerned, the required tasks are very well structured, since all the close-down items have been defined by both the manager and the client or customer. There is a firm time schedule established. Finally, we come to the last measurement, leader-member relations. By now, the team is operating as a well-oiled mechanism, which means that the leader-member relations are at a peak.

In my terms, these three measurements show that uncertainty is fairly low. This means the situation is very favorable for the leader. Thus, leadership behavior should again be task-oriented, according to Fiedler. This describes a return to the bottom-line approach: getting all those forecasted open items done and the project closed down correctly.

According to this model, then, the best leadership sequence would be first task orientation, then relations orientation, and back to task orientation. It's a logical prescription, and, according to Fiedler, it's the typical behavior exhibited by a low LPC leader. A low LPC leader is task-oriented under pressure, such as would be the case when a project is beginning or ending, but is more relaxed when the pressure is lifted, such as in the middle of the project. An obvious potential problem in this changing leadership pattern is not *what* the behavior should be (we know that now) but *when* the leader should be exhibiting which kind of recommended behavior. This problem is now neatly solved: It's whenever the leader is behaving naturally if he or she is primarily a low LPC person. Consider the following:

Task-motivated (low LPC score of 57 or less) leaders place primary emphasis on task performance. These leaders are the no-nonsense people who tend to work best from guidelines and specific directions. If these are lacking, their first priority is to organize and create these guidelines and then assign the various duties to their subordinates. However, under relaxed and well-controlled situations, task-motivated leaders take the time to be pleasant and pay more attention to the morale of their employees. . . . [Conversely,] Relationship-motivated (high LPC score of 64 or above) leaders tend to be most concerned with maintaining good interpersonal relations, sometimes even to the point of letting the task suffer. In relaxed and well-controlled situations, this type of person tends to reverse his or her behavior and become more task conscious. (Fiedler, et al., 1976, p.213)

If my descriptions of projects are typical of your projects, if you feel that Fiedler's theory is applicable to your company, if uncertainty is really a more subjective measurement of Fiedler's three situational measurements, and if you are a low LPC person, you probably have the best of all possible leadership situations. Now, all you need to do is act naturally. There's a lot of ifs to consider.

4.0 ANOTHER CHOICE: ORGANIZATIONAL MATURITY THEORY

With all the ifs in Fiedler's theory, it's possible that his theory might not be as applicable as we would like it to be. The Organizational Maturity theory is a slightly contrasting leadership theory within the Contingency theory framework. It minimizes a major part of the situational leadership interaction, that is, the attitudes of the leader. Fiedler's theory measured a leader's global thinking processes using the Least-Preferred Coworker test (LPC) in addition to situational measurements. The LPC test was basically a measure of self-perception. Hersey and Blanchard, (1982, p.52) say that self-perception instruments tend to measure attitudinal frameworks, that is, how I would like to behave or what I feel is acceptable behavior, rather than actual behavior.

Hersey and Blanchard's theory downgrades the inferred overall thinking processes of leaders and emphasizes the situation. The *situation*, according to their ideas, is measured by the maturity of the subordinates. In summary, *maturity* is defined as the capacity of subordinates to set high but attainable goals, their willingness and ability to take responsibility, and the education or experience of an individual or a group. These variables of maturity should be considered only in relation to a specific task to be performed. That is to say, an individual or a group is not mature or immature in a total sense but only in respect to the job that has to be done. These theorists have various testing mechanisms to determine the degree of maturity, but for our purposes it is sufficient to describe how the leader should behave in dealing with different levels of subordinate maturity. It assumes that everybody can be a good leader if he or she behaves appropriately and these are the descriptions the theorists give for appropriate leadership behavior when leading different kinds of people.

1. *Telling:* for low-maturity individuals.
2. *Selling:* low- to moderate-maturity individuals.
3. *Participating:* moderate- to high-maturity individuals.
4. *Delegating:* very mature individuals.

This theory has the same limitations as others. It's subject to interpretation. However, if there is a definition of the level of the team members' maturity with respect to their project tasks (Goldsmith, 1980, p.40), if the project tasks, at the time, seem to require the degree of maturity that the team members have, and if the manager is able to change his or her behavior to fit the requirements of the situation (such as telling, selling, participating, or delegating behaviors), he or she is again in the best of all possible leadership worlds. Unfortunately, there's a lot of ifs here too.

5.0 USING LEADERSHIP THEORY

How can these ideas be used in building management theory if they are subject to so much interpretation? Let's use the scientific method:

1. Define the situation: According to subjective criteria, is it favorable or unfavorable (assuming one wants to use a Fiedler type of theory)? Are the team members mature or not mature (assuming one wants to use a Hersey-Blanchard type of theory)?
2. Which predictive theory should be used?
3. What leadership behavior should be tried as a testable hypothesis? How will the results of that trial be measured? In other words, what standards will be set to see if the behaviors chosen work well?
4. What is the present "typical" behavior? How will it be modified if there is dissatisfaction with the resulting interactions of leadership behavior and those of the team?

It's a difficult process and it doesn't even have the controls of a laboratory research project. But it provides a subjective and even reasonably logical process for self-evaluation and improvement of one's own leadership behavior. It's subjective because the leader has to analyze what he or she is, for example, a low LPC or high LPC person by Fiedler's tests. This test is available in your library. Then define the situation or the maturity of the group if you use Hersey and Blanchard's theory, and select the optimum behaviors to match that situation. Finally, take the time to measure the end result (perhaps on an ordinal scale from 1 to 10) to detemine if the behaviors were the best for that situation.

So far, we've covered two of the three parts of personal management or leadership style—delegation and leadership behaviors. It's clear, by this time, that these two parts are interrelated. Delegation implies leadership and vice versa. There is another part of personal style—handling conflict, which is just as interactive as the other two parts.

6.0 CONFLICT RESOLUTION: THE THIRD PART OF STYLE

The techniques used for delegation, leadership, and conflict resolution seem to go together. Conflict is inevitable in every organization. There are never enough resources to satisfy the changing and perhaps diverging interests of the various people in the organization. Accordingly, we must consider the various reasons for conflict and learn how to constructively manage it. Conflict will always be there, and it is impossible to eliminate it entirely. In other words, as the manager, conflict resolver, or peacemaker, the task

> might be to ensure that conflict "makes sense" in a desirable state of affairs that involves some "optimum" amount of conflict of the "right" kind rather than try to eliminate the conflict completely. (Butler, 1973, p.317)

We'll start by trying to make sense of the kinds of conflicts that seem to repeatedly appear in projects.

6.1 CONFLICT: IN PROJECTS

Every project has some built-in endemic conflicts (see Fig. 9-3). Some of them are caused by differences between the project structure and that of the parent functional organization. These should have been dealt with in the project manual. In addition, there may be an underlying role conflict. For example, a project manager may require a certain activity from a particular team member that contradicts what was received from his or her functional manager. We've also discussed this previously.

As usual, we start with definition. *Conflict* is a difference or a variance between two or more ideas, directions, or concepts. It can also be defined as the behavior of a group or person that restricts or impedes another group or person from attaining goals. Some conflicts are predictable and can usually be managed through the overall project systems design that we've covered in previous chapters. But the more specific time-based, project-related conflicts require a set of more flexible management strategies. These management strategies, as applied daly in management tactics, are a part of a leader's management style.

One group of researchers began to develop strategies by dividing the typical project into four major stages: (A) start, (B) early phases, (C) main phases, and (D)

CONFLICT

A variance or difference between at least two concepts or ideas.

Conflict resolution with people: closed and supportive situation and,

Conflict resolution with "things": open and investigative

Figure 9-3.   Conflict

end. Then they analyzed the kinds of conflict that typically erupt at these stages and measured their subjective intensity. Following this, they developed some strategies for conflict management according to the conflict intensity. The seven potential conflict sources found are as follows (1 indicates the most conflict and 7 the least):

	Mean Intensity	Start	Early	Main	End
1. Schedules	1	3	2	1	1
2. Priorities	2	1	1	4	4
3. Manpower resources	3	4	5	3	3
4. Technical opinions and performance trade-offs	4	6	4	2	6
5. Administration	5	2	3	5	7
6. Personalities	6	7	6	7	2
7. Cost	7	5	7	6	5

The conflict sources just listed can be defined as follows:

Schedules—conflicts with other areas of project performance.

Priorities—the organization did not have prior experience with a current project undertaking, and there were conflicts with supply departments.

Manpower—no organizational slack available.

Technical opinions and performance trade offs—not enough of a broad overview.

Administrative procedures—poorly outlined authority, responsibilities, and reporting relationships.

Personality—The various team members' behaviors did not support cooperation. Although this is not very high in conflict source, it can be very difficult to deal with.

Cost—This was in negotiations with performing departments. (adapted from Thamhain and Wilemon, 1979, p.339)

According to these findings,

> The three areas most likely to cause problems for the project manager over the entire project cycle are disagreements over schedules, project priorities, and manpower resources. One reason these areas are apt to produce more intense disagreements is that the project manager may have limited control over other areas that have an important impact . . . particularly the functional support departments. . . . In our discussions with project managers who have experienced problems in these areas, almost all maintain that these problems frequently originate from a lack of effective pre-project planning. (Thamhain and Wilemon, 1979, p.353)

Since we have developed the proper overall reporting and authority levels in prior systems, we have minimized most of the disagreements that could be forecasted. This shows the need for a really effective sequence of preproject planning or overall project systems design. But even this process is sometimes inadequate. Defining a problem and designing the overall system for it's solution(s) are necessary but insufficient, because there are always unforeseen situations. Those unforeseen situations involve the day-to-day, specific problems that are not really solved with ''standard'' answers. Now we must be able to deal with unique, specific problems and conflicts. Handling those conflicts as they appear should minimize potential delays and losses that could affect the entire project. Overall systems just provide an overall solution to a forecasted repetitive problem. Now we must get more specific.

6.1.1 Resolving Conflict: Try Logic First, Separate the Problem

Assuming that some disagreement has occurred, the obvious first step is to define the problem or disagreement. This is easier said than done. If the definition of the problem involves any of the topics discussed previously under the topics of systems design, the solution(s) should be relatively straightforward, since they're already built in to the project systems that we've designed. These systems are intended to resolve the typical problems involving authority, reporting relationships, forecasting processes, feedback data, or any of the other topics that we've covered. But now we have the unforeseen, nonrepetitive occurrences that cannot be covered through any general systems design.

We can now develop general guidelines for actions or a personal strategy for conflict management. *Conflict management strategy* begins by separating the conflict into that which is objective (about things or data, using specific tests or information involving the physical world) and that which is subjective (about people or feelings, regarding emotions such as intuition, prejudices, and subjective judgments). Objective conflicts involving things are usually related to the project's technical achievements. These kinds of conflicts respond well to more familiar problem-solving techniques, such as testing and data gathering. The best resolutions are usually obtained by means of some kind of crucial test(s) that provide a digital (i.e., either a yes or no) answer. If this type of test isn't available, the next step might be to develop such a test to provide analog (or interpretable) answers. If neither kind of test (digital or analog) can be defined, you either don't know enough about the problem (if it's really dealing with the physical world) or the problem is subjective, that is, human emotional aspects are the major determinant.

If a lack of knowledge is preventing the development of a crucial test, the next step could be to break up the problem into smaller pieces. For example, is it possible to develop a minitest or a feasibility study that will provide *some* objective data upon which to resolve at least part of the problem? Alternatively, if there isn't even enough data to support breaking up the problem or conflict into smaller pieces for minitesting or feasibility studies, the leader must determine why and what is needed. The problem-solving meeting might be the next strategic step. At least, this kind of a meeting will generate some additional data.

The problem-solving meeting, which we covered in Chapter 7, is started by circulating the best written description of the problem outline or conflict that you have at present to the meeting attendees and requesting that anyone interested in contributing to possible solutions should provide his or her response in writing. The response time is usually inversely related to the magnitude of the problem. The greater the problem, the shorter the response time required for meeting attendees to answer. (For instance, if there is a major fire in the office, it's not the time for lengthy deliberations. Nevertheless, this is a flexible suggestion, and you may not want to apply it.)

When the meeting leader receives the responses, they are summarized according to the various problems outlined. This summary is then circulated as the meeting agenda. If the data noted in the summary on the agenda show that the particular problem can potentially be resolved with an objective test, the meeting chairperson should see that the test is run before the meeting so that the results can be provided to the meeting attendees. Sometimes the responses themselves provide enough data to start the solution process. If the result is digital (that is, yes, it worked, or no, it failed for the following reasons), the agenda should include recommendations for the next objective step. If the results are analog (that is, if the test didn't show a failure, but there were excessive strains), the meeting attendees can now discuss the more limited problem of, say, product safety rather than total product acceptability. The conflict has been somewhat diminished.

Let's look at a simplified example. The question is Does the new hydraulic control device meet all operating and safety requirements? The problem breakdown can be classified as function, operating, and static safety conditions. The responses from the meeting attendees show that the functional specifications have been met, including the operating safety requirements, but there might be occasional transient conditions when there is an overload. Therefore, an objective test might be called for to see if the accumulator will burst when the internal pressure occasionally reaches a load of 3,000 pounds per square inch absolute at an ambient of 500°F. The normal load is only 1,500 pounds, but what happens under very unusual conditions?

If no objective tests are developed, and if we are dealing with subjective criteria—for example, ''Will the appearance of the new machine designs help to increase sales?''—an objective test might not be possible. The problem-solving meeting should be the place to define differences among meeting attendees, evaluate them, and develop an approach to a solution.

Since anyone who was interested in the problem has already submitted his or her solution in writing before the meeting, the only reason to meet is to gain synergy from the total group and iron out differences among attendees. As long as the conflicts seem to be resolvable with a minimum amount of emotional input, these interactive processes should be carried out in public using open discussions and decision making.

But this somewhat ideal process may not be feasible all the time, especially when the conflict seems to involve much human emotion. Even though the manager may consider conflict-management as a game in which everyone gains from the solution, not everyone agrees. They may think of a certain solution as a ''zero-sum'' game, in which there are some participants who lose "everything" if they don't "win." Sometimes, it's difficult to distinguish what kind of a game the individual players are in. Using our familiar idea of "uncertainty," the manager now has to make a decision. If it is decided that this is a conflict about "things," then data gathering, objective tests, and problem-solving meetings are good starting points. They might even be partially useful in emotional conflicts, but there should be a major difference in the arena chosen to handle objective conflicts versus subjective ones. Public forums are great for logic, but emotions are better handled in private.

Of course, it is probable that conflicts involving people can't be separated into logical and emotional parts. People aren't that easily divisible, and logic and emotion are often inextricably tied together. In other words, if we think that it's going to be primarily a subjective conflict, it shouldn't be resolved in front of an audience. With people, the fewer participants, the better, mainly because there are sometimes human "losers" in these conflicts. Conversely, there are no "losers" in resolving conflicts about "things." Mother Nature doesn't care what you do.

6.1.2 Understand the Conflict or You'll Get Nowhere

The process of resolving emotional conflicts starts similarly to that of resolving objective conflicts: Understand the conflict. Understanding involves communication. Try to get the conflict participants to clarify the differences by telling each other what he or she understands concerning the other's position. The other person should listen to what is said and either agree or clarify the differences between what was heard (or said) and what was meant. Then try it again. In other words, person 1 says to person 2, "Let me tell you what you said." Person 2 either agrees or else restates his or her position and lets person 1 try again. When person 2 agrees that what person 1 said is what is meant, the roles are reversed. Now person 2 says, "Now I'll tell you what *you* said." This should go on until there is agreement or at least understanding of the differences between the positions.

With this agreement or understanding, we can go on to a typical next step such as classifying the differences in terms of importance by using an ordinal scale. For example, ask each conflict participant, "On a scale from 1 to 10, where 10 is wonderful and 1 is terrible, what score would you give to each of these differences between your two positions?" In some cases, just the understanding of opposing positions and the consequent classification process will diminish the overall intensity of the conflict or confine it to those differences in which the ordinal scores are very high. The differences receiving a low score can either be removed from further discussion or dealt with after the major differences are resolved. To briefly repeat, first define the problem, then classify or measure it, and finally begin resolution.

Logic requires experimentation and objective testing in an open and public discussion of the results. Objective conflict often responds to public problem solving. Emotional or subjective conflict requires a more private forum. Objective conflict can sometimes be dealt with in situations where there is bargaining and problem solving between the principals and interventions by the third party, whereas emotional conflict requires a restructuring of a person's perceptions and the working through of feelings between the principals, as well as conciliative interventions by a third party. The former processes are basically cognitive; the latter processes more affective." (Hill, 1979, p.372)

Emotional conflict is obviously the most difficult to manage since it is not susceptible to the logical techniques that are openly available. Sometimes it is difficult to separate logic and emotion, just as it's difficult to separate a person's thinking or cognitive processes from his or her emotional or affective processes. A person is just not easily divisible into separated parts that are either cognition or emotion. People are entire entities, and they react as "totalities." Therefore, it might be appropriate to consider other ways to resolve conflict that include both the logical (objective, or cognitive) and emotional (subjective) components. As there isn't "only one way," each leader has to determine the best way. Therefore, it may or may not be appropriate to use the following techniques as part of an overall conflict management style when dealing with people (see Fig. 9-4):

1. **Confrontation**—a complete and open discussion of differences, resulting in a unanimous answer freely arrived at.
2. **Compromise**—giving up less than what you expect to receive.
3. **Forcing**—Be reasonable, we'll do it *my* way.
4. **Smoothing or Withdrawing**—O.K., we'll do it *your* way.
5. **Ignoring**—Write that out in triplicate and send it to me. Then I'll get back to you.

These can be used sequentially as the project goes through its life cycle: first, confrontation, then compromise, forcing, smoothing, and finally ignoring. The amount of time each requires to implement seems to decrease; however, the length of time that the problems *stay* resolved also decreases. For example, it usually takes more time to use confrontation than compromise, more to use compromise than forcing, and so forth. However, time becomes more precious as the project ages; there's less of it. Therefore, try to move sequentially through these techniques. But there is a cost. (Isn't there always?) It may take less time to use forcing than compromise, but the problem may not remain solved for as long as it would if it were

CONFLICT ABOUT PEOPLE

1. Confrontation
2. Compromise   Length of time conflict is resolved
3. Forcing
4. Smoothing
5. Ignoring

Ease of implementation

Figure 9-4. Conflict about people

solved using compromise. Similarly, a problem solved through compromise will be solved for a shorter time period than one handled through confrontation. Choose the technique that's appropriate *at the time*. A minor problem that occurs toward the end of the project might be handled through smoothing, since there's less time left and the problem probably won't reoccur before the project ends.

Therefore, at the project start, when uncertainty is high, the most effective conflict-resolution style is confrontation. Conflicts should be handled through nonpressured discussions of the options available, but always within the project's three "golden limits." A complete and unanimous agreement is required at this stage. The time to use confrontation is during the initial system design. Later in the life cycle, during the middle of the project, the alternatives are forcing or smoothing to resolve technical trade-offs and worker conflicts. As the project winds down and major conflicts are about schedules and personalities, the choices might be smoothing or ignoring.

This is, of course, a very general, subjective sequence. One chooses the management tools that seem to match the subjective definition of the project uncertainty. If some problem has caused project uncertainty to go up in the middle of a project due to something unforeseen, change behavior (for the time being) and readopt confrontation. In my opinion, these techniques are easier to use as one moves down the list from confrontation to ignoring, yet as one moves down that list, the conflicts seem to stay resolved for shorter periods of time. In other words, confrontation is a costly and difficult technique, but the conflict will probably stay resolved forever. But if the project is almost over and ignoring will only be applied for a short while, why not use it?

There are other findings about this sequence that you can consider.

> While confrontation was the most favored mode for dealing with superiors, compromise was most favored in handling disagreements with functional support departments. . . . If project managers are aware of the importance of each potential conflict source by project life cycle, then more effective minimization strategies can be developed. . . . It appears to be more significant that project managers, in their capacity as integrators of diverse organizational resources, employ the full range of conflict-resolution modes. While confrontation was found as the ideal approach under most circumstances, other approaches may be equally effective depending upon the situational content of the disagreement. Withdrawal, for example, may be used effectively as a temporary measure until new information can be sought or to "cool off" a hostile reaction from a new colleague. (Thamhain and Wilemon, 1979, p. 352, 355).

As we have seen, the situation defines the tools or behavior techniques to be used. These techniques are fairly specific and are directed toward constructive conflict resolution. But conflict may not always be constructive. It can open problems for discussion and resolution; it can also unreasonably delay project goals. It is best for a leader to make sure conflicts do not remain hidden. Get them out in the open. Once others become aware that there will be no counterattack or selling, the real source of the conflict can be more easily defined. Stay calm and move slowly; acknowledge conflict's existence. Reinforce the part of of the project culture that says conflict is accepted and understood as a part of the project itself.

6.1.3 Conflict: More Techniques

Conflict resolving techniques can also be broadly classified into those dealing with one's self and those dealing with others. Since one should know one's self better, we'll start there. These are some questions to ask.

Did I sincerely admit and accept the existance of conflict?

Did I accurately describe the nature of the conflict?

Did I develop and adequately examine all the alternative paths of actions available to me?

Did I select the optimal path? Does this path, in the light of current relationships, continue to be optimal?

Did I miss a step in the path?

Did I establish the appropriate standards and controls?

Or, is my evaluation of progress in error? Am I expecting too much? Too little?

Should I go back to assure myself that my assumptions are accurate? (Worthman, 1981, p.94)

Conflict-resolving behavioral style involves dealing with others as well as one's self. With others, a more specific technique of information gathering is useful. It involves the art of asking questions. Depending on the way questioning is done, a discussion can be opened or closed. The following types of questions can be used:

- *Open or direct questions* that utilize the old standbys of who, what, when, where, how, and why and invite the subordinate to express openly what he or she feels or thinks. Example, "How do you think we ought to handle this problem?"
- *Leading questions* which invite the subordinate to give his opinion even though there is an implied direction in which the manager thinks the problem should be effectively tackled. Example, "How about using this approach?
- *Unemotional questions* that appeal to reason and evoke little or no feelings. Example, "What would be your first steps toward solving this problem?
- *Off-the-hook questions* which allow the subordinate to decline a request without losing face. Example, "We have a serious overrun on this project. I don't suppose that you would want to put in two or three extra hours tonight and tomorrow night?
    - Don't settle for an answer to the primary question you ask, regardless of the response. . . . i.e. "Why is that?"
    - Don't slam the door of communication by agreeing or disagreeing flatly with the response given. . . "Um-hum, I see.
    - Still another way to keep the door open is to say nothing until the other person says what is on his mind. (Raudsepp, 1981, p.190)

Management behavior during this kind of interaction is obviously neutral and supportive and can handle most situations involving affective problems that are nonrepetitive and that involve mixed cognitive-affective problems or conflicts. Sometimes, however, we have a different situation—repetitive affective problems with the same individual. How do we deal with an inherently "difficult" person? The following are some suggestions:

> Assess the situation: Be sure you're dealing with someone who displays chronically difficult behavior, not someone temporarily responding in the worst way because of a particular situation. If you answer yes to any . . . [part of the next question] . . ., you're not dealing primarily with a difficult person, but rather with a situation that has elicited impossible behavior.

- Has the person usually acted differently in three similar situations?
- Am I reacting out of proportion to what's warranted by the situation?
- Was there a particular incident that triggered the troublesome behavior?
- Will direct, open discussion relieve the situation? . . .

> Put some distance between you and the difficult behavior. Gain some perspective by separating yourself psychologically and emotionally from the problem. The more objective and detached you can be, the better equipped you are to understand the behavior—and how to deal with it. . . .

> Decide on a way to structure the interaction between the two of you. . . .

> Implement your plan. Decide on a strategy for putting your plan into practice, and time it so that

> **a.** It occurs when the difficult person is not overburdened, and
> **b.** You have the time and the energy for the implementation.
> Monitor your plan and modify it as needed. (Auerbach, 1981, p.178)

There's really nothing very novel about these steps. They follow our classical Scientific method: (a) define the problem, (b) select appropriate theory (strategy), (c) implement a hypothesis (try it), (d) measure the variances between expected and actual results (did it work?), (e) if necessary, go back to (a) and try again. (I'm not minimizing that it is difficult to follow this reasonable approach when your emotions (and other's) seem to indicate that the "tigers are attacking.")

Occasionally, even the most logical and rational approach to conflict resolution fails because the other individual seems to have problems that are not clearly definable. For example, some individuals "reject" an entire idea or report because it is "inadequate" or "not prepared well" or for some other overall reason. If the reason(s) for rejection is not clear, one useful technique is to determine if *any* part of the idea or report is acceptable. Try to segregate the acceptable parts from those that are unacceptable. Then concentrate on revising the unacceptable parts.

Another type of individual is the one who cannot or who refuses to confront problems directly. I call this type of person a *problem avoider*. Conflict arises because differences are delayed or responses are not connected to the inquiry. I suggest that this is dealt with by direct confrontation: "Either give me a direct answer or else we shall move up the organizational ladder until one is forthcoming.'' It is difficult enough dealing with project uncertainty without adding someone else's uncertainty to it.

Finally, for those of us who either cannot or do not agree with defining a general strategy beforehand (because "it's not practical"), the world will provide one anyhow: it's called *firefighting* or solving the "crisis-du-jour". A lack of a general strategy always results in daily crises that must be solved immediately. It is an unsatisfactory way to manage. It does not allow the manager to think about the elements of today's "crisis" that could have been forecasted and eliminated before they occurred. However, I knew one project manager who somewhat facetiously remarked, "I don't like to think about future problems. I have so many today that the future can take care of itself." Of course, "today's" problems often come back in the future; only in different clothes.

7.0 SUMMARY

We started our discussion with the definition of personal management style and then covered how that style might respond in typical project-oriented situations. Style is the application of strategy, manifested by the behaviors (and tactics) that we show to others. Those behaviors depend upon our internal states and our interpretation of external reality. It's a very simplistic definition of a complex subject, since "human behavior is like a centipede, standing on many legs. Nothing that we do has a single determinant, whether conscious, preconscious, or unconscious." (Kets deVries, 1984, p.179) But a simple definition is a beginning. It can help us to understand both the situation and ourselves. It also enables us to use basic ideas as building blocks when constructing our own special, and obviously more complex, management styles.

7.1 DELEGATION

For example, when we understand the standard basic concept of the downward delegating process (defining a task, then providing resources including authority and determining feedback and measurement criteria), we are ready to accept the more complex idea that this type of delegation is actually measured by the amount of trouble that we can get into, as delegators. And this means evaluating both the complex nature of the task and the capacity of the individual to whom that task is to be delegated. It's another application of the "uncertainty" idea, here evaluated with people in mind.

Conversely, when given a task by upper management without sufficient resources (such as authority), we can quickly recognize that the complex idea of accomplishing that task effectively will rarely work. The more realistic solution is to immediately advise the delegator of the task what you believe *can* be done with the limited authority, money, personnel, and time given to you, and therefore why the total task cannot be accomplished. In effect, you're delegating "upward' and are saying, "If you give me this authority, you will get the delegated task done, otherwise this is what you will get, since this is all I have to put into it."

7.2 LEADERSHIP

Leadership behaviors are part of management style. These behaviors change with project life cycles. When there is an optimal fit between the situation and the leader's behaviors, it is more likely that the project will succeed. And one of the more critical parts of the project situation is the willingness of the project team members to accept the project manager's leadership. For example,

> Our findings strongly suggest that decisions made by typical managers are more likely to prove ineffective because of deficiencies in acceptance by subordinates rather than deficiencies in decision quality. (Vroom, 1983, p.508)

Acceptance by subordinates is an important measurement of the situation. When it dooesn't exist, uncertainty is sure to be high. This typically occurs at the beginning of project life cycles. But as time goes on, there is a social exchange between the manager-leader and the project team. The project manager's behaviors affect the behaviors of others and their behaviors affect his or hers. Uncertainty decreases and leadership behaviors become less directive and more supportive. Hersey and Blanchard's theory also outlines optimum leadership behaviors for general situations, but these situations are defined for different stages of the organization's maturity. If you use their ideas, you have to determine the maturity of your project team. It's another theory to use in building one's own leadership behaviors.

I have found little hard research data about the influence of the project manager's boss on the leadership criteria that the project manager uses. Perhaps this is because experienced project managers recognize that this influence may be greatly overrated. In my opinion, functional managers are more influenced by their perceptions of their bosses' leadership criteria than project managers are. In projects, there is less time for that influence to build up. The boss really doesn't know what the project manager is doing in as much detail as would be known in functions. Projects are not as predictable. Of course, it also follows that the project manager doesn't really know what the project team members are doing in detail. It's a very logical situation. If everyone knew exactly what every subordinate was doing, there wouldn't be any uncertainty and, consequently, there would be no need for managers. Finally, when all the unknowns are gone, there is no longer any need for humans. Computers can handle the tasks. But we really don't know or even need to know what everyone is doing, just that the overall tasks are defined and that the people involved in doing them can handle them.

7.3 CONFLICT

Project managers spend a lot of time communicating. Very little of that time is spent with superiors, some of it is spent with peers, and most of it is spent with subordinates. When there are conflicts and differences of opinion, they should be appropriately handled. An overall method begins with logic. Try to separate the facts from the opinions. Facts are relatively easy to determine. Opinion and emotions are much more difficult. The relative organizational position of the individuals in the conflict situation should *not* affect the specific conflict-resolution tools chosen. Of course, normal diplomacy and tact are always necessary.

The next chapter is intended to be both a summary and a beginning. It's a summary because it deals with how-to get things to happen. Designing systems and selecting better management techniques are important, but if we can't implement them, they're useless. And it's a beginning, because it also discusses the very basic considerations at the root of all of our actions—ethics and the stress that management can impose. As both technical specialists and management generalists, we make nonrepetitive decisions, and they have to be solidly rooted in a positive ethical stance.

If managing is defined as a form of decision making, implementing change is the way that we can make it happen, and ethics are why we make it happen the way that we do.

My Suggested Answers to the Case Study

1. Both of these problems are typical, since they are an uneven mixture of situation and person-based problems. We'll start to untangle it within the easier part of the problem first, the situation. Are company procedures being followed, and do those procedures satisfy the present needs of the project? If the answers to either of these questions is no, either follow the procedures or else determine if they should be changed. They were not being followed by May when she spoke to Carol, so that could be May's first step.

    Assuming the behaviors persist after procedures are followed, we can now proceed to handle the person problem. Carol's passivity seems to be saying, "I'll tell you anything you want just so long as you stop bothering me for a while." This passive kind of behavior may occur when the person is not in control of her job or feels that she is at the beck and call of others. Therefore, being passive is a reasonable (although short-term) solution and minimizes immediate conflict.

    Bill's aggressive behaviors might be well adapted to the demands of managing many projects and project managers. In projects, it's results that count; you have to meet the three golden rules—make it to specifications, on time, to budget. This kind of behavior can occur when the person believes that he or she is in control of the situation. It may even seem unreasonable to others, but once understood, it can be dealt with.

    Carol, as a passive, and Bill, as an aggressive, are both difficult to deal with. When Carol agreed to a short delivery schedule, she seemed to be saying, "Why rock the boat unnecessarily? Something might change before long and I might even make my original promise good if it does." Therefore the first step is for May to follow the correct company policy with respect to project operations planning and request manufacturing delivery schedules. If following policy does not result in acceptable responses from Carol, May can be reasonably sure that the situation is not the major problem. Now she must define solutions to minimize the passivity of Carol. The same process of following company policy first can be used with the aggressive behavior of Bill. An initial step for solving both problems could be developing a type of problem-solving meeting with each person. Always start solving management problems using what you know, then move to what you don't know. In this case, at least we know that some company policies and procedures apply.

2. When May was in the Engineering department, the policies and procedures were very clearly defined. She may have reported to both her functional boss, (at that time Ira) and some project manager, but she always knew that her primary limits were those of the Engineering department. As a project manager, her limits are much larger than before. Now she may even deal directly with the user or the customer. In this way, she is quite similar to a general manager. The whole situation in projects is less restricted. There is more personal freedom and more uncertainty.

3. The first general steps of the solution should be to select a neutral area and after defining the problem, May has to tell Carol it is a *total* problem that is being dealt with. When Carol is asked a question or receives a request for a delivery schedule, it cannot be interpreted according to her feelings about the person asking the question. A professional is supposed to provide an unbiased opinion. If Carol's delivery schedule won't meet May's needs, then there is still the option to either get the work done elsewhere or advise the project team (and management) of the delay in time to take corrective action if needed.

    Passive behavior is a kind of protective shield for the person. "If that shield isn't there, I might make a mistake or else somebody else won't like me, etc." This behavior should be countered with a clear and unmistakable message, "I want your opinion and you're the only person who can provide it. Others may offer additional data or other opinions, but your opinion is the only one that I want." Reinforcement is also important here. No acceptance of "wishy-washy" answers allowed!

4. The underlying strategy of open and direct problem solving is the same, but the tactics are different for an aggressive person.
    a. *Select a neutral area:* Ask Bill if they could meet at lunch time in a quiet office where no one will disturb them. Ask for his help in providing supportive criticism directed at solving the problems that he sees.
    b. *Define the problem:* "Bill, is the whole thing unsatisfactory? If it is, can you help me develop a revised project plan outline so that I can develop an acceptable one quickly? If any part is satisfactory, what part is it, because then we can concentrate on the unsatisfactory parts and revise them. Finally, are there procedures or methods that I should be using in order to prevent this from happening in the future?"
    c. *Set up a minisolution:* Find a minisolution that will be to Bill's benefit.

"Bill, could we use a few minutes tomorrow during lunch hour to review what I've done so far? Then, we'll both be sure that I'm on the right track and you'll be able to get what you want from my project."

d. *Reinforce the behaviors that you want:* "Bill, I really appreciate your helping me."

Unless you wish to use superior force (which is an alternative that doesn't seem to be readily available when we most need it), aggressive behavior is often difficult to handle until you know what the other person really wants. By providing "small" solutions of an overall problem, you're breaking the problem into understandable pieces and trying to reinforce successful cooperation rather than continuing the conflict.

BIBLIOGRAPHY

AUERBACH, SYLVIA, "Difficult Workers: Handle with Care," *Computer Decisions*, December 1981, p.178.

BUTLER, A. G., "Project Management: A Study in Organizational Conflict," *Academy of Management Journal*, March 1973, pp. 316–37.

FIEDLER, FRED, *A Theory of Leadership Effectiveness.* New York: McGraw-Hill, 1967.

FIEDLER, FRED, "Relationship of Traits to Leadership," oral presentation at *Academy of Management meeting, San Diego, Ca., Aug. 13, 1985.*

FIEDLER, FRED, "Styles of Leadership," in *Current Perspectives in Social Psychology*, (2nd ed.), eds. E. P. Hollander, and R. G. Hunt. New York: Oxford University Press, 1967.

FIEDLER, FRED E., MARTIN M. CHEMERS, and LINDA MAHAR, *Improving Leadership Effectiveness.* New York: John Wiley, 1976.

HILL, RAYMOND, "Managing Interpersonal Conflict in Project Teams," in *Matrix Organization and Project Management* (Michigan Business Papers, 64), eds. R. E. Hill and B. J. White, Ann Arbor, Mich.: University of Michigan Press, Division of Research, Graduate School of Business Administration, 1979.

KETS DEVRIES, MANFRED F. R., The Irrational Executive, Psychoanalytic Studies in Management. New York: International Universities Press, 1984.

McCALL, MORGAN W., "Leaders and Leadership: Of Substance and Shadow," in *Perspectives on Behavior in Organizations*, J. Richard Hackman, Edward E. Lawler III, and Lyman W. Porter. New York: McGraw-Hill, 1983.

MACCOBY, MICHAEL, "The Corporate Climber Has to Find His Heart," *Fortune Magazine*, December 1976, pp. 100–10.

RAUDSEPP, EUGENE, "Your Career: Questions Can Be Ambiguous, Don't You Think?" *Computer Decisions*, September 1981, pp.190–93.

SAYLES, LEONARD, and MARGARET CHANDLER, "The Project Manager: Organizational Metronome," *Managing Large Systems*, eds. L. Sayles and M. Chandler. New York: Harper and Row, pub., 1971.

THAMHAIN, HANS J., and DAVID L. WILEMON, "Conflict Management in Project Life Cycles," in *Matrix Management and Project Management;* (Michigan Business Papers, 64), eds. R. E. Hill, and B. J. White. Ann Arbor, Mich.: University of Michigan Press, Division of Research, Graduate School of Business Administration, 1979.

VROOM, VICTOR H., "Can Leaders Learn to Lead?" in *Perspectives on Behavior in Organizations* (2nd ed.), eds. Richard Hackman, Edward Lawler III, and Lyman W. Porter. New York: McGraw-Hill, 1983.

WORTMAN, LEON A., *Effective Management for Engineers*. New York: John Wiley, 1981.

OTHER READINGS

BASS, BERNARD, M., *Stogdill's Handbook of Leadership*. New York: Free Press, 1981.

BLAKE, R. R., and J. S. MOUTON, *The Managerial Grid*. Houston, Tex.: Gulf, 1964.

BLAU, P. M., and W. R. SCOTT, *Formal Organization*. San Francisco: Chandler, 1962.

BOWERS, DAVID G., JEROME L. FRANKLIN, and PATRICIA A. PECORELLA, "Matching Problems, Precursors and Interventions in O.D.: A Systematic Approach," *The Journal of Applied Behavioral Research*, 11, 4(1975), 391–409.

DEUTSCH, MORTON, *The Resolution of Conflict*. New Haven, Conn.: Yale Univ. Press, 1973.

GREINER, LARRY E., "Red Flags in Organiation Development," in *Current Perspectives in Social Psychology*, (2nd ed.), eds. Edwin P. Hollander, and Raymond G. Hunt. New York: Oxford University Press, 1967.

HAYAKAWA, S. I., *Language in Thought and Action*. New York: Harcourt Brace Jovanovich, Inc., 1964.

HERSEY, PAUL, and KENNETH BLANCHARD, "Leadership Style: Attitudes and Behvior," *Training and Development Journal*, May 1982, pp.50–52.

———, *Management of Organizational Behavior: Utilizing Human Resources* Englewood Cliffs, N.J.: Prentice-Hall, 1969.

HERSEY, PAUL, and MARSHALL GOLDSMITH, "A Situational Approach to Performance Planning," *Training and Development Journal*, November 1980, pp. 38–45.

KARP, H. B., "Working with Resistance," *Training and Development Journal*, March 1984, pp. 69–73.

KATZ, E., P. M. BLAU, M. L. BROWN, and F. L. STRODTBECK, "Leadership Stability and Social Change: An Experiment with Small Groups," *Sociometry*, vol. 20(1957), 36–50.

KEEN, PETER G. W., "Information Systems and Organizational Change," *Communications of the ASM*, 24, no. 1 (January 1981), 24–33.

KOONTZ, HAROLD, and CYRIL O'DONNELL, *Principles of Management* (3rd ed.). New York: McGraw-Hill, 1964.

KOTTER, JOHN P., and LEONARD A. SCHLESINGER, "Choosing Strategies for Change," in *Current Perspectives in Social Psychology* (2nd ed.), eds. Edwin P. Hollander and Raymond G. Hunt. New York: Oxford University Press, 1967.

McCLELLAN, D. C., *The Achieving Society*. New York: Van Nostrand, 1961.

McGREGOR, DOUGLAS, *The Human Side of Enterprise*. New York: McGraw-Hill, 1960.

———, *The Professional Manager*. New York: McGraw-Hill, 1967.

MARGULIES, NEWTON, and ANTHONY P. RAIA, *Conceptual Foundations of Organizational Development*. New York: McGraw-Hill, 1978.

MILLER, GEORGE A., "The Psycholinguists: On the New Scientists of Language," in *Current Perspectives in Social Psychology* (2nd ed.), eds. Edwin P. Hollander, and Raymond G. Hunt. New York: Oxford University Press, 1967.

NADLER, DAVID A., "Concepts for the Management of Organizational Change," in *Current Perspectives in Social Psychology*, (2nd ed.), eds. Edwin P. Hollander, and Raymond G. Hunt. New York: Oxford University Press, 1967.

NICHOLS, RALPH G., "Listening Is a Ten-Part Skill," in *Readings in Interpersonnal and Organizational Communication* (2nd ed.), eds. Richard C. Haseman, Cal M. Logue, and Dwight L. Freshley. Boston: Holbrook Press, 1971.

OUCHI, WILLIAM G., *Theory Z: How American Business Can Meet the Japanese Challenge.* Reading, Mass.: Addison-Wesley, 1981.

OUCHI, WILLIAM G., RAYMOND L. PRICE, "Hierarchies, Clans and Theory Z," in *Perspectives on Behavior in Organizations* (2nd ed.) eds. J. Richard Hackman, Edward Lawler II, and Lyman W. Porter. New York: McGraw-Hill, 1983.

PASCALE, RICHARD T., and ANTHONY G. ATHOS, *The Art of Japanese Management.* New York: Penguin, 1982.

PERROW, CHARLES, "The Short and Glorious History of Organizational Theory," in *Perspectives on Behavior in Organizations* (2nd ed.), eds. J. Richard Hackman, Edward Lawler II, and Lyman W. Porter. New York: McGraw-Hill, 1983.

RAHIM, M. AFZALUR, "A Measure of Styles in Handling Interpersonal Conflict," *Academy of Management Journal,* 26, no. 2(June 1983), 368–76.

SCHEIN, EDGAR H., *Organizational Psychology* (3rd ed.) Englewood Cliffs, N.J.: Prentice-Hall, 1980.

———, "Does Japanese Management Style Have a Message for American Managers?" *Sloan Management Review,* Fall 1981, pp. 77–90.

SELYE, HANS, *Stress Without Distress.* Philadelphia: Lippincott, 1974.

STEWART, JOHN M., "Making Project Management Work," *Business Horizons,* Fall 1965, pp. 266–85.

WIGGINS, JERRY S., *Personality and Prediction, Principles of Personality Assessment.* Reading, Mass.: Addison-Wesley, 1973.

Chapter 10   Getting Things Changed—Handling Stress and Ethics

PERSONAL MANAGEMENT STYLE: THE INDIVIDUAL SYSTEM

PEOPLE MANAGEMENT: THE PROJECT SYSTEM

PROJECT OPERATIONS: THE PROJECT SYSTEM

MEASURING: THE ORGANIZATIONAL SYSTEM

FORECASTING: THE ORGANIZATIONAL SYSTEM

FOUNDATION: PROJECT MANAGER'S AUTHORITY

THE CASE OF THE FRACTURED PLASTIC

Cast:

Rose Eggen, Project Manager, Store Fixtures Project
Noah Johnson, Project Engineer for Freezer Displays
Arnold Slatman, General Manager, Store Constructors, Inc.
Gale Freed, Materials Scientist

Store Constructors, Inc., develops and manufactures food-preparation equipment and installs it as new fast-food stores are opened by a few major fast-food chains.

It was a lovely Friday morning in April. Rose had just finished the first critical design review meeting and was relaxing for a minute in her office. The design review meeting was intended to determine if all the functional specifications of the food-preparation equipment were met. The meeting seemed to go well. The food-handling conveyors, the cooking tops, the ventilation machinery, and the computers that controlled everything were all on schedule and had passed the first operating tests. The way seemed to be clear for approval for manufacturing, and then Gale appeared at Rose's office.

Gale: Good morning, beautiful day, isn't it? Say, I just got a copy of the agenda for the design review meeting. I guess it must have gotten lost in the company mail because I received it after the meeting was over. I would have liked to attend.

(She then sat down as if preparing herself for a long visit. Rose stole a glance at her watch. She was supposed to meet with Arnold in his office in 15 minutes about another project that she was managing for the company. How long would this take?)

Rose: Why? We've worked together for years now, and you always tell me that those meetings are a waste of your time.

Gale: You're right. Usually they're very boring affairs as far as I'm concerned, because my specialty, "materials", has a low priority around here. But these store fixtures interest me.

Rose: Well, as you know, the meeting was a problem-solving meeting. Everybody was supposed to get their inputs in beforehand, so the agenda could be distributed. By the way, the meeting went very well. Noah even was on time with the new design of the refrigeration unit that keeps the food preserved before they are ready to cook it and serve it. That must be a "first." Usually, we have to delay things until he finishes something or other. I don't know why he has trouble getting his jobs done on time. (Rose noticed that the time was fast approaching for her meeting with Arnold. She had better get Gale to come to the point quickly.) Anyhow, what did you want to talk about?

Gale: Well, I know that it's really not my job since I'm only responsible to be sure that the materials we use are the best for the design. To get to the point, I happened to see the assembly design for Noah's new refrigeration unit, and I think that there's a problem with it. I noticed that the throttle valve for the high-pressure refrigerant is right next to the front plastic panel of the unit. The design standards for that plastic require that it be used only in room temperatures and conditions. If some employee happens to accidentally run one of the food handling carts into the front of the unit and it severely dents the plastic just at the point where the valve is, the valve could rupture.

If that happens, the high-pressure refrigerant will escape, and the flow will directly impinge on the plastic panel. The refrigerant itself is nontoxic, so it won't hurt anybody, but that plastic becomes brittle at low temperatures. If someone bumps it again before it has a chance to warm up, it will fracture and the pieces could fly around and hurt someone.

Rose: Let me see if I understand you. This is what must happen in your "worst case" scenario. Someone hits the plastic front hard enough to break the throttle valve, while the refrigerant is under pressure. Then, they do it again as the refrigerant escapes and are hit by flying plastic. Is that right?

Gale: Right.

Rose: It's a bit difficult to imagine all these things happening at once, isn't it? But, have you discussed this with Noah or anybody else in his engineering design group? What do they say?

Gale: They say that the possibility of all this happening is so remote that they're not concerned. They also said that there isn't really enough room inside the refrigeration cabinet now, and if they had to move things around, it would delay the project for several weeks while they completed some more space and func-

tion tests. I was going to bring the point up in front of the whole project team during the design review meeting but, as I told you, I didn't know about it till today.

Rose: Let me look into it, and I'll get back to you. I have a meeting with Arnold in five minutes about another project.

During the meeting, Arnold brought up several items that were holding up two other projects. Rose became involved in working them out, and the plastic problem that Gale had mentioned that morning never got discussed. Several weeks went by. Rose received a note from Gale about the possible interaction of cold refrigerant and the plastic. Rose decided to take it up with Noah. She quickly went over to Noah's office. He was in.

Rose: Hi, Noah, got a minute? I want to discuss a small problem with you.

Noah: Sure, what can I do for you?

(Rose quickly outlined the problem as Gale had presented it. Noah had an exasperated expression on his face when she finished.)

Look, Gale has been a pain in my side every since we took on this store fixture job. I think that she's just bored because she doesn't have enough to do. Everytime some new development or potential problem is outlined in those technical magazines she reads, she comes running down here to tell us about it. She's wasted more of my time since this job began than I can afford. I've looked into this so-called problem. The chain of events that have to happen before a problem occurs is almost impossible. And even if we wanted to change the design at this time, there's too much work done already. The cost would be prohibitive.

Rose thanked Noah, went back to her own office, and phoned Arnold. She asked if she could meet with him. He told her to come over since he had a few minutes right then. She went over to his office. She quickly reviewed the situation.

Arnold: O.K., now that I understand the situation, what do you want to do? Consider these points, First: we're two weeks late on this store fixture job. Second, this is the first time that we've broken into this new market and our success in meeting the schedule within cost will probably determine whether we stay in this business or not. Third, if you succeed as the project manager this time, who knows what kinds of promotions you'll get for the next job?

Rose: Thanks a lot, you're making my job a lot tougher with these points. I really don't know what to do. Noah seems to be so sure about his design and Gale does occasionally get overexcited, but I don't want to approve something that could hurt somebody.

Arnold: It's your decision, you know. Have you really determined the cost-benefits in this decision? There's lots of other questions that I could ask you, but I've got to go. I have a board meeting in five minutes and I'll be late if I don't go now.

The next day, Rose met with both Noah and Gale. Arnold joined the group.

Rose: I have a suggestion that might solve the problem for everybody. Why not put a decal on the front of the fixture that cautions against bumping into the unit when it is operating?

Noah: That's not bad. Even if the warning is not needed, and I don't think that it is because of the impossibility of all those things happening, this suggestion is easy to use and it won't hold up production.

Gale: I don't like it. A warning decal can come off. I think that the whole design should be redone.

Rose: Well, why not include another warning in the operating instructions in addition to the decal. Then the store operators will be informed about this very remote problem.

Arnold: I don't think that you want to point out a problem that may never happen. How about this suggestion? We'll ask the training people to emphasize that store operators have to be very careful about the front of our refrigerating units. We can tell them that it will cost them a lot of money to repair them if anybody bumps into them. That's true and since those owners don't like to spend money on repairs, it should do the trick.

Gale: Arnold, I still think that we should redesign the units.

Arnold: No way, it's too costly, but I'll go along with putting a warning into the training manuals.

Everyone seemed to agree with that suggestion. Gale agreed too, but she was very quiet during the rest of the meeting. As time went on, the refrigeration units went into production and many hundreds were successfully installed in stores. The success of this project supported the growth of Store Constructors, Inc., into other kinds of renovations. Rose was promoted to Chief of Project Management. About a year later, Gale left the company to take another job with a competitor.

Soon after that, Arnold received a phone call from the Vice President for Engineering of the chain of food stores. He was informed that a refrigerating unit in one of their fast-food stores had broken and wanted to know if Store Constructors would look into it.

Arnold immediately sent out a field inspection team. They reported back in several days. It seemed that the employees of this store had been using the front of the refrigerating unit as a backboard against which they bounced small coins for gambling during lunch breaks. That was not approved by the store management. Then, several of the employees had gotten into a disagreement and pushed against the refrigerating unit when it was operating. The plastic front had been displaced enough to break the value, the refrigerant had quickly sprayed out, freezing the plastic against which it impinged. Then the plastic broke, sending slivers into one employee's eye. The employee had lost the sight of that eye. While Arnold was reading this report, he received a phone call from the firm's attorneys.

A criminal and civil suit had been filed against the company by the employee who had lost his eye. There was more news; the newspapers had found out about it and had interviewed Gale. She had blamed the company for negligence.

Questions

1. Can you list several alternatives that were available to Rose when she first heard about this problem? What procedure would you use to compare them?

2. Who do you think should be held personally liable for this? The company, Noah, Rose, Gale, Arnold? Why?
3. What can the company do about preventing or eliminating this kind of situation? What about individuals? Is this action necessary? Why?
4. Assuming the worst situation, what should you do as a project manager? If you were Gale, what system would you use to protect yourself? What do you think the results would be?
5. Do you believe that a decal on the front of the equipment and a warning note in the operating manual are sufficient? Why?
6. Can you think of any situations in your experience where this kind of thing could or has happened? What would you do now? Do you think that it would work?

1.0 INTRODUCTION: IMPLEMENTATION IS DIFFICULT

Designing systems is one thing, implementing them is another. It's a much tougher job. A system is almost like an intent, outlining general rules and procedures for people to follow. But intentions are not always followed in practice. Implementing any design means that some people will have to act or behave differently, and there are more problems with that than there ever have been in designing systems. It's one thing to describe how people should behave, it's quite another to get them to do it. Consider the obvious example of the simple traffic light system. A red light means "stop" and green means "go." It's logical, very useful, and apparently quite easy to understand as a system. However, it doesn't always achieve the correct behavior. Systems implementation requires new behaviors to meet some predicted, and novel, requirement *consistently*. Intellectual understanding may be a good beginning, but like the green and red traffic lights, understanding doesn't always result in consistent behavior. In addition to understanding, implementation depends upon the individual's acceptance, that is, his or her realization that it's better to follow the system than not to. When the traffic fine for passing a red light costs more than the few minutes gained, the person accepts the rules and will probably stop.

In other words, passing those first hurdles to get the organization and the project team members to intellectually accept the forecasting or the measuring or the other systems designs is only part of the task. It's undoubtedly an important beginning, but when the time comes for people to actually follow the system procedures, there are often pitfalls, because each individual (either cognitively or unconsciously) determines how he or she will actually behave.

This isn't as much of a problem for functions as it is for projects, since functions are long-lived. Projects are not. The relatively limited project life provides less time to develop consistent behavior. There's less time for everyone to learn who has what information, how they must share it, the rules about how to do it, and then behave appropriately. But there are some compensating advantages in projects. Since project organizations are usually small, behavioral variances from those pre-

dicted by the system procedures are more obvious and can be dealt with promptly, either by changing that behavior or else changing the system. Sometimes variances lead to improvements. Variances are not necessarily bad, just indications that there is a difference between forecast and measurement.

The management techniques in the systems we have been designing are expected to provide pragmatic solutions to repetitive project problems. If valuable time is not to be wasted during the critical initial project phases, the systems including these techniques must be designed (or modified if they have been designed for previous projects) and implemented quickly. The time at the beginning of a project is very valuable, and using it to develop and implement systems is expensive. Most projects also require a high level of social and interpersonal skill. Since every project is somewhat unique, the importance of designing and implementing project systems that result in predictable behaviors should be apparent to everyone. When we know how to handle repetitive jobs, we can concentrate on learning how to handle nonrepetitive problems. Project systems are intended to help solve repetitive problems. But project systems design and implementation can be difficult in addition to being expensive; therefore, as much as possible should be done before the project itself has to appear.

2.0 THE NEVER-ENDING IMPLEMENTATION PROCESS

However, even when the importance of successful systems implementation is obvious to the project manager, it's not always equally obvious to the equivalent functional managers who supply the team members. The following comments are typical: "After all, those people worked well together in their original departments, why is it necessary to change things just because they have been assigned to your project?" "Why is it necessary to spend so much time at the beginning of a project talking to each other? Why not just get out and do the job?" There can be real gaps in understanding. Furthermore, an initial successful implementation of project-related systems doesn't end the implementation process. Often project-oriented systems are under pressure from the functional organization. They therefore require almost constant support.

> For the functional manager, a matrix organization is often experienced as involving a loss of status, authority and control. He becomes less central and less powerful as parts of his previous role as initiator move from the function to the business manager . . .

> Because the structural element of the matrix is so fiendishly difficult to many, we observe organizations trying to shed the form while maintaining the substance. Our diagnosis is that it can be done successfully only where appropriate matrix behavior is so internalized by all significant members that no one notices the structural shift. Even then, however, we anticipate that through the years the structural imbalances will increase. (Lawrence, Kolodny and Davis, 1977, p. 511, 517)

Thus, we need to consider an almost never-ending design, support, and maintenance process against organizational forces that are often in opposition to these

systems. The opposition can continue because of what the change represents in terms of lowered prestige for equivalent functional managers, more complex cost reporting for accounting departments, and less standardization across the company for top management. These are possible sources of destructive friction. This is often a fair description of the situation that exists, so we have to deal with it as it is, not as we would like it to be.

2.1 WHEN DO WE HAVE AN IMPLEMENTATION?

Systems implementation means change, and change is the basis for a modification of human behavior. When the actual behavior and the systems predictions coincide more or less consistently, change subsides. The system is implemented. In other words, we design systems because we want some expected behavior (which is probably different from present behavior), and if we perceive a change from present to expected behavior within expected variances, the system has been implemented well. We are not concerned with behavior that is presently satisfactory, just that which is presently unsatisfactory.

On the other hand, if the system's predictions have not been met, either the system is unsatisfactory (back to the drawing board to do another design iteration) or else it hasn't been implemented well (back to the person who is supposed to be following the system's procedures). In either of these cases, the intended behavior and the expected results have not coincided. This defines how to measure the end result or the success of an implementation process. In general, it is how well the forecast and measurement agree. Does this sound familiar to you? But how do we develop the process itself?

2.2 AN APPLICABLE CHANGE THEORY

There is a fairly well known, general theory about change that we can use as a beginning. (Lewin, 1951) It's one of the few universalist theories that can be adapted to specific people. In very general terms, Lewin suggests that change consists of three sequential stages. He calls them *Unfreezing, Changing or Developing New Behaviors,* and then *Refreezing.* He also suggests some methods to accomplish these stages. These suggestions make his theory adaptable to fit various situations. For example,

> Stage 1: *Unfreezing—Creating Motivation to Change*—Present behaviors must actually be disconfirmed or must fail to be confirmed over a period of time. . . . There must be a creation of psychological safety. . . . No matter how much pressure is brought to bear on a person to change, nothing will happen till that person feels that giving up old responses is safe and learning something new is equally safe.
>
> Stage 2: *Changing—Developing New Behaviors*—Identifying with a role model or some other person and learning to see things from that other person's

point of view. . . . Viewing the situation for information about one's particular problem and selecting that which is applicable. . . .

Stage 3: *Re-freezing*—Testing new behaviors to see if they are really appropriate. . . . Will significant others accept the new behavior patterns?

This theory provides relatively straightforward processes. The unfreezing occurs with the initial project charter approvals by top management. The changing begins with the participation in the bottom-up revision of the charter. Re-freezing occurs when the expected behaviors and the systems requirements match. However, changing is a very personal business and participating in the charter revision process is only part of it. There must be learning and motivation.

1. Any change process involves not only learning something new, but unlearning something that is already present and possibly well integrated into the personality and social relationships of the individual.
2. No change will occur unless there is motivation to change, and if such motivation to change is not already present, the induction of that motivation is often the most difficult part of the change process.
3. Organizational changes such as new structures, processes, reward systems, and so on occur only through individual changes in key members of the organization; hence organizational change is always mediated through individual changes.
4. Most adult change represents attitudes, values, and self-images, and the unlearning of present responses in these areas is initially inherently painful and threatening.
5. Change is a multi-stage cycle, ... and all stages must be negotiated somehow or other before a stable stage can be said to have taken place. (Schein, 1980, p. 208)

2.3 QUESTIONABLE WAYS TO IMPLEMENT CHANGE

There are other approaches to implementing change. Some of them that have been popular in the recent management literature include sensitivity training (or T groups), managerial grids, survey feedback interventions, organizational development, quality circles, and management by objectives.

Very briefly, *sensitivity training* allows a leaderless group to highlight what they think are necessary changes in one's behavior. (Perrow, 1983). *Managerial grids* provide training in an attempt to have the individual exhibit an ideal leadership set of behaviors. (Blake and Mouton, 1964). *Survey feedback* provides summarized information from the group back to the individual with the intent of restructuring the person's cognitive views (Nadler, 1967). *Organizational development* provides a supportive environment in which the person will try to improve his or her performance (Trice and Beyer, 1984). *Quality circles* is a modification of organization development that is primarily applied to improving the quality of products and processes (Hackman, et al, 1983). And *management by objectives* is designed to provide individualized goals and rewards (or penalties), within an overall organizational framework, for people who achieve (or don't achieve) these goals (Hackman, et al, 1983).

In my opinion, the major difficulty with all of these approaches is their underlying assumptions about people. They assume that everyone inherently wants to improve social interaction on the job and expects that the change from this interaction will benefit them. It's almost a theory Y view of people (McGregor, 1960). Most of these techniques don't provide the benefits expected because their basic assumptions don't hold. While some people may have a need for improved social recognition, some may just want more money, a bigger office, to be left alone, or whatever. The research is clear.

> Management should be advised that the attempt to produce change in an organization through managerial grids, sensitivity training, and even job enrichment and job enlargement is likely to be fairly ineffective for all but a few organizations. The critical reviews of research in all these fields show that there is no scientific evidence to support the claims of the proponents of these various methods, that research has told us a great deal about social psychology, but little about how to apply the highly complex findings to actual situations. (Perrow, 1983, p. 96)

Or, in even more direct language, these authors feel that these change-implementation processes are based on a myth, not even an assumption.

> Rites of renewal apply to a variety of elaborate activities intended to refurbish or strengthen existing social structures and thus to improve their functioning. Prominent current examples include most so-called organizational development (OD) activities, management by objectives (MBO) programs, job redesign, team building, sensitivity training, survey-feedback interventions, quality-of-work-life (QWL) programs, quality circles, and so on. . . .

> OD techniques appear to be based on a combination of humanistic and positivistic scientific ideologies. For example, an activity such as team building is justified by the myth that a family-like bond exists within work groups that can be nurtured and used for the company's benefit. . . . Most OD activities use certain standardized sets of techniques that can be viewed as rituals, in that few of them have demonstrated their intended, practical effects. (Trice and Beyer, 1984, p.660–661)

I think that we have avoided many of the unwarranted assumptions about people and the myths that the project team is just like the person's family. Of course, we require motivation (as evidenced by achievement-oriented behavior), cooperation (operating as a team member), and commitment (exerting one's self for project goals) in the prior People Management system, and this system uses techniques of reward or penalty to get them. These rewards include many things in addition to social recognition, such as money or promotions. We also designed our systems with the advice and consent of the project team members. Self-interest, no matter how manifested, is the best motivation force for change, and every team manager has a chance to exhibit his or her self-interest during the design processes.

When implementing systems, there will be variances. These variances in actual versus predicted behaviors require that we use several of the conflict-resolving tools covered in Chapter 9—confrontation, compromise, forcing, smoothing, and ignoring. We also have the guidance of a well-established change theory (Lewin's unfreezing, change, and refreezing) as an overall plan. And finally we are going to be cautious about adopting current management fads or myths that use universalistic implementation solutions. It's better to take the more direct approach of analyzing the specific situation that we face as the project manager today.

3.0 HOW TO IMPLEMENT CHANGE SUCCESSFULLY

In my opinion, the first variable to consider is the speed of change. It is rare that long-lasting change takes place *quickly* without trauma of some kind. Trauma can happen in very unusual and temporary circumstances such as in an emergency that changes things quickly. For example, ''We've got to make that delivery by September or else we are bankrupt, so our regular working hours will include Saturday overtime for the next six months!'' This trauma will change behavior quickly, but these changes will not last when the pressure is removed. The kind of systems implementation that we want, the kind that builds a supportive, adaptable project culture, occurs more slowly because we are concerned about individuals and their motivation, cooperation, and commitment. This doesn't occur under duress. In projects, the use of directive authority is generally undesirable, since it would inhibit the creativity of individuals. So we make sure that everyone is allowed maximum freedom, removing unwarranted stress.

The project charter is, of course, given to each team manager when he or she is initially recruited. Then, during the bottom-up meeting, each team manager is able to provide both his or her own charter and comment on revisions to the overall project charter. Usually this provides sufficient time for any resistance to surface and be handled in any of the appropriate ways that we discussed in Chapter 9. When resistance is thus dealt with very openly and quickly, the change process of unfreezing is supported, and intellectual understanding begins. But even intellectual understanding and acceptance of a change don't always result in equally acceptable behaviors. It takes practice, and that may take time. For example, let's look at the fifth system—Personal Style. Just consider how long it will take you to easily assume the behaviors required of the leader if you follow Fiedler's leadership model.

However, it should be possible to implement the first four systems for others in a general format and get a good start on the fifth system for one's self. When this is done, the next project managed will be that much easier to handle. In a very general way, it's almost like raising a family. The first child is sometimes more difficult than those who follow, because the parent has learned from the first one how to generally handle the rest. In a very general sense,

> An old truism [exists]: introducing a product or process to the world is like raising a healthy child—it needs a mother (champion) who loves it, a father (authority figure with resources) to support it, and pediatricians (specialists) to get it through difficult times. It may survive solely in the hands of specialists, but its chances for success are remote. (Quinn, 1985, p.79)

Translating this extract into project terms, the project manager is the "champion," the boss is the "authority figure with resources," and the "pediatricians" or "specialists" are the team members who support the project. Getting change implemented (or learning how to complete the child-raising project) cannot be done overnight. Presenting the charter to the team members and supporting their bottom-up reviews coincide with the implementation process called *the small win*. It is a good way to start.

> The author's research has focused on . . . "the theory of the small win." Patterns of consistent, moderate size, clear-cut outcomes—patterns of small wins—are a special subclass of managerial activity patterns influencing future change. . . .
>
> Frequency and consistency are two other primary attributes of effective pattern shaping. A pattern of frequent and consistent small successes is such a powerful shaper of expectations that its creation may be worth the deferral of ambitious short term goals. . . . Support of completed actions typically generates further actions consistent with the rewarded behavior. . . . By varying his patterns of reinforcement he can substantially influence people's behavior over time, often several levels down in the organization. (Peters, 1983, p.16–29)

"The small win" implementation process can be compared with Lewin's change theory. Unfreezing can be done in small sections. It doesn't have to be done all at once. Pascale and Athos (1982), describe this very effectively in their book about Japanese management techniques. They say that "The best course may be to move toward the goal through a sequence of tentative steps rather than by bold stroke actions." Accepting a lack of clarity in the situation of the specific team managers might even be desirable. The overall systems covering the project may be clear, but it may not be necessary to request equal clarity from team managers with respect to the way they handle their own responsibilities. Let them manage their own areas. When one considers that very few projects are completed exactly according to an initial forecast, it is the overall project goals of technical achievements, time, and finances that will probably be re-evaluated as the project progresses through successive interactions. If the overall project system has to be changed or if a failure within a team manager's system requires corrective and detailed action, a redesign can be done and evaluated. The repetitive, successive evaluation process involving the team managers might, in itself, be an aid in obtaining a common culture and the consequent predicted behaviors.

3.1 PARTICIPATION

The theory of "the small win" or the lack of complete clarity usually supports another activity called *participation*. Participation can be a powerful tool in the change process if used wisely, since "it is inviting someone into the situation and that also invites them to accept the premises of that situation. And thus to plan within the same constraints that they may have previously opposed. . . ." (Salancik, 1982, p.218) This is *not* the same thing as proposing participation merely to undercut opposition. This kind of participation allows the opposition to join the team and contribute. Participation cannot be requested and then ignored by the manager receiving it. It must be responded to openly, otherwise the level of conflict will escalate. The manager cannot promise and then not deliver. As a practical consideration, because projects have so few managers involved, it becomes very easy to detect management subterfuge. Don't even think about it.

3.2 GETTING STARTED

Using the theory of "the small win" doesn't mean that all alternatives except the one under consideration are discarded. Just as a feasibility study is used to reduce uncertainty in technical areas, a tentative system may be partially implemented for a short time period and then evaluated or improved by the project team. Or, if time is very short, several concurrent approaches can be tried and the best one kept. For example, dual reporting may be tried and evaluated on a "time" basis, for example, when one's ordinal score is an overall 9, which is multiplied by the time he or she has spent on the project. The total scores are then divided by the total time spent. Or, the dual reporting system can be tried on a specific-goal basis. For example, specific goals must be accomplished by a specific time. One's appraisal will depend on how close he or she comes to achieving these goals. This is just like trying several concurrent design approaches when time is short and unknowns are high. In other words,

> The most common mistake that managers make is to use only one approach or a limited set of them *regardless of the situation*. . . . A second common mistake that managers make is to approach change in a disjointed and incremental way that is not part of a clearly considered strategy. (Kotter and Schlesinger, 1983, p.548)

3.3 TRAINING THE INDIVIDUAL: WHAT HAPPENS?

Management training can educate people to help implement change. However, this kind of formal training is *not* usually a major responsibility of the project manager. It is the functional manager's responsibility. However, the project manager is expected to recommend people for further training if they perform inadequately or to deal with them after training. Therefore, it might be wise to understand what training is and what it accomplishes.

By definition, *training* is intended to provide new and useable behaviors through providing different information and then allowing the trainee to practice what has been learned. It starts by restructuring cognitive processes, thereby improving resulting job performance. But it is almost an act of faith. The resulting change (or lack thereof) in job performance is contingent upon many factors that are outside the training experience. Therefore, most formal training is inadequate by itself. It requires practicing new behaviors, and some of the opportunities for this practice are presented by the project.

> To transfer to the job what has been learned during training, trainees need to receive continuing opportunities to practice what has been learned. This, in turn, needs to be coupled with feedback or self-reinforcement about the trainees' practice efforts. (Bass, 1981, p.580)

Practice and feedback do occur in projects. The project manager develops goals for each team manager and provides assessment and feedback as those goals are achieved. But there are other considerations.

> Sykes (1962) conducted a case study of a firm in which participants in a management development program regarded the training as unsuccessful because top management was unwilling to correct grievances and unsatisfactory conditions toward which they had become sensitized during the discussion sessions. (Bass, 1981, p.581)

In my experience, the case described in the extract is typical. If the training of team members in management isn't supported by top management, when those team members try out their newly learned ideas and skills, there is a *decrease* in job performance. This may happen in training team members in new project-related behaviors if the functional managers provide no support. It makes project management systems implementation more difficult. This potential decrease in job performance is only partially alleviated by the participative environment of the bottom-up approach. It must be totally supported to be optimally effective. If upper management is unwilling to support project management systems, the probable success of any project will drop. In addition to this,

> Most important to whether training will modify behavior back on the job is the trainee's immediate supervisor. . . . Haire (1948), Fleishman (1953), and many others have suggested that for management training to be effective, the entire management of the organization should be subjected to the same or similar program. It is self-defeating to train lower-level managers in a style of supervision incompatible with that of their superiors. (Bass, 1981, p.582)

Open resolution of common problems cannot flourish in a semi-closed environment. When different organizational levels use incompatible supervision styles, the signals produced are confusing. When people being recruited for a project have been trained to operate with one set of signals and then receive an incompatible set, implementation will fail and so will any projects. With no consistent systems in place, chaos is endemic.

For example, conflict could simmer under the surface until it finally erupts at the most inopportune time. Since projects are, by definition, novel, they need creative solutions. These solutions are occasionally questioned. The systems provide a fair and open forum. Not all answers are self-evident; some must be tested. And tests must be conducted in a non-punitive atmosphere. An incorrect answer is not "wrong," it is another piece of data that helps the project when the correct answer is discovered. If we already knew all the answers, we wouldn't need a project.

In other words, when the project environment is open and supportive and the particular team manager's functional organization is not, there is a built-in conflict. The team manager is getting different input signals. When this happens, successful recruiting and integration into a team are almost impossible, because the project manager is trying to "undo" behavior that has served the team manager well before. There's often not enough time to effect new behavior, even if the project was assigned the retraining task.

One obvious solution to minimize the potential problems in differences between the viewpoints of team members and the project manager is to define the overall environment very clearly. A clear definition of the project manager's authority and the approval of the first draft of the charter by top management helps. Any potential conflicts will appear at the next stage when the team managers review the project charter. Resolving conflict during that review requires confrontation. The confrontation technique is also an excellent informal training tool.

3.4 SUMMARIZING CHANGE IDEAS

In all cases, the willingness to change is tied to the individual's self-interest. The following are action steps to motivate change through using an individual's self-interest:

1. Identify and surface dissatisfaction with the current state. . . .
2. Build in participation in the change. . . .
3. Build in rewards for the behavior that is desired both during the transition state and in the future state. . . .
4. People need to be provided with the time and opportunity to disengage from the present state. (Nadler, 1983, p.556)

These action steps can help the transition from the unfreezing stage, through the change stage and on to the refreezing stage by providing a clear image of the future, its benefits, and opportunities. These benefits should include rewards for changed behavior that meets predictions, during both the transition period and thereafter.

Finally, time and opportunities must be provided to extinguish the present behavioral patterns. This implementation of change also fosters project culture. Discussing common problems encourages the development of cohesive groups and the complex, informal communication channels that are so necessary to projects.

Few of the change ideas that we've discussed can be neatly separated into ideas that you apply to groups, to other individuals, or to one's self. They are all interactive.

4.0 STRESS

Managing projects can be very stressful. Stress occurs in some people because the multiple tasks involved in taking on a new job each time and having to organize the whole thing from the foundations up seem to be greater than one's mental resources can support. It is like a psychological overload. These are some practical suggestions to reduce the overload.

> Don't strive for perfection (it doesn't exist), rather strive for the best in each category; genuine simplicity in life earns much goodwill and love; keep your mind on the pleasant aspects of life and on actions which can improve your situation; forget ugly events. . . . When frustrated, take stock of your past successes and rebuild your confidence; when faced with a painful task yet very important, don't procrastinate—cut right into the abscess to eliminate the pain instead of prolonging it by gently rubbing the surface; love your neighbor and work hard to earn your neighbor's love. (Selye, 1974, p.134–135)

Selye's suggestions may seem to be similar to an act of faith, but they work. Here are some other suggestions that might be useful.

1. Is that disaster certain or probable, and if it happens, what's the worst that can happen to you? Remember, slavery is no longer socially acceptable. You may not like the idea, but it is still only a job; you and the world will go on.

2. Avoid excessive or temperamental language. Do you *really* have the worst project in the world? Define exactly what is bothering you. Perhaps it's not, "I can never meet a schedule," but it's "The schedule was set too optimistically by my boss. It will never get done, and I'm hesitant about telling anyone. Maybe it will work out."

3. What do you really know, as opposed to what you *think* you know? Shakespeare said it very well, "Present fears are less than horrible imaginings." Deal with what is and less with what you interpret.

4. There is a big difference between the symptoms and the disease. Would that temporary sickness disappear if that design review meeting would too? Well, the meeting won't.

5. You are responsible for most of it, and you can't do anything about the rest of

it. You can't blame everything that goes wrong on someone else. The uncooperative team member or your boss are all nice candidates for blaming. If you can change them, fine, but it's not likely. Therefore, do what you can and accept the responsibility for it.

6. You're not the only one who's been in this situation. You are unique, but someone else, somewhere, has been through a similar situation. I and many others survived as project managers; you can too.

Assuming that you have the usual tolerance that most of us have for the minor (or major) problems of everyday life, try the following technique to minimize the physical effects of stress when change is required. I've found that by (privately) tensing every muscle in my face and hands as hard as I could for 30 seconds, and then suddenly letting them go, there is an immediate feeling of complete relaxation. Then, if time and space permit, I change my thoughts by reading a poem, enjoying the scenery, or perhaps going for a short walk. The first step is to calm down. Then I'm able to diagnose or define the problem causing the stress more clearly. I may not be able to control my feelings, but I can control my own thinking processes. When I do this, I find that the stress decreases. The stress could be from not wanting to be in a certain situation anymore. But I am as much in control of the situation as anyone can be. I can't allow the situation to control me because I have choices.

5.0 ETHICS

Projects always deal with unknowns; that's why they are created. And there are occasions during every project life when the unknowns *cannot* easily be classified into neat categories that are typical of the sciences. The level of uncertainty is no longer primarily concerned with objective things like resources, people, and money. It's concerned with subjective things like personal values and integrity. They apply to the important kinds of decision that involve ethics. As Peter Drucker points out,

> There is only one set of ethics, one set of rules of morality, one code, that of individual behavior in which the same rules apply to everyone alike. . . . What difference does it make whether a certain act or behavior takes place in a business, in a non-profit organization, or outside any organization at all? The answer is clear: none at all. (Drucker, 1985, p.235–254)

Most of the industrial world holds those of us who make technical decisions personally responsible for those decisions. And that applies no matter what organization we work in. This is called a *medical model* because it is most applied to physicians. In a very simplified sense, the physician is expected to exhibit only a level of competence that is consistent with his or her immediate peers in medicine. It works something like this for the rest of us: when you make a technical or professional decision, you are responsible for the effects of that decision and your competency is determined by your technical peers. Your peers, of course, may now be the whole world, but it is a start. In other words,

> The professional can usually be held liable for damages only if he or she performed below the level of expertise in the immediate community. Increasingly "community" has come to mean the entire inhabited world. (Kolb and Ross, 1980, p.85)

5.1 LEGAL LIABILITIES

There is an interesting situation with respect to personal legal liability that has been developing for a number of years. For example, consider the obvious example of product safety and assume that your project has just been completed and it is out on the market successfully. Kolb and Ross point out that you may be liable for damages if

- Your product is defective in its design (not suitable for its intended use).
- Your product is defective in its manufacture (inadequate inspection and testing).
- Your product is inadequately labeled as to proper use and possible warnings.
- Your product is packaged in such a way that safety-related shipping and handling damage can result.
- Your product is packaged in such a way that parts can be sold separately or instructions can become detached before sale, allowing sale in an incomplete and dangerous form.
- You fail to maintain proper records of product sales, distribution, and manufacture.
- You fail to maintain proper records of failures and customer complaints.

> Not only manufacturers, but parts and component suppliers, wholesalers, retailers, lessors, and bailors have been held liable for defective products, even though the wholesaler, for example, might not have uncrated the product before passing it along. . . .
>
> With the privity requirement abolished the courts have allowed recovery by the family of the product buyer, by other members of the buyer's household or by guests therein, by innocent bystanders hurt by someone else's product, or by children who couldn't understand written danger warnings, and even by Hispanics who couldn't understand warnings in English. . . .
>
> The burden of proof had been removed from customers and placed upon businesses.
>
> The passage of time matters little. There is effectively no "statute of limitations" on how long a manufacturer is responsible for his products. There are cases where the makers of a drill press have been held liable for injuries caused by their product, even

though the press had been resold many times in its 30 year life, and had been modified extensively by the various owners. . . .

If the risk of a significant hazard cannot be eliminated, it is equally the design engineer's responsibility to ensure that the product carries proper warning and that operators are educated in its use. Thus, design engineering staff must:

1. Understand the priorities.

2. Develop systems of hazard identification that assume nothing is safe.

3. Promulgate countermeasures. . . .

Where safety is concerned, those personnel responsible may discover that some of their recommendations are greeted less than enthusiastically by management. As one product liability lawyer put it, ''. . . the engineer's responsibility to implement his safety recommendations often exposes him to virtually limitless explanation, rationalization, and, in certain instances, management harassment.'' At one memorable product safety panel in Washington, D.C., in 1971, not one of the speakers still worked for the companies or government agencies at which they had raised objections to management safety practices. (Kolb, John and Steven S. Ross, *Product Safety and Liability: A Desk Reference.* New York: McGraw-Hill, 1980, p.83. Reproduced by permission.)

Therefore, it is very clear that both the project manager and every team member can be held personally liable for their decisions. But safety is such an obvious topic. How can violations of safety considerations knowingly occur, and what are the remedies that can be used? Answering these questions should help in answering similar questions about decisions that are not as obvious, such as contractual, financial, or promotion-transfer decisions.

5.2 HOW WE CAN BECOME LIABLE
WHEN DIRECTED BY OTHERS

One of the most well-known experiments in social psychology was performed by Stanley Milgram (Milgram, 1974). The intent was to discover the reasons why apparently otherwise normal people performed various atrocities during World War II when told to do so by their superiors. The experiment went like this.

Subjects were informed that they were to be part of a memory and learning experiment. When they arrived at the laboratory, they were introduced to the experimenter, dressed in a laboratory coat, and were informed that he was an important person in the scientific community. The subjects were ordered by the experimenter to apply electric shocks to the ''learner'' whenever he or she failed the memory test. The shock was supposedly increased with each memory failure. But it was a sham, because there were no real shocks and the apparent ''learner'' and ''experimenter'' were only acting their parts in the experiment. The experiment was intended to see how far the subject would go in following orders from the experimenter ''authority'' figure.

The results were unusual. When the "subjects" were in an adjoining room separated from the "learner" by a shaded glass window, over half were willing to follow orders including giving the maximum electric jolt of 450 volts. This happened even though the "learner" could be seen strapped in a chair, writhing in (apparent) agony. The same results happened when the subjects were allowed to hear the (apparently) painful screams and the protests of the "learner," which became intense from 130 volts on. However, there was a striking difference when subjects were placed in the same room within touching distance of the "learner." Then the subjects willing to continue to the maximum shock dropped by one-half.

Milgram explained these results by citing a strong psychological tendency in people to be willing to abandon personal accountability when placed under authority. He saw all his subjects ascribing all initiative, and thereby all accountability, to what they viewed as legitimate authority. And he noted that the closer the physical proximity, the more difficult it becomes to divest oneself of personal accountability. (adapted from Martin and Schinzinger, 1983)

While there may be little similarity between the Milgram experiments and the management direction of project team members, the possibility of this situation occurring in projects is very real. There must be systems to preclude it, since there are many unknowns, and who is to determine, for example, which safety factor to use in a design? However, response to management or legitimate authority are not the only causes of potentially incorrect or potentially unethical decisions.

It may happen even in the sciences when we know so much more about the subject. I recall designing a steel truss many years ago in one of my first jobs. I got a reminder from the design supervisor that I hadn't put in enough of a "safety" factor. I said that I had used the load factors in my engineer's handbook, and if the materials were right, why did I need a safety factor? The supervisor quite rightly said, "Because we never really know." As the designer, I was responsible. In this example, I was not told what "safety factor" to use but even then, I was still responsible. The unknowns are always there, and they are even more evident in projects. Therefore each of us must decide, ourselves, how to handle them.

5.3 PREVENTION TECHNIQUES SUPPORTING ETHICAL DECISIONS

Irrespective of the sources, either in unauthorized management directions or else in the unknowns of project operations, there are several techniques that can be used to prevent becoming involved in unethical decision making. We start by stating that the engineer, project manager, or team member should always protect themselves *first*.

The situation should be carefully documented with formal, written (and witnessed, if necessary) statements, backed up by all pertinent data. Some legal experts have argued that such files are time bombs, ticking away until detonated by plaintiff's attorney in a liability suit. But the absence of such information is also a bomb of sorts. It suggests to a jury that the company never even considered safety to be an issue in design, or is trying to hide something.

The written record should show as impartially and as fully as possible the positions of management and the other parties to any disagreement. (Kolb and Ross, 1980, p.84)

Written records should include the usual records of any well-run operation, such as design reviews, calculations, drawings, test data, quality information, production specifications, service instructions, correspondence, installation directions, field feedback reports, and so forth. As a general rule, about three years should be sufficient for everything except financial data. The length of time those records are kept depends upon current tax and legal regulations.

Maintaining accurate records should take care of most decisions, but there are some that do not fit into any of these categories. Where there is no public documentation, there is another answer: the engineering notebook. Every engineer and team member should have one. The notebook has fixed and numbered pages. All decisions that were not distributed publicly in company documentation should be noted in this notebook. Periodically someone *not* connected with the project should sign the bottom of the last page upon which some notations were made. This would mean that those notes were made between the date of that signature and the one before it. It would then be possible to determine in what time period any decision in the notebook was made.

Then, assuming that one has protected one's self first, there are other alternatives when confronted with an ethical problem. Some are:

1. Don't think about it.
2. Obey.
3. Leave.
4. Conscientiously object.
5. Secretly blow the whistle.
6. Publicly blow the whistle.
7. Negotiate and build conscientiousness for a change in the unethical management. (Nielsen, 1985, p.416)

Whether it is done secretly or publicly, "blowing the whistle" is a real alternative to consider. It's not an easy choice and it's not even protected very well legally. These are some suggestions.

1. Be very analytic and careful in assessing the facts on which your protest would be based. Can you document company wrongdoing? . . .
2. Determine what kind of company conduct you are protesting. (Is it clear illegality or unsafe conduct or is it lawful business and social policies?)

3. Inform yourself fully of the procedures in your company for appealing the policy . . . up through the chain of command. . . .

4. Before and while moving through the internal channels, document your position at every step of the way with records, letters, and other hard evidence. . . .

5. When moving into either internal or external channels of complaint, seriously consider getting an attorney, or contacting a public interest group, professional society, labor union, or other organization active in your area of concern or industry. . . .

6. If the problem is one regulated by law, . . . learn what the requirements and procedures are for lodging a complaint with the relevant government agency. . . .

7. If you are fired or forced to resign because of your protest, be aware that your right to discuss the case in public is strong but not unlimited. . . .

8. If you cannot win damages or reinstatement through an appeal to a government regulatory agency with jurisdiction over the issue you have raised, be ready to consider a lawsuit alleging that your discharge or punitive treatment violates public policy. (Westin, 1981, p.161)

Obviously any kind of protest involves a difficult moral choice which should be taken only after management has been informed appropriately that someone feels that something illegal, unethical, or dangerous is happening and that he or she feels strongly that something should be done about it. But sometimes there's no way to avoid making this kind of decision. The law states that someone who knows that a crime has been committed or some safety regulation is knowingly being violated and does *not* come forward can also be guilty. It provides one with no alternatives.

When thinking about these kinds of decisions, remember we don't ever know all the answers. There may be something that we're not aware of. Our job as managers is to make nonrepetitive decisions by absorbing uncertainty, and this means we don't really "know" all the answers. (If you recall, the uncertainty curve never dropped to the horizontal axis of zero uncertainty, even when the project was closed). As Mitroff points out, "Newer methods are being developed that realize that if completeness, closure, and certainty were required before action was warranted, then action would never be warranted. One would have to wait forever." Therefore, we still have to make decisions, not knowing all the answers, and we are definitely held responsible for those decisions. Society (and that includes the company and the law) holds us personally responsible for the decisions we make as a professional.

Consequently, any decision that is made must be one that could be approved by a jury of our professional peers acting at the time the decision was made. It's almost as if those decisions are made in an open courtroom. It's not a case of making the "right" decision, since we're never really sure of that, just making it in a way that our peers would accept in accordance with the standards of our professional environment.

6.0 SUMMARY

We have covered some of the major areas in implementation and maintenance of project systems. Those systems were generally implemented as they were designed by following the life cycle of the project itself. They started with Forecasting and ended with Personal Management Style. The Personal Management Style system is probably the most difficult one to implement, because it involves restructuring one's own thinking patterns, and this is not easy. Learning how to delegate, to lead, to resolve conflict, to handle stress, and to document one's decisions requires discipline and self-control. It's very difficult, but when considering the alternatives, it's easy to see that it's worth it.

To briefly review some of the concepts in this chapter, we start with implementing change. For example, one author suggests that you use

1. Breakthrough projects—start with a small, easily achieved project.
2. Traditional information for top management—provide traditional kinds of information voluntarily.
3. Retention of power—allow administrators to retain their power.
4. Policy recommendation—propose only policy recommendations that can easily be accepted.
5. Slow down—don't push for too much change.
6. Schedules aren't the end of the world—keep schedules and other tools in the background. Results are what count.
7. Decode all information—produce reports in the style of the receiver. (Stuckenbruck, 1981, p.171–175)

The suggestions noted above basically apply when others are involved. However, when you are the only variable in the equation such as when implementing your own management style, there are other things to think about. Reduction of stress is one of them. It's important to remember that there's probably no one better qualified than you, at this moment, in this place. You may not always win the first time. After all, the basic premise of project operations is the *iteration* or re-evaluation of what you have done and what has to be done next. This means you have to rethink how you're going to manage. You may not hit the target the first time, and there may be some disappointment there. But you can also learn from that.

If the goals were unrealistic, partial failure will inevitably result. Face this fact squarely. Sometimes the pain of self-examination supports growth for the next attempt. One learns from the past, but it's pointless to try to live there.

There is irony in all human experience, and no less in the solutions to the problem of disappointment. The deepest irony of all is to discover that one has been mourning losses that were never sustained and yearning for a past that never existed, while ignoring one's own real capabilities for shaping the present. (Zeleznick, 1967, p.245)

We have depended heavily on the ideas inherent in Contingency or Situational theory. These ideas point out that there are no simple answers about human behaviors in organizations,

> but that if one can spell out enough of the prior situational conditions, enough about the human actors in the situation, and enough about the properties of the task and the environment within which the task is being carried out, one can then specify hypotheses or propositions. . . .
>
> In the area of how to organize work, similar kinds of contingency theories are being developed pertaining to how to divide labor, how to integrate effort, how much to decentralize, how to control the organization, and so on. . . . While it is still worthwhile to ask basic questions such as, What is human nature? or What is an organization? we must recognize that *we will not find simple answers*. (Schein, 1980, p.44–49)

Basically, we have started with the management artist and tried to spell out as many situational conditions as the appropriate research, my experience, and the experience of others could provide. This is partially a clinical model of learning. There are no exact, simple, this-is-the-only-way-to-do-it answers. We're dealing with an art form, which means using a novel kind of learning method—a clinical learning method. The clinical learning method involves observing the processes and the behavior of other, successful professionals in somewhat similar (but not exactly the same) situations. The artist who learns brushwork by duplicating the fine lines of a Rembrandt painting, or the medical intern who assists in complex operating procedures are both using the clinical method, that is, learning from others. The artist, of course, will never paint *exactly* the same way as Rembrandt, nor will the intern operate *exactly* the same way as others. Nevertheless, both will have learned enough to act *similarly* in similar situations.

Most of us are more comfortable with the more objective Scientific learning method of learning, the intellectual, define-theorize-test-replicate model that I initially used to start our thinking and systems development processes. And although the clinical learning model may not be as familiar, it has a long and honorable history, much older than the Scientific method that was developed within recent history. In the past, most learning was done through example, by observing others, not from any rigorous theorizing posture. When dealing with people, this model still applies. I'm sure that you will agree that

> Managing others is most effectively done by example rather than by preaching or policy. If the example is lacking, the most moving sermon and the wisest policy rarely work. "Do as I tell you and not as I do" is the motto of the outsider, the consultant. Effective executives know that their associates will do as the boss does, and not necessarily as the boss says. (Drucker, 1985, p.3)

7.0 THE END OR THE BEGINNING?

Well, we have come full circle. We started by saying that management was an art and the project manager had a very difficult but fascinating job. His or her job is to become the best management artist possible. It is a never-ending job, because there are always ways to improve one's management talents. It is very rewarding, because as one becomes a better management "artist," he or she is able to handle more complex and larger management tasks. This is even more obvious in project operations, since the higher levels of uncertainty that are absorbed during the beginning of project operations (and when the project gets into trouble) often require that the project manager operates at a higher level than equivalent functional managers. Thus, the project manager might be operating at a general manager's level in terms of the level of uncertainty.

If the challenges of a project are met successfully, the project manager's future is almost unlimited. Therefore, the sooner he or she begins to develop unique systems, the sooner that success will occur. But care must be taken to use the available tools objectively and selectively. They are powerful and are intended to operate in an open environment. But they do work, and they work well. Working well does *not* mean achieving every impossible goal, meeting every deadline, and staying within every budget. This could only happen if the project manager possessed the ability to see into the future and have the philosophical strength to understand what was seen. Fortunately, one can't foresee the future perfectly, but one can have the philosophy to deal with it when it comes. For example,

> All his important decisions must be made on the basis of insufficient data. It is enough if a man accepts his freedom, takes his best shot, does what he can, faces the consequences of his acts, and makes no excuses. It may not be fair that a man gets to have total responsibility for his own life without total control over it, but it seems to me that for good or for bad, that is just the way it is. (Kopp, 1976, p.187)

If this seems reasonable to you, you are undoubtedly enlightened (by now, anyhow) and are probably well on your way to success as a project manager. You may not have total control over your life, but I think that it is possible to improve some parts of it, especially those parts that are spent as a project manager. In any event, by trying to improve, you can't lose, because winning is in the mind of the beholder, and that is you. In the end, you are the only one that judges if you have succeeded or not, and *today* is a good day to begin that personal success. *What do you want to do now?*

My Suggested Answers to the Case Study

1. I can think of four alternatives, as follows:
   A. Request Gale to run some fracture tests using the same design configuration as the front of the refrigeration unit and determine the actual results. Use those for further action.

    B. Accept Gale's recommendations and send a change-of-scope directive to Engineering for a redesign.

    C. Follow the actions described in the case study but add additional field service costs to supervise the training of store employees. Set up a training qualification program and make sure that all trainees understand the possible problems that could result if the equipment is not used correctly. Include video demonstrations of the possible effects of misusing the equipment.

    D. Buy additional product liability insurance.

In all cases, I suggest that a statistical analysis be made of the likelihood of failure and a cost-benefit analysis be made to determine potential failure modes and possible corrective actions if needed.

2. Everyone can be held liable. They may be held liable if the product is defective in design, in manufacture, improperly labeled, or improperly packaged so that safety-related damage can result.

3. The company can provide a supportive environment for the technical staff including training in product and personal legal liability requirements. Individuals can maintain the TV rule; that is, imagine that you are making decisions with a TV camera over your shoulder and the whole world is watching you. They can also document their decisions using internal memos, letters, and engineering notebooks to record their decisions in the event no widely distributed memo or letter is written. Yes, it is necessary because individuals, as well as companies, can be liable for technical decisions.

4. Make sure that no design decisions are made without complete disclosure. Professionals can be held liable only if he or she performed below the level of expertise of the immediate community. Gale should be able to do a mathematical analysis of the likelihood of a product failure occurring. She should then suggest several alternative approaches to a solution. In an extreme example, she might even consider the risks or gains in "blowing the whistle" on Store Constructors, Inc., if she felt strongly enough about it. The risks may be very high if she chooses that alternative, but she must think of herself first.

5. In my opinion, decals and warnings in manuals are insufficient unless there is a knowledgeable user. Always consider the product user or the client first. If they are knowledgeable, that is, equipped by training or knowledge to handle the equipment, decals and printed warnings may be O.K. For example, a qualified buyer of dynamite understands what the DANGER sign on the box really means. Therefore, the first step is to define the user, then take appropriate design steps to assure safe and successful product use.

BIBLIOGRAPHY

Bass, Bernard M., *Stogdill's Handbook of Leadership*. New York: Free Press, Copyright © 1981.

Drucker, Peter F., *The Changing World of the Executive*. New York: Times Books, 1985.

KOLB, JOHN, and STEVEN S. ROSS, *Product Safety and Liability: A Desk Reference.* New York: McGraw-Hill, 1980.

KOPP, SHELDON B., *If You Meet the Buddha on the Road, Kill Him!.* New York: Bantam, 1976. Originally printed by Science and Behavior Books, Palo Alto, Calif. U.S.A., 1972.

KOTTER, JOHN P., and LEONARD A. SCHLESINGER, "Choosing Strategies for Change" in *Current Perspectives in Social Psychology* (2nd ed.), eds. Edwin P. Hollander, and Raymond G. Hunt. New York: Oxford University Press, 1967.

LAWRENCE, PAUL R., HARVEY F. KOLODNY and STANLEY M. DAVIS, "The Human Side of the Matrix," *Organizational Dynamics,* Summer 1977, pp. 56–59, New York: American Mangagement Association.

MARTIN, MIKE W., and ROLAND SCHINZINGER, *Ethics in Engineering.* New York: McGraw-Hill, 1983.

MILGRAM, STANLEY, *Obedience to Authority.* New York: Harper & Row, Pub., 1974.

MITROFF, IAN I., *Stakeholders of the Organizational Mind.* San Francisco: Jossey-Bass, 1983.

NADLER, DAVID A., "Concepts for the Management of Organizational Change," in *Perspectives on Behavior in Organizations* (2nd ed.), eds. Richard Hackman, Edward Lawler III, and Lyman W. Porter. New York: McGraw-Hill, 1983

NIELSEN, RICHARD P., "Alternative Managerial Responses to Unethical Strategic Management," in *Proceedings of the 45th Annual Meeting of the Academy of Management,* San Diego, Calif., 1985, p. 410–416.

PERROW, CHARLES, "The Short and Glorious History of Organizational Theory," in *Perspectives on Behavior in Organizations* (2nd ed.), eds. Richard Hackman, Edward Lawler III, and Lyman W. Porter. New York: McGraw-Hill, 1983.

PETERS, THOMAS J., "Symbols, Patterns and Settings: An Optimistic Case for Getting Things Done," in *Perspectives on Behavior in Organizations* (2nd ed.), eds. J. Richard Hackman, Edward Lawler II, and Lyman W. Porter. New York: McGraw-Hill, 1983.

QUINN, JAMES BRIAN, "Managing Innovation: Controlled Chaos," *Harvard Business Review,* May-June 1985, p.73–84.

SALANCIK, GERALD R., "Commitment Is Too Easy!" in *Readings in the Management of Innovation,* eds. Michael L. Tushman, and William L. Moore. Marshfield, Mass.: Pitman Pub., and Ballinger Pub. Co., 1982.

SCHEIN, EDGAR H., *Organizational Psychology* (3rd ed.) Englewood Cliffs, N.J.: Prentice-Hall, 1980. Reprinted by permission of Prentice-Hall, Inc.

SELYE, HANS, *Stress Without Distress.* Philadelphia: Lippincott, 1974.

STUCKENBRUCK, LINN C., ed., *The Implementation of Project Management: The Professional's Handbook.* Reading, Mass.: Addison-Wesley, 1981.

TRICE, HARRISON M., and JANICE M. BEYER, "Studying Organizational Cultures Through Rites and Ceremonies," *Academy of Management Review,* 9, no. 4 (October 1984), 653–69.

WESTIN, ALAN F., *Whistle Blowing! Loyalty and Dissent in the Corporation.* New York: McGraw-Hill, 1981.

ZELEZNICK, ABRAHAM, "Management of Disappointment," *Harvard Business Review,* November-December 1967.

OTHER READINGS

ANDREWS, K.R., *The Effectiveness of University Management Development Programs.* Cambridge, Mass.: Harvard University Press, Graduate School of Business Administration, 1966.

BLAKE, R.R., and J.R. MOUTON, *The Managerial Grid.* Houston, Tx.: Gulf, 1964.

BOWERS, DAVID G., JEROME L. FRANKLIN, and PATRICIA A. PECORELLA, "Matching Problems, Precursors and Intervention in O.D.: A Systematic Approach," *The Journal of Applied Behavioral Research,* 11, no. 4 (1975), 391–409.

FIEDLER, FRED E., MARTIN M. CHEMERS, and LINDA MAHAR, *Improving Leadership Effectiveness.* New York: John Wiley, 1976.

FLEISHMAN, E.A., "Leadership Climate, Human Relations Training, and Supervisory Behavior," *Personnel Psychology,* 6 (1953), 205–22.

GLASER, B.G., and A.L. STRAUSS, *The Discovery of Grounded Theory,* Chicago: Aldine, 1967.

GREINER, LARRY E., "Red Flags in Organization Development," in *Current Perspectives in Social Psychology* (2nd ed.), eds. Edwin P. Hollander, and Raymond G. Hunt. New York: Oxford University Press, 1967.

HACKMAN, RICHARD J., EDWARD LAWLER III, and LYMAN W. PORTER, eds., *Perspectives on Behavior in Organizations* (2nd ed.). New York: McGraw-Hill, 1983.

HAIRE, M., "Some Problems in Industrial Training," *Journal of Social Issues,* vol. 4 (1948), 41–47.

HAYAKAWA, S.I., *Language in Thought and Action.* New York: Harcourt Brace Jovanovich, Inc., 1964.

HOLMES, FENWICKE W., "IRM: Organizing for the Office of the Future," *Journal of Systems Management,* January 1979.

KEEN, PETER G.W., "Information Systems and Organizational Change," *Communications of the ASM,* 24, no.1 (January 1981), 24–33.

KELLER, ROBERT T., "Project Group Performance in Research and Development Organizations," *Academy of Management Proceedings,* Aug. 11–14, 1985, San Diego, Ca., p. 315–318.

LEWIN, KURT, *Field Theory in Social Science.* New York: Harper & Row, Pub., 1951.

MCGREGOR, DOUGLAS, *The Human Side of Enterprise.* New York: McGraw-Hill, 1960.

MARGULIES, NEWTON, and ANTHONY P. RAIA, *Conceptual Foundations of Organizational Development.* New York: McGraw-Hill, 1978.

MILLER, GEORGE A., "The Psycholinguists: On the New Scientists of Language," in *Current Perspectives in Social Psychology,* (2nd ed.), eds. Edwin P. Hollander, and Raymond G. Hunt. New York: Oxford University Press, 1967.

NADLER, DAVID, A., "Concepts for the Management of Organizational Change," in *Current Perspectives in Social Psychology,* (2nd ed.), eds. Edwin P. Hollander, and Raymond G. Hunt. New York: Oxford University Press, 1967.

OUCHI, WILLIAM G., *Theory Z: How American Business Can Meet the Japanese Challenge.* Reading, Mass.: Addison-Wesley, 1981.

OUCHI, WILLIAM G., and RAYMOND L. PRICE, "Hierarchies, Clans and Theory Z," in *Per-*

spectives on Behavior in Organizations (2nd ed.), eds. J. Richard Hackman, Edward Lawler III, and Lyman W. Porter. New York: McGraw-Hill, 1983.

PASCALE, RICHARD TANNER, and ANTHONY G. ATHOS, *The Art of Japanese Management.* New York: Penguin, 1982.

PIRSIG, ROBERT M., *Zen and the Art of Motorcycle Maintenance.* New York: Morrow, 1974.

SCHEIN, EDGAR H., "Does Japanese Management Style Have a Message for American Managers?" *Sloan Management Review,* Fall 1981, 77–90.

SYKES, A.J.M., "The Effect of a Supervisory Training Course in Changing Supervisors' Perceptions and Expectations of the Role of Management," *Human Relations,* Vol. 15 (1962), 227–43.

Appendix 1 One Way to Evaluate Projects

This evaluation procedure is one that was developed for evaluating various information systems projects.* While it may not be completely applicable for your situation, the concepts used in the questionnaire could be applied to almost any industry, company, and situation. It requires thinking about the major project areas and developing a questionnaire that would fit. The difficult part of developing this kind of questionnaire for your own organization is in determining where the proposed project can *go wrong*. This is a bit different from the usual investment justification process, since there the proposal emphasizes the future benefits to the organization. It is very difficult to do a similar future-oriented economic justification on computer systems or management information systems, because we cannot predict the savings that will occur when we have this better information. How can you measure the difference between your "new" and "better" system and the "present" system. How can you measure it since you won't be making all those "old" mistakes anymore? This evaluation process is divided into four parts:

1. *User's impact*: To be completed by the senior user manager, who may be the program manager.
2. *Size*: To be completed by the program manager.

*Adapted from Dr. Clayton Harrell, Jr., *Investment Risk Assessment for Information Management Projects*, Xerox Corp. (Webster, N.Y.) (used with permission)

3. *Technology*: To be completed by the project leader (engineer).

4. *Structure*: To be completed by the information systems manager.

The program or project manager will ensure all four parts are completed, scored, and evaluated. The questionnaires should be completed as objectively as possible.

The objective of this evaluation is to assist the project team by suggesting various management approaches and technical methodologies that are known to decrease the risk of project failure. The results of this evaluation and the action plan of the program manager must be included in the management part of the next report.

USER'S IMPACT

What is your *overall* estimate of the risk of this project relative to (1) meeting the schedule, (2) accomplishing all your objectives, (3) remaining within budget, or (4) becoming implemented across all planned sites? Place a single X on the line at the appropriate point where you feel the total risk of the project falls. The results of this question will not be included in the risk score. It is included only for an assessment of your intuitive judgment concerning the risk of this project.

Level of
Risk Scale

0	1.0	2.0	3.0	4.0	5.0
No risk	Low risk	Average risk	High risk	Very high risk	Excessive risk

QUESTIONS

1. If this system is not correctly implemented, on time, within budget, to your requirements, what are the consequences?
   _____ **A.** Drastic, with an impact on the company profit and loss.
   _____ **B.** Heavy, although a profit and loss will probably not occur, certain management plans will be impacted.
   _____ **C.** Heavy, because there are legal or like requirements that must be considered.
   _____ **D.** Moderate; the implementation is a key ingredient in meeting our objectives.
   _____ **E.** Slight.
   _____ **F.** None that I know of.

2. Will the implementation of this system require reorganization?
   _____ **A.** Yes, it will require a major reorganization.
   _____ **B.** Yes, it will require some personnel movements, but not a major reorganization.
   _____ **C.** Yes, but only minor changes.
   _____ **D.** No, at this time it is not anticipated that changes will occur.

3. Is there a history of frequent user department reorganizations?

_____ **A.** Yes.

_____ **B.** No.

4. Is there opposition from your management to this project, its justification, need, or cost?

_____ **A.** Yes, there is considerable discussion as to the validity of the project plan.

_____ **B.** Yes, but only from the standpoint of costs or schedule.

_____ **C.** No, all management is behind this project.

_____ **D.** Not applicable, or visibility to management is limited.

5. How would you assess the technical approach being discussed?

_____ **A.** I am not in a position to understand the proposed technical solution.

_____ **B.** I understand the concept of the technical solution and believe it is moderate or low-risk and within the capabilities of the organization to implement it.

_____ **C.** We really have not discussed the technical approach.

_____ **D.** I am fully confident the approach is feasible.

_____ **E.** I am not sure the approach is valid.

6. The proposed tangible benefits for this system are:

_____ **A.** Fully achievable and trackable.

_____ **B.** May not be fully realized for some time after implementation.

_____ **C.** Are very "soft," and I am not sure can ever be realized.

7. The proposed costs for this system

_____ **A.** Seem reasonable for what I know now.

_____ **B.** May be very challenging to control.

_____ **C.** Are probably low, but the project would not "sell" if they go any higher.

_____ **D.** If costs start going up, we can always pull some functions out of the system for a later implementation to keep it within cost target.

8. Is new or unfamiliar hardware required in the user area to process the data for the proposed system?

_____ **A.** No.

_____ **B.** Hardware appears to require little training.

_____ **C.** Hardware will require extensive training.

_____ **D.** Hardware is too complicated for this installation.

9. Is relocation of personnel or facilities necessary for full implementation of this system?

_____ **A.** No.

_____ **B.** Only to another part of the building.

_____ **C.** Yes, to a new facility being planned in this local area.

_____ **D.** Yes, to another city that has a company office.

_____ **E.** Yes, to another city with no existing company office.

10. Will the data from the present system have to be converted physically for the new system?

_____ **A.** No.

_____ **B.** Yes, but it is understood that it can be done routinely.

_____ **C.** Yes, files will have to be converted and audited prior to full turn-on.

_____ **D.** Yes, but no one has thought of conversions so far.

_____ **E.** Not applicable.

11. What is your impression of your management's commitment to this project?

_____ **A.** Very committed and supportive.

_____ **B.** Not applicable, or the system is too small to be concerned about.

_____ **C.** Supportive, but cautious.

_____ **D.** Not supportive.

12. What is the classification of the data (e.g., secret, confidential, etc.) being used in this system?

_____ **A.** Classified data will be used within the system and held on the data base.

_____ **B.** Classified data will be used for reporting only and not resident on this system's data base.

_____ **C.** No classified or sensitive data will be used.

13. If this system is not implemented on time, or to the requirements that are specified, the impacts on interfacing systems are:

_____ **A.** Significant.

_____ **B.** Moderate.

_____ **C.** Slight.

_____ **D.** None.

14. What is the impact of data controls (i.e., internal and external balancing, standard and nonstandard file coordination, etc.)?

_____ **A.** Critical.

_____ **B.** Routine.

_____ **C.** None.

15. Has the user worked with the project leaders or project manager on previous projects?

_____ **A.** Yes.

_____ **B.** No.

16. Has the end user participated in the requirements definition of this system?

_____ **A.** Yes.

_____ **B.** No.

_____ **C.** Not applicable.

PROJECT SIZE

1. Total project team worker hours estimated for development and implementation: Using the graph in Figure A1-1, find the intersection of elapsed worker hours (including job shoppers, vendors, technical, and users) and the estimated calendar time from concept proposal through implementation. Please check appropriate letter

_____ **A.**

_____ **B.**

_____ **C.**

_____ **D.**

_____ **E.**

2. What is the average number of people to be assigned to each identified project leader or subsystem manager (e.g., data-base design)?

_____ **A.** 1 to 3.

_____ **B.** 4 to 6.

_____ **C.** 7 to 12.

_____ **D.** 13 or more.

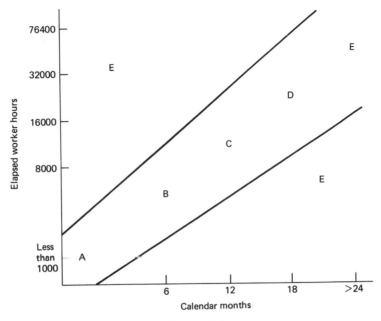

Figure A1-1

3. What is the average calendar time (i.e., elapsed) per subsystem (or total project if no subsystems)?
   _____ **A.** Less than 6 months.
   _____ **B.** 6 to 12 months.
   _____ **C.** 12 months or more.

4. Who will do the development work?
   _____ **A.** System support group associated with the primary user and at the initial implementing site.
   _____ **B.** Significant portions by on-site personnel with assistance by other technical groups or vendors.
   _____ **C.** Mostly by off-site personnel (staff organizations not at the initial site).
   _____ **D.** System design group or vendor personnel representing the users with limited user contact other than periodic meetings and with no prior relationship.

5. How many user departments are involved?
   _____ **A.** 3 or more.
   _____ **B.** 2.
   _____ **C.** 1.
   _____ **D.** No identifiable user.

6. Approximately how many user department people will be necessary to operate the system once it is implemented fully?
   _____ **A.** Fewer than 10.

———— **B.** 10 to 20.
———— **C.** 21 to 50.
———— **D.** Over 50.

7. At how many different geographic locations (buildings, cities, etc.) will this system be implemented that will require some level of support (training, equipment installation, etc.)?
———— **A.** 1.
———— **B.** 2.
———— **C.** 3 or more.

8. How many existing data processing systems must the new system interface with?
———— **A.** None.
———— **B.** 1.
———— **C.** 2.
———— **D.** More than 2.

9. How many new data processing systems now under development must the new system interface with?
———— **A.** None.
———— **B.** 1.
———— **C.** 2.
———— **D.** More than 2.

10. Does the project leader(s) have experience with projects at least as large as this one?
———— **A.** No, none do.
———— **B.** Some do and some do not.
———— **C.** We plan to use consultants or experienced persons for guidance.
———— **D.** Yes, with comparable projects.
———— **E.** Yes, with much larger projects.

11. Does the project leader(s) have experience with projects at least as complex as this one?
———— **A.** No, none do.
———— **B.** Some do and some do not.
———— **C.** We plan to use consultants or experienced personnel for guidance.
———— **D.** Yes, with comparable projects.
———— **E.** Yes, with much complex projects.

TECHNOLOGY

1. Is additional hardware required for this system?
———— **A.** None.
———— **B.** Yes, devices other than terminals and controllers.
———— **C.** Yes, terminals and lines in the user's area.
———— **D.** Yes, mini or microcomputers in the user's area.
———— **E.** Yes, more than one of the above.

2. Which hardware is new to the Information Systems department?
———— **A.** Not applicable.
———— **B.** Devices other than terminals and controllers.

_____ **C.** Terminals and lines.

_____ **D.** Mini or microcomputers.

_____ **E.** More than one of the above.

3. Is system success dependent upon the new hardware?

    _____ **A.** No, or no hardware is involved.

    _____ **B.** Somewhat.

    _____ **C.** Very heavily.

4. Is the hardware the first installation for the vendor?

    _____ **A.** Not applicable.

    _____ **B.** No vendor involved.

    _____ **C.** No, it has been done before.

    _____ **D.** No, but some features are new or customized to our requirements.

    _____ **E.** Yes.

5. Is the software the first installations for the vendor on the equipment?

    _____ **A.** Not applicable.

    _____ **B.** No vendor involved.

    _____ **C.** No, it has been done before.

    _____ **D.** No, but some features are new or customized to our requirements.

    _____ **E.** Yes.

6. How many hardware or software vendors (suppliers) are involved with this project?

    _____ **A.** None.

    _____ **B.** 1.

    _____ **C.** 2.

    _____ **D.** 3 or more.

7. Does the system include network applications?

    _____ **A.** No.

    _____ **B.** Yes, new application on existing utility network.

    _____ **C.** Yes, uses new network but is being developed by another project team.

    _____ **D.** Yes, this project includes a new network.

8. Will the system be modeled first for the user?

    _____ **A.** No.

    _____ **B.** Yes, using APL or a language other than the one used for production.

    _____ **C.** Yes, using the language of the production team.

9. What programming (non-data-base) language(s) is being planned?

    _____ **A.** Report generator.

    _____ **B.** COBOL, BASIC, or FORTRAN with or without the languages noted above.

    _____ **C.** APL alone, or in combination with any of those noted in A or B above.

    _____ **D.** Other, specify.

10. Is any system software (nonoperating system) or programming language new to the project team?

    _____ **A.** No.

    _____ **B.** 20 percent new.

    _____ **C.** 50 percent new.

    _____ **D.** Over 50 percent new.

11. Are software packages being planned for use on this project?

_____ **A.** No.
_____ **B.** Yes, to some degree.
_____ **C.** Yes, to a large degree.

12. How good is the anticipated vendor or supplier support for the software?
_____ **A.** Not using software packages.
_____ **B.** Unknown.
_____ **C.** Adequate.
_____ **D.** Good.

13. What is your opinion of the proposed software package(s)?
_____ **A.** Stable, as reported by other users.
_____ **B.** Untested, new software.
_____ **C.** Not applicable.

14. Are modifications to the proposed software planned?
_____ **A.** No.
_____ **B.** Yes, modifications to existing package or system, and will be supported by originator.
_____ **C.** Yes, modifications to existing package or system, and will be done by this project team.
_____ **D.** Not applicable.

15. What is the system complexity?
_____ **A.** Straightforward or average.
_____ **B.** Unknown.
_____ **C.** Complex, with many interactions.
_____ **D.** Intermediate or above average.

16. How knowledgeable is the technical team in the proposed business application?
_____ **A.** Limited.
_____ **B.** Understands concept, but no experience.
_____ **C.** Has been involved in prior implementation efforts.
_____ **D.** None in this area.

17. Is a data dictionary being used?
_____ **A.** No.
_____ **B.** Yes, but with no experience.
_____ **C.** Yes, by some of the experienced analysts.
_____ **D.** Yes, by all the analysts and they have experience.
_____ **E.** Use of this technique is not possible because of the nature of the project.

18. Are structured analysis techniques being planned?
_____ **A.** No.
_____ **B.** Yes, but with no experience.
_____ **C.** Yes, by some of the experienced analysts.
_____ **D.** Yes, by all the analysts and they have experience.
_____ **E.** Use of this technique is not possible because of the nature of the project.

19. Will reusable programming code be used?
_____ **A.** Yes.
_____ **B.** No.

20. Are the key people experienced in the proposed data-base technology?
_____ **A.** Yes.
_____ **B.** No.

PROJECT STRUCTURE

What is your overall estimate of the risk of this project relative to (1) meeting the schedule, (2) accomplishing all your objectives, (3) remaining within budget, or (4) becoming implemented across all planned sites? Place a single X on the line at the appropriate point where you feel the total risk of the project falls. The results of this question will *not* be included in the risk score. It is included only for an assessment of your intuitive judgment concerning the risk of this project.

Level of
Risk Scale

0	1.0	2.0	3.0	4.0	5.0
No risk	Low risk	Average risk	High risk	Very high risk	Excessive risk

1. The system is best described as:
   _____ **A.** Totally new.
   _____ **B.** A replacement of an existing manual system.
   _____ **C.** A replacement of an existing automated system.
   _____ **D.** A combination—new functions with replacements of automated or manual system components.

2. If a replacement system is being proposed, what percent of its functions simply replace existing ones (as opposed to being new functions)?
   _____ **A.** Not a replacement system.
   _____ **B.** Less than 25 percent.
   _____ **C.** 25 to 50 percent.
   _____ **D.** Over half.

3. What is the degree of procedural changes in the user department caused by the proposed system?
   _____ **A.** Low.
   _____ **B.** Medium.
   _____ **C.** High.
   _____ **D.** None.

4. Proposed systems design methodologies and/or procedures, such as structured analysis are
   _____ **A.** First to be used here, but we have had good training.
   _____ **B.** First to be used here and we have no help with their use.
   _____ **C.** First of kind for us in this group (although other company teams have used them).
   _____ **D.** Standard procedures will be followed.

5. What degree of flexibility and judgment can be exercised by the system organization in the area of design?
   _____ **A.** Very little.
   _____ **B.** Average.
   _____ **C.** Very high.
   _____ **D.** None at all, the core system is being developed by another system development center or vendor.

6. How confident are you in the team's ability to use the data-base technology being planned?

    _____ **A.** High.

    _____ **B.** Medium.

    _____ **C.** Low.

    _____ **D.** Question does not apply.

7. How cooperative do you feel the user is to work with on this project?

    _____ **A.** Uncooperative.

    _____ **B.** Adequate.

    _____ **C.** Very cooperative.

8. What is the general attitude of the user?

    _____ **A.** Poor, anti–data processing.

    _____ **B.** Fair, some reluctance.

    _____ **C.** Good, understands the value of a data processing solution.

9. How committed is group level user management to the system?

    _____ **A.** Unknown.

    _____ **B.** Somewhat reluctant.

    _____ **C.** Adequate.

    _____ **D.** Extremely enthusiastic.

10. How do you feel about the level of user support commensurate to the size of the project?

    _____ **A.** No problem anticipated.

    _____ **B.** Could use more support than is planned.

    _____ **C.** User support is lacking or uncertain at this time.

11. How knowledgeable is the user in the area of data processing?

    _____ **A.** None.

    _____ **B.** Adequate.

    _____ **C.** Very knowledgeable.

12. How knowledgeable is the user representative in the proposed business function?

    _____ **A.** Limited.

    _____ **B.** Understands concept, but no experience.

    _____ **C.** Very knowledgeable.

    _____ **D.** None.

13. What form of project management will be used?

    _____ **A.** Informal.

    _____ **B.** Formal, but one that I have used successfully before.

    _____ **C.** Formal, but never used before in this group.

    _____ **D.** Formal, and has been used before, but not by me.

14. To what extent do you plan to follow company standards?

    _____ **A.** As specified in corporate policy, including documentation and check-out points.

    _____ **B.** Follow the intent of the standards, with some modifications where necessary.

    _____ **C.** I probably will not use them.

15. Will a formal change-control procedure be used?

_____ **A.** No.
_____ **B.** Yes.

16. How well are the user's requirements defined?
_____ **A.** Too vague, further definition required.
_____ **B.** Adequate, for coping purposes only.
_____ **C.** Adequate, for systems definition.
_____ **D.** Well defined.

MY GENERAL NOTE AND EXPLANATION

It should be obvious that the mechanism of scoring depends upon the particular organization using this type of evaluation documentation and procedure. Usually there is a score for each answer, and the total for all the scores is divided by the maximum total for all scores. This is done for each of the four sections. This results is a *risk score*, or a percentage for each section. By adding all the risk scores together and dividing by four, you can get a total risk score for the project.

It seems to me that this evaluation procedure and the scoring that accompanies it is a very valuable process for two reasons:

1. Developing measureable project scores: The actual process of selecting applicable questions for your kind of projects and then setting up the scoring mechanisms is similar to outlining the network in a PERT diagram. It is a tremendous assist in defining what you want to do and what you think is important. Even if you take the extreme position of not using the form after it is developed (for whatever resons), the development process itself is valuable because it helps you to define what you feel are the actual content and value of the project.

2. Comparing how different people think about the project: As I mentioned before, we should use ordinal scoring to measure our management decisions. Without some kind of a measurement, we cannot compare alternatives and an ordinal scale is about the best that we can get under these conditions. The scoring of this project is done on an ordinal basis. It is therefore possible now to compare how different people feel about the problems and potential successes in the project by comparing their scores after different people have completed the four sections of this evaluation procedure.

This completed form (or an equivalent one that you develop to fit your situation) should be the basis for a project-definition meeting to outline an action plan to minimize the risk of project failure. If necessary, after appropriate corrective action has been taken, the form should be completed again (without referring back to the original completed forms, of course) and then comparisons between the original forms and these can be made to determine if sufficient progress has been achieved, as measured by the improved project total risk score.

Appendix 2   Project Management: A Goal-Directed Approach

The questions noted under each heading below are intended to be indicative of the general types of questions to be answered during each phase of a project. The authors have a much more extensive series of queries, and for further information, I refer you to the original source material.*

- *Idea or Concept*
  Why establish the project?

  What is it that the project is expected to accomplish?
- *Feasibility Study*
  Is the project feasible? (Usually this phase is done as a sort of miniproject to minimize later unknowns or determine the limits of several major variables such as marketing studies or physical tests. The output should be some kind of specification with a listing of the major problems evaluated and a probability as to potential solutions.)
- *Project Definition*
  How will the project itself be managed?

*Adapted from Michael G. Assad, and G.P.J. Pelser, "Project Management: A Goal Directed Approach," *Project Management Quarterly*, June 1983, pp. 49-58. Project Management Institute, P.O. Box 43, Drexel Hill, Pa. 19026

(The usual questions of what, when, where, who, and how much should now be answered.)
- *Vendors' Quotations*
This step is dependent upon a preliminary make-or-buy analysis. It includes:
What will we need to buy?

How reliable will our major venders be?

Where will we buy it?

How much will it cost?

When do we get it?

How will we make sure that we get what we ordered?
- *Final Design*
Will it meet the three "golden limits," i.e., meet the specifications, on time, within budget?
- *Contracts*
Can we now sign the purchase contracts that were so carefully evaluated in the vendors' quotations step above?

What is the probability that the vendor will meet the requirements of the purchase order?

What have we done to minimize the risk to the project?
- *Physical Execution*
Have we completed the planning for production satisfactorily?

Is the product or process going to be manufactured or constructed properly?

What does that mean, and how will we know it?
- *Product Testing*
What kind of in-process, incoming, subassembly, and completed assembly tests will have to be designed?

Can these tests be completed within the budgets set for the project?

Do we know enough about the potential product or service to devise tests that will be satisfactory?

And now that all this is done, has the product or service passed?
- *Final Tests*
How close do these tests match that environment in which the product or service is expected to perform?

Have we checked all the applicable physical and legal constraints?

And then, has the product or service passed or not?

Is there a feedback from all testing to the source of error, and has corrective action been taken?
- *Project Close Down and Summary*
Has the customer or user been consulted and satisfied?

Has the plan for sequential close down of the project and documentation of results been satisfactorily completed?

- *Services and Maintenance*

    Is there a service or maintenance contract with the customer or user?

    If so, have plans been made for training, spare parts, and other requirements?

    Have the services or maintenance managers been consulted about managing this part of the project?

Appendix 3   The Technique of Design Review

by Richard M. Jacobs, P.E. Consultant Services Institute, Inc., Livingston, N.J. 07039*

INTRODUCTION

As society demands more products and services of increasing complexity which are safer, the manufacturers and designers are under intense pressure to satisfy those demands. Those pressures stem from groups generally classified as "consumerists" who are publicly voicing their demands, often without technological knowledge. Additionally, related pressures are being applied by the courts through the evolution of strict product liability suits. The normal competitive pressures also require that products be designed, produced and operated safely and economically.

The combination of these pressures and the increasing sophistication of products have resulted in the need to further develop the techniques of Formal Synergistic Design Review. Certainly, Formal Synergistic Design Review is not a new concept. On the contrary, almost every engineering department will claim, upon inquiry, that Formal Synergistic Design Review is being employed.

Frequently the designs are "reviewed" by engineering, manufacturing, marketing, purchasing, quality control, etc. in sequence or with a few interested parties meeting informally at one time. Usually, a review takes place under the pressure of

*Reproduced with the permission of the American Society for Quality Control, from the Product Liability Prevention Conference, 1979.

a delivery date, a release to manufacture date, or other non-arguable demands. The result is often a cursory "review" at best and not infrequently a sign-off without review. The intent of this paper is to offer a method which could improve upon the results of sequential reviews which may be prevalent in the reader's organization.

Most engineering department practices vary considerably due to product, company, and industry. However, many Formal Synergistic Design Review programs have been successfully implemented when:

a. The product is competitive.

b. Engineering manpower is at a premium.

c. The schedule is very tight—frequently trying to beat the competition to the marketplace.

d. Company development funds are limited.

e. Competition has a product on the market and the company is attempting to meet the challenge.

f. Regulatory bodies are imposing arbitrary requirements.

Design Review is a term that may not accurately describe the action taken, especially when no design exists. Therefore, for the purpose of this paper the term Design Review is a process of communication between personnel of different technological persuasions which is concentrated on a single subject.

Formal Synergistic Design Review is a method that has found application in process industries as well as in product manufacture, services, office routines, banks and insurance operations. It is primarily a formal system forcing the different technical disciplines to communicate. Many new developments over the years have adopted Formal Synergistic Design Review as a key function toward achieving the established goals. For example, Formal Synergistic Design Review was used in the early 1950s when a new multidisciplined complex product was developed for the Air Force. Since then, the Formal Synergistic Design Review has matured into the most viable technique available to assure that the product is as reliable as designed, has the capability of being produced in quantity at the required quality level within cost constraints, and is as free from hazards as is possible when time, funding, and anticipated use are concerned.

Formal Synergistic Design Review (FSDR) also has the ability to minimize costs, speed delivery, etc. These attributes of FSDR are due primarily to the fact that all interested parties can and do communicate collectively, concurrently and not sequentially, with the responsible engineers.

The value of FSDR had not been ignored by overseas competition in its effort to build and sell better products. FSDR has been in use in most Western countries as well as in Japan, Hong Kong, Israel, Czechoslavakia, U.S.S.R., South America, Australia, and South Africa. The challenge is ever present; the rewards enormous and the opportunities to reduce costs, frequency of product failure, complaints, and returns abound.

I. WHAT IS A DESIGN REVIEW?

A Design Review is an aid in product development activities, which serves to guide and control the quality of a product so that it meets the needs and demands of the market. Its purpose is to utilize and exploit collective expert knowledge and experience within a company to ensure that intended qualities are incorporated in a product during the design phase; in other words, an assurance that demand specifications have been complied with.

A Formal Synergistic Design Review is not, therefore, a commonplace, informal activity, but is:

> A formal, documented, and systematic study of a design by specialists not directly associated with its development (1). (Numbers pertain to references at the end of this article.)

Design Reviews, in the form of formal gatherings, occur at predetermined stages of development and design processes. Participants at these meetings comprise the designers in question and appointed representatives for various functions within the company inclusive; e.g., Marketing, Service, Production Engineers, Buyers, Quality Technologists, Laboratory Personnel, and other kinds of personnel and their inputs are restricted to actual matters under discussion.

Within this definition are several words or phrases, the first of which is "formal." The term "formal" implies that the FSDR is "organized and controlled," and these two characteristics are of utmost importance. The organization of the Review helps assure that all significant topics are considered and appropriate personnel are in attendance: the "control" characteristic helps to assure that time is not spent on non-pertinent matters. Thus, formalizing the Design Review procedure will increase the effectiveness of the tool while holding down the cost of invested man hours.

The second key word in the above definition is "documented." In an environment of "Caveat Vendor" (let the seller beware), it behooves any manufacturer to fully document the design and production processes so that in the event of a liability litigation, he can adequately defend his design. This can be accomplished only by fully documenting the design decisions and the rationale that supports them.

The third key phrase in the definition is "systematic." The overworked term "system" is applicable here. It can be further explained, however, by the need to logically examine the design or process by following its evolution or by examining the subdivision; such as steering system, brakes, power train, or by separating the item into electrical, mechanical, and materials parts for discussion purposes.

The fourth key item in the definition of FSDR is "by specialists not directly associated with the product's development." By choosing people not associated with the design, an unbiased objective and *new* point of view are obtained. As a person works on a project, he may become somewhat myopic; that is to say, he becomes locked into his own train of thoughts relative to the particular design. Outsiders can examine the design and ask unrestricted questions concerning specific decisions. The suggestions resulting from or implied by the questions asked can provide the investigative designer with an additional perspective in viewing his work, thus increasing his overall effectiveness.

The fifth and most significant term is a word that is in the title but not in the definition, "synergistic." This word is why FSDRs are successful investments. Webster defines synergism as "the simultaneous action of separate agencies which, together, have greater total effect than the sum of their individual effects." "Brainstorming," a popular method of generating opinions and obtaining solutions, is dependent upon synergism to spark the process design.

SCOPE OF FSDR

The scope of the FSDR must encompass the total product picture. To provide maximum system effectiveness, the product must be produced economically; make a profit; perform safely; and meet the customer's expectations for performance capability, quality, and economy. To insure that the above requirements are satisfied, the design is reviewed at each stage of its development from the product definition at the conceptual stage, up to and including the start-up of the first units in the operational stage. A thorough program of this nature will provide the product integrity demanded by tomorrow's marketplace.

Essentially anything associated with the product as delivered to the customer should be subject to FSDR. The following is a list of those items which have been found defective in Product Liability suits during the past 20 years. A document search was conducted which resulted in these terms.

- Pre-sales documents: Brochures, catalogue data, price lists, advertising (mail and periodicals).
- Post-sales documents: Care and use books, warranty cards and explanations, instruction manuals.
- Shipping and display packages, labels, nameplates, decals, and tags.
- Field assembly and/or installation services.
- Service and maintenance.
- Spares.

DESIGN REVIEW OBJECTIVES

The objective of the Design Review is to mature the design the first time. (2) In other words, Design Review seeks to age the product more rapidly, to "debug" the product and its manufacturing process *before* it goes into production. This is accomplished by considering the following four guides in FSDR:

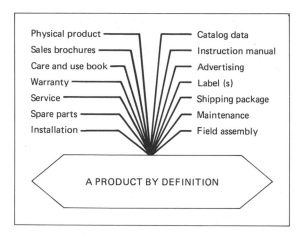

Figure A3-1.   Definition of a product

1. *Perspective*: Observe or review the design from viewpoints other than the designer's to minimize the risk of unforeseen contingencies; in other words, seek to plug the soft spots or holes.

2. *Experience*: Use personnel who have the widest range of experience outside the assigned design group within the company or, if necessary, outsiders. They are to provide guidance in identification and elimination of potential problem areas.

3. *Compatibility*: Assure compatibility with other mating designs, environment, personnel, and foreseeable application details.

4. *Economy*: Achieve the most economical design for the required performance, including reliability and safety. (3) In considering the economy factors, the decisions of what costs are to be minimized must be established and require the considered opinion of the accounting operations. In some instances, desirable cost data may not be available because the accounting system is not programmed to accumulate such information. In these situations, the accounting system may have to be altered or the data estimated.

II. WHAT ARE THE BENEFITS OF *FORMAL SYNERGISTIC DESIGN REVIEW?*

All designs are compromises between different demands and requirements. The difficulty lies in optimizing the priorities, particularly in the case of complex products. A Formal Synergistic Design Review provides the necessary forum for a comprehensive and simultaneous evaluation of different demands and requirements. It is similar to so-called simultaneous brainstorming whereby a number of specialists work collectively to produce a better and often a safer product. Five of the most significant benefits of FSDR are:

1. A DESIGN CAN MATURE MORE RAPIDLY

The frequency of drawing changes after an FSDR is often reduced by 40 to 60 percent. This reduction stabilizes and "matures" the design by eliminating design and manufacturing revisions.

2. COSTS WILL OFTEN BE LOWER

At an FSDR, proposals and suggestions are put forward for simplifying the design, which consequently could lead to lower costs for tools, manufacturing methods, inspection and quality control equipment, materials, and test equipment. At the same time, one obtains a critical examination of specific and intangibly defined market demands, making it possible to obtain a clear picture of unforeseen demands, which otherwise could result in time delays and subsequent related costs.

The number and extent of post FSDR design modifications or changes are reduced. The cost of each change is reported at a minimum of $75 to a high of $250 for the paper work only, making FSDR a cost reduction vehicle.

3. LESS RISK OF DELIVERY DELAYS

The total time for development, tooling, and manufacture is shortened through uniting and combining knowledgeable personnel from different spheres and activities to the design work early in the cycle. (See Fig. A3-2.)

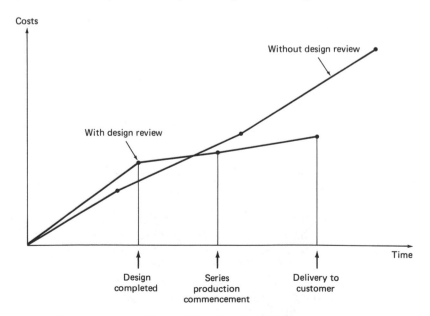

Figure A3-2. Results of FSDR

Intermediate Design Review

Intermediate Design Reviews are held when the following are completed: block or function diagram, energy flow design, mechanical design, styling, and development model tests. They are carried out in conjunction with the presentation of technical solutions, design, and prototype testing. These reviews (Intermediate Design Reviews) provide a recheck of product performance requirements in the light of development and design experience. Several Intermediate Design Reviews are frequently necessary to measure the progress of the design, and are determined by the complexity of the product in question and if needed, one to consider anticipated long lead-time items.

Final Design Review

Final Design Reviews are held when material lists and drawings are complete or when pre-production units are tested and analyzed. This is the last opportunity to effect changes to achieve the design objectives without seriously affecting schedules and at a cost significantly less than field changes or model changes. The Design Review should concentrate on final performance requirements, critical cumulative tolerances, and instruction books, in addition to all items previously studied. Action should be taken to dispose of all questions still outstanding before the design is released to production.

Manufacturing Design Review

This is a special form of Intermediate Design Review. Manufacturing Design Reviews are concerned with the method(s) needed to produce the items designed and consider the tooling, handling, flow, jigs, fixtures, sequence of operations and assembly conveyor system, and test and packaging procedures.

Installation Design Review

Installation Design Reviews are primarily used to consider the installation and/or field assembly procedures with the intent of doing the task less expensively and more rapidly. However, of late it has been used as a Product Improvement Design Review.

DEFINITION OF FSDR TOPICS

Topics subject to discussion at FSDR frequently have different meanings in different organizations. To eliminate misinterpretations, further explanation is offered for some of the items listed above.

Specifications.   This category includes specifications and regulations at all

4. PRODUCT SAFETY INCREASES

FSDRs offer the opportunity of discussing accident risks in conjunction with foreseeable usage of the future product and the elimination of such risks. By engaging field experts in possession of extensive product-environmental experience, who know the usage, maintenance, and handling of products by different customer categories and markets, the designer can obtain forewarning of likely safety risk features in the design as well as a clear understanding of demands or stipulations from other authoritative or official bodies. He may also obtain information about abnormal or unexpected methods of use which could result in direct personal injury or material damage.

Every conscientious designer has product safety in mind in his work and a common desire to make a product as safe to use as possible. Through supplementing this awareness with specialist information, he can further increase product safety which, from a product liability aspect, can prove vital to a company's business, reputation, and good will.

5. PERSONNEL CAPABILITY IMPROVEMENT

All participants in an FSDR exchange knowledge and take advantage of the expertise from different specialists, thus increasing their own knowledge and value to the company and to their careers. A project or product employing FSDR often finds that the design work takes longer and may cost more. Even so, this has proven profitable at subsequent product development stages, such as tooling and initial production, through fewer costly modifications and in the long run accounts for considerable savings in time and expenditure.

A series of FSDRs have been undertaken in the United States, and data collected from over 600 such reviews indicate that a return on investment exceeds twenty to one. An experiment conducted in Europe where multiple facilities produced the same product for local markets was undertaken in the period 1972-1974. Each facility was given a number of basic designs, samples, and test results from a central research operation. Half were introduced to the FSDR concept; the others were not. Costs of design, tooling, and manufacture were closely monitored for over two years and resulted in comparable returns on investment: (15 to 20 to 1). Figure A3-2 provides an overview of the European results of FSDR.

The data based on uncontrolled experiments and manufacturing estimates had the design phase with FSDR less costly than without FSDR.

III. WHEN SHOULD A DESIGN REVIEW TAKE PLACE?

As shown in Fig. A3-3, the product Life Cycle, . . . from the basic concept to the finished product, is divided up into a number of phases. Formal Synergistic Design Reviews commence at the concept or proposal phase and continue until the product

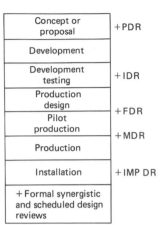

Figure A3-3. Product life cycle and design review

specifications in the design phase are completed. It is important that the FSDR activities are emphasized as early as possible in the development process, as this is where costs are most likely to be influenced. These costs reduce as time progresses, whereas project costs increase because one is committed through decisions on investments of various kinds.

DIFFERENT TYPES OF DESIGN REVIEWS

In order to achieve the desired effect with this type of activity, Formal Synergistic Design Review must be held at a number of decisive occasions or "milestones" during the product's development. Selecting these milestones may vary from company to company depending upon the type of product. Normally, however, one can discern between three or four types of Design Review (see Fig. A3-3):

- Proposal, or Concept, Design Reviews, (PDR)
- Intermediate Design Reviews, (IDR)
- Final Design Reviews, (FDR)
- Manufacturing Design Review, (MDR)
- Improvement or Installation Reviews, (IMP DR)

Preliminary Design Review

Preliminary Design Reviews are held at (see Fig. A3-2) product concept and planning, proposal, bid, or request for funds, and when contracts or authorization is received. It must be clearly determined that the concept or proposal corresponds to market requirements, that there are few deviations or interpretations and all agree. The purpose of the Preliminary Design Review is to establish early communication among Marketing, Engineering, Purchasing, and Manufacturing and to confirm the concept of the product design as truly representative of the customer's requirements. It is intended to expose divergent requirements and interpretations. The conditions and technology currently prevailing satisfy the market demands. In such occasions, it is up to the designer to supply and furnish information and experiences

from similar products, give accounts of problems connected with materials, manufacturing processes, quality control, product safety, ergonomics, transport, installation, maintenance, usage, and working environment, etc. This review is conducted to establish most of the ground rules and goals for the design. Frequently these are the product definitions discussed previously. Some of the items considered and reviewed during the conceptual or later Design Reviews include:

1. Function to be performed by product.
2. Market and sales volume.
3. Design sequence (working elements and artistic appearance).
4. Subsystems concept (if applicable).
5. Make or buy considerations.
6. Subsystem interfaces.
7. Design parameters (which are required in order of importance to function; some are mandatory, whereas others are only nice to have).
8. Test considerations.
9. Documentation required.
10. Critical parts to be used.
11. Environmental considerations.
12. High risk areas (including product liability and safety problems).
13. Reliability and maintainability requirements.
14. Redundancy requirements.
15. Availability.
16. Human factors.
17. Specifications.
18. Safety—personnel.
19. Safety—property.
20. Cost of manufacture.
21. Schedule considerations and cost alternatives.
22. Establishing rank of importance for all requirements.

In this way, and at one meeting, all persons concerned with the design and program planning are a party to and are informed of the reasons for decisions made. In the event of a question that cannot be answered at the Design Review meeting, an "action item" is established and assigned to a specific person for study and a detailed recommendation. The Design Review is not considered complete until all action items are resolved. In this way, the Chief Design Engineer—or in a large program, the manager of the program—keeps watch of who is doing what, when it will be completed, and how much it will cost.

The conceptual Design Review is followed at appropriate periods by the Intermediate Design Review, the Critical (or final) Design Review, and, if required, a Manufacturing Design Review, and as appropriate the Installation Review.

levels; such as company, military, and governmental (legislation and regulatory). Conflicts between these specifications and regulations and/or their interpretation must be resolved in light of the performance and safety requirements established for the product.

Reliability.   Few reasons will dissuade a customer from purchasing an item as effectively as previous experience with a company's product that is a ''part-time performer'' or does not meet his expectations with respect to useful life. Numerical estimates or requirements are needed for each product and the alternative designs.

Safety—Personnel.   Safety must consider anyone who can come into contact with the product. This includes manufacturing personnel, the user, and even third-party bystanders.

Safety—Property.   How will the failure of the product affect the surrounding property? For example, a smoldering appliance power cord may give off smoke of a strong odor or toxic fumes which destroy.

Human-Engineering Factors.   Does it fit the people who will use it? Few companies can afford to put out a product which is ill-proportioned and does not ''fit'' its user.

Cost of Manufacture.   Value engineering factors critical to the product system should be defined at the Preliminary Design Review. For example, producing an expensive design and then changing it to something more economical may provide impressive ''savings'' on a monthly report; however, the Design Review can add economy from the outset by the early definition of these requirements at a lower cost.

Environmental.   The product should be able to survive dust, heat, cold, vibration, corrosion, fungus, shipping, and other natural and human outrages to which it will be foreseeably subjected. The early identification of weaknesses aid the designer in his planning effort. Additionally, the product's effect on the environment must also be defined. The degree of survival and/or resistance to the environment must be defined for the designer and the user.

Maintainability.   The maintenance philosophy of each assembly, as well as that of the whole system, must be defined, projected, planned, and applied from the outset. This philosophy must consider whether the company is in the service repair business and, if so, the profitability of that activity. Maintainability, like reliability, must be designed in: it cannot be added on. Therefore, the designer must have firm parameters toward which to work.

Availability.   It is not infrequent that a product is part of an overall system which needs to be available for service on an instantaneous or momentary notice. An example is a standby lighting or power system. The designer must consider the various trade-offs of a number of outages for short durations or vice versa. Maintainability, reliability, and repairability are all involved in these decisions.

FUNCTIONAL REPRESENTATION
AT THE DESIGN REVIEW

When selecting personnel for the Design Review team, certain individual characteristics are desirable for all participants, the most important being the ability to remain objective. The purpose of an FSDR is to aid the designer. A participant who allows himself to be biased in his attitude will only serve to seriously compromise the success of the Design Review. By the same token, the designer must also be objective in that he must understand that the team is not assembled to criticize or evaluate his work but to assist in improving the design.

The first member of the committee to be chosen, usually by top management, is the Chairman. The function of the Chairman is to handle the procedural aspects and to stimulate the questions and ideas during the Design Review. For this reason, the most important quality in the person selected is management capability. The Chairman must be able to capture the respect of each team member and especially the designer; he must be capable of controlling the progress of the meeting to insure that all pertinent areas are discussed while preventing the discussions from dwelling on insignificant details; and finally, he must have the full support of personnel from management. The Chairman must also be sensitive to the "tone" of the meeting. Statements such as "The design is garbage" and "You don't know what you're talking about" must be avoided at all costs. Only by being very sensitive to the tone and attitudes of the participants can the Chairman prevent the meeting from becoming an execution of the design engineer.

The Chairman should have a broad understanding of the technical requirements of the product. Coupled with the managerial ability and technical capability and not being associated with the design effort will enable the Chairman to maintain his objectivity. However, in a small company, this may not be totally possible; the rapport is different, and the effect is entirely dependent upon mutual respects. The selection of the team is made by the Chairman, often with the assistance of the senior manager of the facility, which is then given Formal Synergistic Design Review responsibility for that product or project. The personnel selected for each review will vary depending upon the product and the agenda for the particular review. A general schedule for the various functions, their responsibilities, and the particular review(s) they should attend can be found in Fig. A3-4.

Those attending Design Reviews include people with special knowledge or experience from different spheres. The composition should be such that the participants collectively cover a sufficiently broad and detailed field of knowledge. When selecting participants, equal importance should be given to knowledge and experience qualities as well as personal attributes. They should be persons who can independently represent their own particular field and function and be able to present viewpoints, demands, and requirements in a constructive manner. The functions that ought to be represented at a review, like the composition aspect, vary according to the type of review. However, one should always attempt to obtain as broad an opinion and representation as possible. At the Preliminary Concept Review, for ex-

Group Member	Responsibilities	Type of FSDR				
		PDR	IDR	FDR	MDR	IMP DR
Chairman	Calls, conducts meetings of group, and issues interim and final reports.	X	X	X	X	X
Design engineer(s) (of products)	Prepares and presents design and substantiates decisions with data from tests or calculations.	X	X	X	X	X
Quality Control Manager or Engineer	Ensures that the functions of inspection, control, and test can be carried out efficiently.		X	X	X	X
Manufacturing Engineer	Ensures that the design is producible at minimum cost and schedule.	X	X	X	X	X
Field Engineer	Ensures that the installation, maintenance, and operator considerations were included in the design.	X		X		X
Product Safety Engineer	Concerned with any/all regulations that affect product safety including warnings, data collection, corrective action, and testing results.	X	X	X	X	X
Procurement Representative	Assures that acceptable parts and materials are available to meet cost/delivery schedules.		X			
Materials Engineer	Ensures that the materials selected will perform as required.		X			
Tooling Engineer	Evaluates design in terms of the tooling costs required to satisfy tolerance and functional requirements.		X	X	X	
Packaging and Shipping Engineer	Assures that the product is capable of being handled without damage.	X	X	X	X	
Marketing Representative	Assures that the requirements of customers are realistic and fully understood by all parties.	X		X		X
Design Engineers. (not associated with unit under review)	Constructively reviews adequacy of design to meet all requirements of customer.	X	X	X	X	X
Consultants, Specialists on Components, Value, Human Factors, etc. (As required)	Evaluates design for compliance, with goals of performance, cost, and schedule.	X	X	X	X	X
Customer Representative (optional)	Generally voices opinion as to acceptability of design and may request further investigation on specific items.			X		X

Figure A3-4. Formal synergistic design review group responsibilities and membership schedule

ample, it is important to invite persons with extensive technical know-how. At this stage, one has not as yet made any commitments or limitations through investments but is in the process of seeking the best, impartial solutions from all aspects and points of view. In later stages, people with specialized knowledge should be selected for the details of the design.

In selecting specific individuals to serve on the review team, the question often arises as to what level should be selected; for example, manager, senior engineer, or engineer. Some firms in their policy statement specify that the functional manager or, perhaps, senior engineer shall serve on the review committee. This method of designation is questionable primarily because it robs the review process of much of its flexibility. By locking in the review personnel in this manner, personality conflicts which can destroy the value of the review are extremely difficult to avoid. The personnel attending any review should be chosen on a case-by-case basis, with the individual personalities of the participants being included as a factor. One consideration will be the viewpoint of the designer. If, for instance, the manufacturing manager attends a review of a product designed by a young engineer, the engineer may be unduly cautious and afraid to disagree with the manufacturing manager. As a consequence, a design may be compromised just for some manufacturing expediency. Conversely, if the manufacturing representative is a young inexperienced engineer, the designer may run "rough shod" over any questions asked by the engineer; or he (the engineer) may even be hesitant in asking questions of the experienced design engineer. In either case, the manufacturing function's input to the design process can be jeopardized.

In summarizing the selection process, the participants must maintain both objectivity and mutual respect for the designer and other participants. At no time can personal criticisms be tolerated.

PLANNING FOR AN FSDR

As with any major undertaking, the development program of a new product requires detailed planning to assure success. Part of this planning should be the inclusion of a Formal Synergistic Design Review Program in the development schedule. In planning for FSDR, several questions should addressed: What products require FSDR, how many and what type of reviews are required, when should the reviews be held, and who should attend? Other requirements are the preparation of data to be given attendees prior to the reviews and the type of documentation required from the FSDR.

Selection of Product for Initial Design Review

Selecting a product is probably the most critical and difficult task in the entire implementation process for the first Design Review. It is wise to select a product which is not (a) behind schedule, (b) overly expended, (c) overly simplified nor overly complex, or (d) in danger of being cancelled. Obviously, the first few re-

4. PRODUCT SAFETY INCREASES

FSDRs offer the opportunity of discussing accident risks in conjunction with foreseeable usage of the future product and the elimination of such risks. By engaging field experts in possession of extensive product-environmental experience, who know the usage, maintenance, and handling of products by different customer categories and markets, the designer can obtain forewarning of likely safety risk features in the design as well as a clear understanding of demands or stipulations from other authoritative or official bodies. He may also obtain information about abnormal or unexpected methods of use which could result in direct personal injury or material damage.

Every conscientious designer has product safety in mind in his work and a common desire to make a product as safe to use as possible. Through supplementing this awareness with specialist information, he can further increase product safety which, from a product liability aspect, can prove vital to a company's business, reputation, and good will.

5. PERSONNEL CAPABILITY IMPROVEMENT

All participants in an FSDR exchange knowledge and take advantage of the expertise from different specialists, thus increasing their own knowledge and value to the company and to their careers. A project or product employing FSDR often finds that the design work takes longer and may cost more. Even so, this has proven profitable at subsequent product development stages, such as tooling and initial production, through fewer costly modifications and in the long run accounts for considerable savings in time and expenditure.

A series of FSDRs have been undertaken in the United States, and data collected from over 600 such reviews indicate that a return on investment exceeds twenty to one. An experiment conducted in Europe where multiple facilities produced the same product for local markets was undertaken in the period 1972-1974. Each facility was given a number of basic designs, samples, and test results from a central research operation. Half were introduced to the FSDR concept; the others were not. Costs of design, tooling, and manufacture were closely monitored for over two years and resulted in comparable returns on investment: (15 to 20 to 1). Figure A3-2 provides an overview of the European results of FSDR.

The data based on uncontrolled experiments and manufacturing estimates had the design phase with FSDR less costly than without FSDR.

III. WHEN SHOULD A DESIGN REVIEW TAKE PLACE?

As shown in Fig. A3-3, the product Life Cycle, . . . from the basic concept to the finished product, is divided up into a number of phases. Formal Synergistic Design Reviews commence at the concept or proposal phase and continue until the product

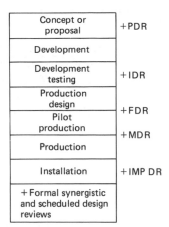

Figure A3-3. Product life cycle and design review

specifications in the design phase are completed. It is important that the FSDR activities are emphasized as early as possible in the development process, as this is where costs are most likely to be influenced. These costs reduce as time progresses, whereas project costs increase because one is committed through decisions on investments of various kinds.

DIFFERENT TYPES OF DESIGN REVIEWS

In order to achieve the desired effect with this type of activity, Formal Synergistic Design Review must be held at a number of decisive occasions or "milestones" during the product's development. Selecting these milestones may vary from company to company depending upon the type of product. Normally, however, one can discern between three or four types of Design Review (see Fig. A3-3):

- Proposal, or Concept, Design Reviews, (PDR)
- Intermediate Design Reviews, (IDR)
- Final Design Reviews, (FDR)
- Manufacturing Design Review, (MDR)
- Improvement or Installation Reviews, (IMP DR)

Preliminary Design Review

Preliminary Design Reviews are held at (see Fig. A3-2) product concept and planning, proposal, bid, or request for funds, and when contracts or authorization is received. It must be clearly determined that the concept or proposal corresponds to market requirements, that there are few deviations or interpretations and all agree. The purpose of the Preliminary Design Review is to establish early communication among Marketing, Engineering, Purchasing, and Manufacturing and to confirm the concept of the product design as truly representative of the customer's requirements. It is intended to expose divergent requirements and interpretations. The conditions and technology currently prevailing satisfy the market demands. In such occasions, it is up to the designer to supply and furnish information and experiences

Intermediate Design Review

Intermediate Design Reviews are held when the following are completed: block or function diagram, energy flow design, mechanical design, styling, and development model tests. They are carried out in conjunction with the presentation of technical solutions, design, and prototype testing. These reviews (Intermediate Design Reviews) provide a recheck of product performance requirements in the light of development and design experience. Several Intermediate Design Reviews are frequently necessary to measure the progress of the design, and are determined by the complexity of the product in question and if needed, one to consider anticipated long lead-time items.

Final Design Review

Final Design Reviews are held when material lists and drawings are complete or when pre-production units are tested and analyzed. This is the last opportunity to effect changes to achieve the design objectives without seriously affecting schedules and at a cost significantly less than field changes or model changes. The Design Review should concentrate on final performance requirements, critical cumulative tolerances, and instruction books, in addition to all items previously studied. Action should be taken to dispose of all questions still outstanding before the design is released to production.

Manufacturing Design Review

This is a special form of Intermediate Design Review. Manufacturing Design Reviews are concerned with the method(s) needed to produce the items designed and consider the tooling, handling, flow, jigs, fixtures, sequence of operations and assembly conveyor system, and test and packaging procedures.

Installation Design Review

Installation Design Reviews are primarily used to consider the installation and/or field assembly procedures with the intent of doing the task less expensively and more rapidly. However, of late it has been used as a Product Improvement Design Review.

DEFINITION OF FSDR TOPICS

Topics subject to discussion at FSDR frequently have different meanings in different organizations. To eliminate misinterpretations, further explanation is offered for some of the items listed above.

Specifications.   This category includes specifications and regulations at all

from similar products, give accounts of problems connected with materials, manufacturing processes, quality control, product safety, ergonomics, transport, installation, maintenance, usage, and working environment, etc. This review is conducted to establish most of the ground rules and goals for the design. Frequently these are the product definitions discussed previously. Some of the items considered and reviewed during the conceptual or later Design Reviews include:

1. Function to be performed by product.
2. Market and sales volume.
3. Design sequence (working elements and artistic appearance).
4. Subsystems concept (if applicable).
5. Make or buy considerations.
6. Subsystem interfaces.
7. Design parameters (which are required in order of importance to function; some are mandatory, whereas others are only nice to have).
8. Test considerations.
9. Documentation required.
10. Critical parts to be used.
11. Environmental considerations.
12. High risk areas (including product liability and safety problems).
13. Reliability and maintainability requirements.
14. Redundancy requirements.
15. Availability.
16. Human factors.
17. Specifications.
18. Safety—personnel.
19. Safety—property.
20. Cost of manufacture.
21. Schedule considerations and cost alternatives.
22. Establishing rank of importance for all requirements.

In this way, and at one meeting, all persons concerned with the design and program planning are a party to and are informed of the reasons for decisions made. In the event of a question that cannot be answered at the Design Review meeting, an "action item" is established and assigned to a specific person for study and a detailed recommendation. The Design Review is not considered complete until all action items are resolved. In this way, the Chief Design Engineer—or in a large program, the manager of the program—keeps watch of who is doing what, when it will be completed, and how much it will cost.

The conceptual Design Review is followed at appropriate periods by the Intermediate Design Review, the Critical (or final) Design Review, and, if required, a Manufacturing Design Review, and as appropriate the Installation Review.

views should be performed on products or projects that can be looked upon as being representative of the organization's effort. Selecting a product that is not representative can bias the reaction and curtail the acceptance of the Design Review tool. When the idea of formal Design Review is first being introduced, it may be advisable, because of the greater impact upon higher levels of management, to hold the first design review on a product which has been carried through the development stage. In other words, an Intermediate Design Review may produce more emphatic results and thus "sell" the procedure to management. Just when this should be held varies with the product, its size, complexity, cost, and manufacturing process. The closer this Formal Design Review can be held to the completion of the design details and before an investment is made in tooling, molds, etc., the better the results. This suggestion may not be applicable to products with extremely long design cycles. For products in this category, a subassembly Intermediate Design Review may be the answer. In any event, the intent is to minimize the investment of time or equipment and have the design reviewed frequently enough to get the benefit of the collective thinking which occurs at a Formal Design Review.

Scheduling of Reviews

The Chairman and Engineering Manager should schedule all Design Reviews with the appropriate attendees and advise them of time, place, and subject. Specific time for the Design Reviews will be incorporated into the product design and development schedule, and approved so that all participating functions can plan their efforts efficiently.

The Engineering Manager and the Chairman should jointly have the responsibility for scheduling and calling Formal Synergistic Design Reviews. The Chairman of a FSDR should be given the authority to request that investigations and related reports be carried out. It is of particular importance to stress that *final authority for design decisions will rest with the Engineer and his Supervisor.*

Establish a schedule for:

(a) Preliminary Orientation Meeting. (Used only to introduce FSDR to personnel as they are selected for the task.)

(b) Formal Synergistic Design Review.

(c) Follow-up Dates.

(d) Issuance of preliminary meeting notice.

(e) Issuance of agenda for FSDR.

(f) Issuance of minutes.

(g) Issuance of report on each FSDR.

Advance Information

At least 10 days before the actual date of the review, information should be distributed or made available to the designated participants. This may be in the form of specifications, a competitive cost exhibit, preliminary layouts, etc. Distributing this information beforehand will help to assure that the participants are well prepared to contribute constructively to the objectives of the review.

The results of an FSDR are most substantial when the material to be reviewed is distributed along with the agenda. This may include customer originated requirements (request for quotes, specifications, standards); anticipated customer needs developed and supported by market surveys and competitor's activities; engineering prepared proposals; photographs of similar products, competitive product data; cost estimates; specifications; drawings, test reports, analyses and requirements; field failure or malfunction summary reports; Quality Control analyses of processes and supplies; inspection reports; and any other pertinent data. This enables the participants to have ample time to study the circulated information, become familiar with the product and its requirements, and prepare their questions. The effort thus spent by participants prior to the meeting reduces considerably the time needed at the FSDR to cover all major parts of the product.

Agenda

Although discussing the need for an agenda seems to be an elementary point, it has been found that a formal advance notification of the agenda has worked well because of the personal recognition effect and the quasi-public assignment of a task to perform. The agenda should:

a. List and be sent to *all* participants.

b. Be sent to all participant's supervisors.

c. Identify projects and provide account charge numbers when appropriate.

d. Identify type of review and what section of the project will be discussed.

e. Date, time to start, and anticipated finish.

f. Place and special assignments, if any.

g. List who will be Chairman and who will be Secretary.

h. Show who will make the various presentations on the various topics to be discussed.

i. List the attachments being distributed as Advance Information.

Suggested general agendas for the three types of Design Reviews are shown in Fig. A3-5. The usual discussion of trade-offs between reliability needs and other important factors is most difficult to settle before or during a Design Review. However, experience has shown that when these items are carefully explained by the Chairman, it reduces discussion time on each suggestion. Establish the relative importance of each factor on the product design early in the Design Review session. Some factors are reliability, cost, performance, delivery date, production release date, appearance, incentive clauses in the contract, weight, volume, size, capability to withstand peculiar environment, etc. Each product is expected to be different,

Preliminary	Intermediate	Final
1. Introduction to Formal Design Review.	1. Restate Design Review Purposes.	1. Restate Design Review Purposes.
2. Marketing—Describes customer product requirements.	2. Chairman reviews product requirements.	2. Chairman reviews product requirements.
3. Discuss Marketing requirements.	3. Review factor priority established for product.	3. Review factor priority established for products.
4. Establish factor priority (reliability, cost, performance, safety, appearance, etc.)	4. Discuss Design—Electrical. Discuss test results, if available.	4. Discuss Design—Electrical. Discuss test results, if available.
5. Design Design and Development Plan.	5. Discuss Design—Mechanical. Discuss test results, if available.	5. Discuss Design—Mechanical. Discuss test results.
6. Discuss schedule.	6. Discuss details affecting safety.	6. Discuss details affecting safety.
7. Summarize.	7. Discuss tooling and packaging.	7. Discuss tooling and packaging.
	8. Discuss installation.	8. Discuss installation.
	9. Discuss instruction book.	9. Discuss instruction book.
	10. Discuss maintenance manual.	10. Discuss maintenance manual.
	11. Discuss schedule.	11. Discuss schedule.
	12. Summarize.	12. Summarize.

Figure A3-5.   General agenda for three types of Formal Synergistic Design Reviews

and the combination may change depending upon the schedule, market conditions, etc. In any event, the priority of items should be recorded at each Design Review held.

Conduct of Design Review Meetings

a. Introductory comments by Chairman: Introductory comments should set a constructive tone and climate for the meeting. The specific objectives of the Design Review should be stated and should relate to the overall objectives: namely achieving optimum product design from the standpoint of reliability, function, cost, and appearance.

b. Appointment of Secretary: An attendee will be appointed secretary and will take notes on useful ideas submitted, and other pertinent comments. He will also record when additional action is required and by whom.

c. Presentation of the subject: The design engineer or product manager should describe adequately the product being reviewed and include a comparison of customer requirements versus the expected performance of the product.

d. Systematic Analysis: The Chairman should make sure that the discussion follows a systematic plan so that no major subject areas are omitted. The discussion should follow the prepared agenda. Checklists should be used to prevent omission of important design considerations.

Since Design Reviews involve people, their ideas, and their productivity, complete understanding is essential to a successful program. All comments should be constructive. Derogatory comments by any attendee or flat refusal to consider change should not be permitted. All comments should be in the form of questions for further information, or inquiring as to whether an alternate design, part, or material was considered. Care should be taken to assure the engineer that the questions raised, subsequent investigations, and opinions on the various results are for his benefit and are not questioning his ability or integrity. He has and retains the full responsibility for the design. The Design Review Group is *advisory* in nature and the prime purpose of Design Review is to assist the design engineer in developing the optimum product design by providing a collective source of specialized knowledge.

The secretary should be able to understand the discussion and record highlights and synopses of the sense of the comments. Verbatim transcriptions should be discouraged as should tape or wire recordings. Tape or wire machines have been used successfully when the secretary or chairman recorded the summary of the comments said. Care should be taken to avoid associating a comment to a person or a group. In each case where some action is required or expected, the name of the person who is to act plus the date for completing the action should be recorded.

The Design Review minutes are to be prepared by the secretary and be distributed as soon as possible. The minutes should

a. Be sent to all attendees and their supervisors.

b. Be sent to managers of Engineering, Manufacturing, Marketing, Purchasing, Reliability, Quality Control, Field Service, etc. and Project Manager or Product Manager.

c. Show references to all documents discussed with dates or issue number.

d. Describe the product.

e. Record all questions raised.

f. Record all responses in summary fashion, particularly action items.

g. Show who will act and by what date.

h. Show priority of factors used in making decisions for this product; reliability, cost, performance, appearance, delivery schedule, incentive clauses in contract, etc.

Fig. A3-6 could be a typical management implementation form for an FSDR.

Post Review Investigations usually lead to the most important results of Design Reviews. The engineer evaluates suggested alternative designs, materials, or

1. Facility Manager _____   Approval Date _____

2. Engineering Manager _____   Approval Date _____

3. Product Selected is _____

	General Manager _____	Date _____
Approved by	Engineering Manager _____	Date _____
	Designer _____	Date _____

4. Chairman of Design Review is _____

5. Secretary of Design Review is _____

6. Participants (and Observers, if any are:

    (a) _____   **(g)** _____

    (b) _____   **(h)** _____

    (c) _____   **(i)** _____

    (d) _____   **(j)** _____

    (e) _____   **(k)** _____

    (f) _____   **(l)** _____

	Goal Date	Actual Date
7. Schedule:		
Orientation Meeting	_____	_____
Design Review	_____	_____
Design Review Minutes	_____	_____
Design Review Final Report	_____	_____
8. Issuance of Orientation Notice	_____	_____
Issuance of Design Review Notice— Agenda	_____	_____
Issuance of Minutes	_____	_____
Issuance of Final Report	_____	_____

9. Results of Design Review (for Management Reporting) are:

    (a) Number of items questioned _____

    (b) Number of action items assigned _____

    (c) Cost of preparation _____

    (d) Cost of Review _____

    (e) Cost of investigation _____

    (f) Change in reliability _____

    (g) Change in performance _____

    (h) Change in manufacturing costs _____

    (i) Change in material cost _____

    (j) Change in operating and maintainability costs _____

    (k) Change in schedule _____

Figure A3-6. Reminder list and data record for implementation of Formal Synergistic Design Review

parts to assure that the best choice is made considering the reliability, cost, schedule, and performance characteristics of the end product. During this evaluation procedure, the engineer frequently discovers that the ideas that he had originally plus the suggestions he heard result in a design embodying the best of all the ideas. It may even trigger a completely new method for doing a particular task. It is, therefore, imperative that the engineer be given every opportunity to measure the usefulness of the ideas offered. Whatever the decision, it should be reported to the Design Review group via the final report. The report to the Engineering manager showing each investigation recommended and each subsequent decision made by the designer should be issued five days after the Review whenever possible. Resolution of differences between the group and the designer should be recorded and brought to the attention of the designer's supervisor and, if necessary, the manager of the department. It is rare that decisions have to be made by the General Manager, although the procedure should provide for that eventuality.

Follow-Up Action

Documentation.   The secretary is responsible for preparing and distributing minutes of the meeting, indicating the ideas generated and the action to be taken. The minutes should also note by whom the action is to be taken and when.

Utilization of Ideas.   The designer is responsible for investigating and incorporating into the design those ideas which will aid in achieving optimum product design. The Design Review Chairman will have the responsibility of following up on the utilization of the ideas proposed and of the assigned action items.

Final Report

A final report of the Formal Synergistic Design Review for a product or project must always be compiled. This report should contain a summary of the results achieved, also a complete list of the reports issued or other forms of documentation distributed, e.g., reports of testing results, activities, etc. In other words, the report should give all details concerning investigations conducted and suggested proposals and concepts, also the reasons for decisions made or refusals. It should also contain information about changes or modifications concerning time factors and costs incurred during the materialization period, plus any other experiences of significance.

Personnel, other than the designer, assigned an action item at the Design Review are to advise participants of their findings (and recommendations, if appropriate); but the designer has the responsibility to make the final design decisions and adaptability of the suggestions made. His decisions are included in the Final Report.

A Final Report is a document of exceptional value for the resultant product and for elucidating design problems or faults which may later be experienced by the user, having previously been discussed, and the reasons for the design solutions selected.

From the product liability aspect, it is important to know that reports from Design Reviews are regarded as extremely valuable documentation of a product and are frequently admissible in Court as evidence of diligence. It is equally important that any proposed measures are followed up and the reasons for selecting one solution in favor of another are clearly explained and stated. The Final Report should:

a. Be sent to all attendees.

b. Be signed by the Chairman.

c. Summarize all items raised at the Design Review showing action taken.

d. Summarize cost of preparation for the Design Review and the investigation time.

e. Show increases or decreases in product reliability, maintenance, and operating and manufacturing costs.

f. Show schedule changes.

CHECKLISTS

Checklists can be developed to remind participants of all items which should be considered in each Review. Such a list also aids the designer by reminding him of things that should not be overlooked, but these become quickly stereotyped.

Checklists are important tools as they serve as a reminder during the design process. It frequently takes a number of reviews before a checklist becomes complete. Checklists are explored and presented in articles previously published. *They are useful, but FSDR is not dependent upon checklists.* They are often used as a crutch and stifle original thinking.

SUMMARY

Design Review changes the attitudes of engineers in at least a small way, and this modification in thinking results in a greater willingness to consider improvements on succeeding designs. This obviously results in a better product at a lower cost earlier in the design cycle. The more Design Reviews, the greater the impact on attitude and the better the product. However, the total value accrued in this complex process cannot be ascertained since increased sales, due to customer acceptance of future products, are difficult to estimate. Customer satisfaction is the most meaningful measure of a product's reliability and safe operation. Formal Synergistic Design Review contributes to his satisfaction and is of primary importance in the operation of the company. It has been found that

1. Maturity of the design is accelerated: The application of expert opinion and seasoned experience results in achieving better (safer) design earlier.

2. Cost improvements are frequently achieved: Design details, tooling,

fabricating methods, materials, finishes, etc., which affect reliability and the cost of manufacture, operation or maintenance, are questioned during a Design Review. Some of the reductions made in total customer costs result in increases in others. These must be questioned and accepted.

3. Delivery dates are frequently improved: The design, development, and manufacturing cycle is frequently shortened by using the combined talents of specialists.

4. Customer design approval is sometimes obtained: Where appropriate, customer design approval has been secured, based upon the formal Design Reviews and their subsequent reports.

5. New product lines may be generated: During Design Reviews, questions on the performance criteria of a product may result in conceiving a new product, model, or market.

6. Staff capabilities are improved: Designers exposed to the constructive criticism of specialists become more valuable to the organization.

7. Design changes are reduced: The cost of developing a product is lower when the frequency of design changes are minimized. Formal Synergistic Design Reviews have aided in the reduction of these changes after the release of the design to manufacturing. Figure A3-2 shows the relative average cost of making design changes versus the design, development, and manufacturing cycle for products using conventional design methods and for products using the Formal Synergistic Design Review technique. It has been found that FSDR successfully combats the philosophy "We'll fix it later! Let's get it into production first!"

8. Formal Synergistic Design Reviews when records are available have formed the basis for successful defenses of product liability cases. Success means that the designing company was either dismissed from the litigation, settlement was reached by the plaintiff and defendant at a very satisfactory figure with which everyone was happy, or judgment awarded to the plaintiff that took care of his immediate medical expenses was substantially lower than anticipated. Formal Synergistic Design Review has thus been useful in limiting the cost of product liability claims.

REFERENCES

1. Jacobs, R.M., "Implementing Formal Design Review," *Industrial Quality Control*, Vol. 23, No. 8, p. 398, February 1967.

2. McClure, J.Y., and E.S. Winlund, "Design Review: A Philosophy, Survey and Policy," General Dynamics Corp., San Diego, Ca. 1963, p. 2.

3. Winlund, E.S., "Design Review," General Electric Corp., Phoenix, Ariz., 1959, p. 3.

4. Jacobs, R.M., "Implementing," *op. cit.*, p. 400.

5. PHILLIPS, W.E., "Guidelines on Planning and Conducting Design Reviews," Sylvania Electronic Systems—East, 1962, p. 3.

6. SIMONTON, D.P., "The Design Review Program," *Machine Design*, November 24, 1960, p. 113.

7. JACOBS, R.M., "Minimizing Hazards in Design," *Trial Lawyers Quarterly*, Vol. 7, no. 2, Summer-Fall 1970, p. 185.

8. JACOBS, R.M., *Design Review Proceedings of Product Liability Prevention*, August 1975.

9. SMITH, MARION P., "Design Review as a Tool for Product Liability Prevention," *Proceedings of PLP-72*, Newark, N.J., August 1972.

10. MIL-STD-1521 (USAF), *Military Standard, Technical Reviews and Audits for Systems, Equipment, and Computer Programs*, September 1, 1972.

11. BOQUIST, E.R., "Tutorial on Formal Design Review," *PLP-73*.

12. JACOBS, R.M., and MIHALASKY, J., "Minimizing Hazards in Product Design Via Design Review," *JEMC*, St. Petersburg, Fla., 1973.

13. JACOBS, R.M., "Design Review as a Liability Preventer," *Winter Annual Meeting A.S.M.E., New York, November 1974.*

Appendix 4   A Brief Overview of PERT (Program Evaluation Review Technique)

Program Evaluation Review Technique (PERT) is a variably complex once-through forecasting technique designed to handle uncertainties in producing a single item or design.* It has two major components:

1. A network of tasks or events showing the events and how those tasks are related to each other, and
2. An estimate of the time between events.

Like the Gantt or Bar charts, PERT makes use of tasks. It show achievements. Those terminal achievements are called *events*. Arrows are used to represent tasks; circles to represent events. You start a PERT forecast by making a diagram of events (see circled letters on Fig. A4-1) which are joined by activities. The direction of the activity is noted by an arrow.

As an example, we can use the development of a fluid valve. The activities that are of concern are tasks a through k, as follows:

a. Complete drawings

*Adapted from M. Silverman, *Project Management: A Short Course for Professionals*, Atrium Associates Inc., (Cliffside Park, N.J., 1984), pp. 68-75, with permission.

b. Complete quotations

c. Complete make-or-buy

d. Complete operation sheets and tooling

e. Complete assembly toolroom lot

f. Complete test

g. Complete detail drawing revision

h. Complete test fixture changes

i. Complete first production lot

j. Complete final test

k. Ship

Each task, then, is limited by a definable event. The event has no duration; it simply serves to tell you that an activity has ended or begun. Except for the first tasks, all activities must have a beginning and an ending event (i.e., a milestone). No task can start until all tasks upon which it depends are completed. There can be no "looping back" of activities in PERT, for a sequence of tasks that leads back to an activity will never end. Your list of tasks and events is the basis for the PERT network diagram.

Arrow length is not significant, but the sequence and interconnections must give a true picture of the tests that must be completed before the next task becomes possible. The numbers shown on the activity lines between events correspond to an estimate of the time required between events. In this example, the time is noted in weeks. According to Fig. A4-1, it takes four weeks from the end of the detailed drawing revision until the first production lot is completed—the time period from g to i.

The longest path through the network is the *critical path*. Since all other paths through the network take less time, the critical path is the one that determines the project completion date. Any delay in the critical path will affect the schedule; any delay in the other paths could possibly be corrected without affecting the schedule.

The planning of a project, using networking, can be applied to almost every action, including determining how and when to paint a living room so that it will be finished during a vacation. In some cases, the network and its control can be extended into a situation that could require expensive computer inputs; however, if

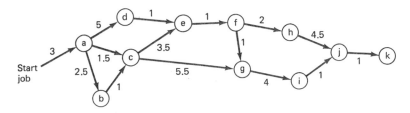

Figure A4-1 Pert chart

kept within a reasonable number of events, the manager can control it by hand notations at minor administrative costs.

When a manager uses PERT, it quickly becomes apparent that networking is a rigorous job. The laying out of activities themselves becomes beneficial, because it requires a logical basis to achieve a goal. In fact unless the problem to be solved is a tremendously large one, most of the benefits come from the networking task alone. Networking requires a try-to-think-of-everything attitude. The manager may otherwise overlook potential problem areas. (By the way, the critical path is Start to a, then b, then c, then e, then f, then h, then j, and finally k, i.e., $3 + 2.5 + 1 + 3.5 + 1 + 2 + 4.5 + 1 = 18.5$). . . . Another way a manager can gain the cooperation and enthusiasm of the project personnel is to allow the people responsible for their own segments of the work to lay out their respective portions. Later these smaller nets can be "stitched" together into an overall project. . . .

PERT-COST is a cost projection status technique based on a PERT activity chart with costs for each activity. It is possible, therefore, to forecast total cost against contract cost in the same manner as total delivery time in PERT against contract delivery time. Another forecastable attribute is manpower, handled similarly to time in PERT or money in PERT-COST.

The variations of PERT (or Critical Path Method, as the building industries call it) are numerous. It is not the manager's function to become an expert on PERT and its variations. He or she is a management expert, and PERT is just another tool to help perform the primary job, that of managing. If the project is large enough, he or she can get an expert to handle the PERT chart.

Appendix 5   Line of Balance

The transition from the conception and definition phases to the acquisition phase requires a different control tool—Line of Balance.* This technique includes four separate components:

1. A *flow chart* showing how the manufacturing process will occur over the *actual time* on the production shop floor. Laying out this flow chart provides similar benefits to those that happen when you lay out a PERT network. It shows discrepancies and inconsistencies in processing, and assists the forecaster by showing possible simultaneous operations that will save manufacturing time (see Fig. A5-1).

2. A cumulative delivery schedule of the final product over the calendar time period that this schedule should occur. The total number of units to be delivered is shown as the Y axis, and the calendar shown on the X axis. (See Fig. A5-2)

3. A Line of Balance chart showing the required number of parts and the actual number of parts. The cumulative number of unit parts to pass a checkpoint in

*Adapted from M. Silverman, *Project Management: A Short Course for Professionals*, Atrium Associates Inc. (Cliffside Park, N.J.) 1984, pp. 75-80, with permission.

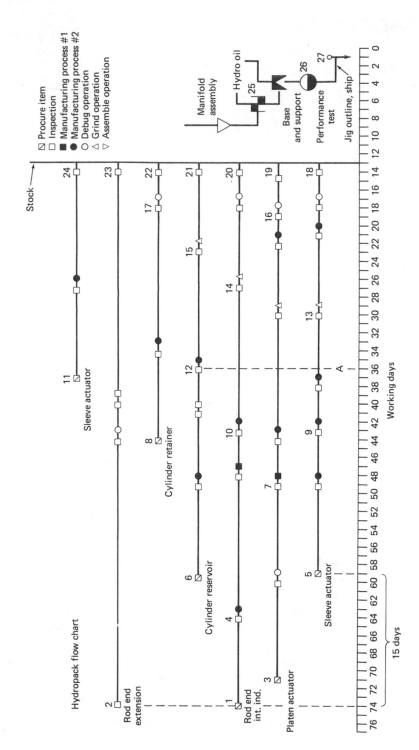

Figure A5-1

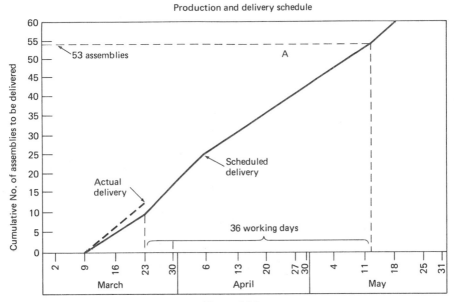

Figure A5-2

the flow chart above (see "A" above) is the *Y* axis, and the *X* axis is the actual checkpoint on the flow chart (see Fig. A5-3).

4. A production report explaining the variances that appear on the Line of Balance. The corrective action report that shows the reasons for either too many units passing a specific checkpoint or too few having been done (see Fig. A5-4).

This technique provides both forecasting and measurement of the result against that forecast. It is a powerful but very easy-to-use technique. It is powerful because it fulfills most of the requirements of a good forecasting technique. It requires the participation and the commitment of the team manager responsible for production; it objectively reports the actual status against that forecast; and it requires corrective action to be reported. In other words, it (a) forecasts *today* what the future situation is supposed to be, (b) measures *in the future* the actual against the forecast, and (c) demands *corrective action* to bring the actual measurement into line with the forecast.

As an example, we will set up the Line of Balance for a hydraulic control known as Hydropack. Our intention will be to forecast production quantities of required components which will allow us to ship the required number of assemblies at the right time in the future.

As a first step, the appropriate team manager for production accepts the responsibility for laying out a flow chart to show when the parts are to be made or bought, and how they will flow together to make subassemblies and assemblies.

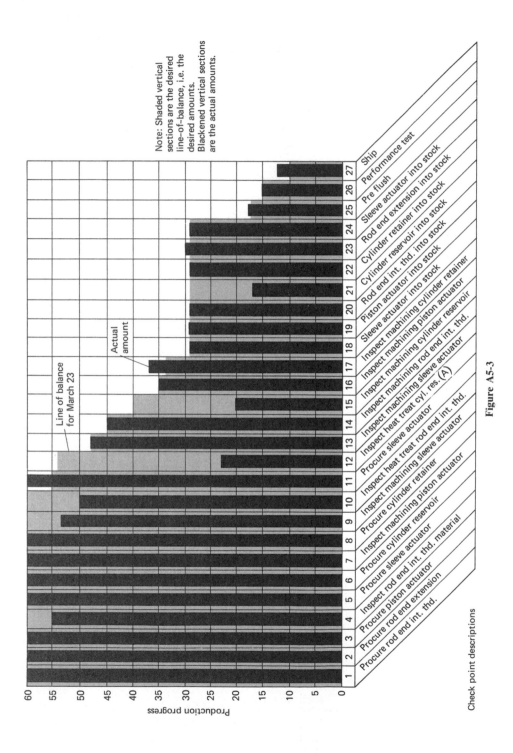

Figure A5-3

Check point descriptions

338

PRODUCTION REPORT — 23 March				
Check point	Description	LOB Req'd	LOB Actual	Corrective action
2	Procure rod end extension	60	58	2 pieces due in tomorrow
4	Inspect rod end internal	60	55	5 pieces in process
9	Inspect machined sleeve actuator	60	53	7 pieces in process
10	Inspect H. T. rod end	60	50	Vendor delay — 15 pieces tomorrow
12	Inspect H. T. cylinder reservoir (A)	53	23	Vendor delay — 40 pieces tomorrow
15	Inspect machined cylinder reservoir	40	20	Holdup — checkpoint 12
21	Cylinder reservoir into stock	28	17	Holdup — checkpoint 12

Figure A5-4

The flow chart will show parts being purchased, manufactured, or assembled in a specific lot size. For example, if it takes one hour to machine a hole, consequently when we are processing a 100-piece lot, the processing time will be 100 hours plus the setup, delay, and movement time. The time for machining a hole then may be, say, 145 hours, not 100 hours. The finished flow chart time generally has little relationship to the actual processing time. The chart may show months while the unit processing time may be in hours. Obviously, just laying out a chart like this will highlight some potential improvements in manufacturing.

We will (as usual in every forecast) work backward from the finished product (zero time) to the major and then the minor parts. "Major" and "minor" here refer to how long the part is in the processing stream, not to how large or critical the part is or to how long the processing actually takes.

Measure *actual working time* horizontally on the X axis. The horizontal line is important here as opposed to the PERT network chart, where it is not. The flow chart goes from zero (shipping) time on the right to the maximum (ordering of long lead items) time on the left.

Now the important benchmarks or measuring gates which we will be monitoring are numbered. Various supervisors will report progress on all of these points at monitoring intervals. Number from left to right in time sequence, so that a lower number is always to the left of the higher number. The shipping date has the highest number. Keep the number of benchmarks to a minimum needed to give you a clear idea of what's going on—the more benchmarks, the more laborious the measuring job becomes.

In the example of Fig. A5-1, the sleeve actuator (beginning with event 5) need not be started until 15 working days after the purchase order for the rod end extension (2) has been placed, i.e., 74 days minus 59 days (event 2 time minus event 5 time). Figure A5-2 shows a cumulative contractually required delivery schedule.

The cumulative delivery schedule is then developed, usually by the contract administrator based upon the contracts for the product. It shows cumulative shipments as the *Y* axis (in accordance with the contract) and horizontal *calendar* time as the *X* axis. The change of the slope of that charted cumulative shipment indicates variations in production rate. But using Fig. A5-2 to determine future production needs requires a calculation of *working*, not calendar, days in order to develop a Line of Balance.

The next chart, Fig. A5-3 is simply a bar chart on which is noted both the forecasted number of parts, subassemblies, and assemblies that should have passed each benchmark according to the cumulative delivery schedule and the actual number that did so at that date. We draw the Line of Balance (LOB) on this chart. Suppose that today is March 23. Benchmark 12 on the Hydropack flow chart must be passed 36 days before final delivery. On the delivery schedule, 36 *working* days forward from March 23 (today) brings us to May 13. A vertical intersection at May 13 to the cumulative delivery schedule shows that 53 assemblies must be delivered by then. On the bar chart (Fig. A5-3), we draw the March 23 LOB to cross benchmark 12 at 53. It also crosses benchmark 15 at 40 assemblies and benchmarks 18 to 24 at 28 assemblies. (Benchmark 18 is 13 working days ahead, and that intersects the cumulative delivery schedule on April 12, taking today as March 23. Therefore on April 12, 28 assemblies must have been delivered, and that means 13 days ahead of April 12, which is today (March 23), 28 sleeve actuators should be in stock.

The LOB tells us exactly how many parts and assemblies should have passed each benchmark by the date under consideration, if the scheduled delivery is to be met. (See the lightly shaded portion on Fig. A5-3, the LOB chart.) An in-process report covering how many parts or assemblies actually passed each benchmark on March 23 is plotted on the LOB. (That's the darker vertical bar.) The desired actual is one that just meets the plotted line. Too many parts completed ahead of time ties up capital unnecessarily. Too few parts may mean a missed delivery schedule. Figure A5-4 has the periodic production report.

Finally, a periodic production report lists each benchmark that lags behind or is ahead of schedule and notes the corrective action that has or will be taken. This last report is another confirmation of the management-by-exception idea.

The LOB can also be used to project funds required at any point in the production cycle. By counting the required number of purchased items at a given time, one can determine the accounts payable. By counting the number of manufactured items required multiplied by the standard manufacturing hours per item and the standard rate for each of those hours, one can determine the labor costs required.

The production LOB (Fig. A5-3) shows the actual requirement of each component that goes into an assembly to meet a future delivery at some point in time. It projects trouble areas and allows time for correction prior to a crisis stage.

Index